Pitman Research Notes in Mathematics Series

Submission of proposals for consideration
Suggestions for publication, in the form of outlines and representative samples, are invited by the Editorial Board for assessment. Intending authors should approach one of the main editors or another member of the Editorial Board, citing the relevant AMS subject classifications. Alternatively, outlines may be sent directly to the publisher's offices. Refereeing is by members of the board and other mathematical authorities in the topic concerned, throughout the world.

Preparation of accepted manuscripts
On acceptance of a proposal, the publisher will supply full instructions for the preparation of manuscripts in a form suitable for direct photo-lithographic reproduction. Specially printed grid sheets are provided and a contribution is offered by the publisher towards the cost of typing. Word processor output, subject to the publisher's approval, is also acceptable.

Illustrations should be prepared by the authors, ready for direct reproduction without further improvement. The use of hand-drawn symbols should be avoided wherever possible, in order to maintain maximum clarity of the text.

The publisher will be pleased to give any guidance necessary during the preparation of a typescript, and will be happy to answer any queries.

Important note
In order to avoid later retyping, intending authors are strongly urged not to begin final preparation of a typescript before receiving the publisher's guidelines and special paper. In this way it is hoped to preserve the uniform appearance of the series.

Longman Scientific & Technical
Longman House
Burnt Mill
Harlow, Essex, UK
(tel (0279) 26721)

Longman Scientific & Technical
Churchill Livingstone Inc.
1560 Broadway
New York, NY 10036, USA
(tel (212) 819-5453)

Titles in this series

1 Improperly posed boundary value problems
A Carasso and A P Stone

2 Lie algebras generated by finite dimensional ideals
I N Stewart

3 Bifurcation problems in nonlinear elasticity
R W Dickey

4 Partial differential equations in the complex domain
D L Colton

5 Quasilinear hyperbolic systems and waves
A Jeffrey

6 Solution of boundary value problems by the method of integral operators
D L Colton

7 Taylor expansions and catastrophes
T Poston and I N Stewart

8 Function theoretic methods in differential equations
R P Gilbert and R J Weinacht

9 Differential topology with a view to applications
D R J Chillingworth

10 Characteristic classes of foliations
H V Pittie

11 Stochastic integration and generalized martingales
A U Kussmaul

12 Zeta-functions: An introduction to algebraic geometry
A D Thomas

13 Explicit *a priori* inequalities with applications to boundary value problems
V G Sigillito

14 Nonlinear diffusion
W E Fitzgibbon III and H F Walker

15 Unsolved problems concerning lattice points
J Hammer

16 Edge-colourings of graphs
S Fiorini and R J Wilson

17 Nonlinear analysis and mechanics: Heriot-Watt Symposium Volume I
R J Knops

18 Actions of fine abelian groups
C Kosniowski

19 Closed graph theorems and webbed spaces
M De Wilde

20 Singular perturbation techniques applied to integro-differential equations
H Grabmüller

21 Retarded functional differential equations: A global point of view
S E A Mohammed

22 Multiparameter spectral theory in Hilbert space
B D Sleeman

24 Mathematical modelling techniques
R Aris

25 Singular points of smooth mappings
C G Gibson

26 Nonlinear evolution equations solvable by the spectral transform
F Calogero

27 Nonlinear analysis and mechanics: Heriot-Watt Symposium Volume II
R J Knops

28 Constructive functional analysis
D S Bridges

29 Elongational flows: Aspects of the behaviour of model elasticoviscous fluids
C J S Petrie

30 Nonlinear analysis and mechanics: Heriot-Watt Symposium Volume III
R J Knops

31 Fractional calculus and integral transforms of generalized functions
A C McBride

32 Complex manifold techniques in theoretical physics
D E Lerner and P D Sommers

33 Hilbert's third problem: scissors congruence
C-H Sah

34 Graph theory and combinatorics
R J Wilson

35 The Tricomi equation with applications to the theory of plane transonic flow
A R Manwell

36 Abstract differential equations
S D Zaidman

37 Advances in twistor theory
L P Hughston and R S Ward

38 Operator theory and functional analysis
I Erdelyi

39 Nonlinear analysis and mechanics: Heriot-Watt Symposium Volume IV
R J Knops

40 Singular systems of differential equations
S L Campbell

41 N-dimensional crystallography
R L E Schwarzenberger

42 Nonlinear partial differential equations in physical problems
D Graffi

43 Shifts and periodicity for right invertible operators
D Przeworska-Rolewicz

44 Rings with chain conditions
A W Chatters and C R Hajarnavis

45 Moduli, deformations and classifications of compact complex manifolds
D Sundararaman

46 Nonlinear problems of analysis in geometry and mechanics
M Atteia, D Bancel and I Gumowski

47 Algorithmic methods in optimal control
W A Gruver and E Sachs

48 Abstract Cauchy problems and functional differential equations
F Kappel and W Schappacher

49 Sequence spaces
W H Ruckle

50 Recent contributions to nonlinear partial differential equations
H Berestycki and H Brezis

51 Subnormal operators
J B Conway

52 Wave propagation in viscoelastic media
F Mainardi
53 Nonlinear partial differential equations and their applications: Collège de France Seminar. Volume I
H Brezis and J L Lions
54 Geometry of Coxeter groups
H Hiller
55 Cusps of Gauss mappings
T Banchoff, T Gaffney and C McCrory
56 An approach to algebraic K-theory
A J Berrick
57 Convex analysis and optimization
J-P Aubin and R B Vintner
58 Convex analysis with applications in the differentiation of convex functions
J R Giles
59 Weak and variational methods for moving boundary problems
C M Elliott and J R Ockendon
60 Nonlinear partial differential equations and their applications: Collège de France Seminar. Volume II
H Brezis and J L Lions
61 Singular systems of differential equations II
S L Campbell
62 Rates of convergence in the central limit theorem
Peter Hall
63 Solution of differential equations by means of one-parameter groups
J M Hill
64 Hankel operators on Hilbert space
S C Power
65 Schrödinger-type operators with continuous spectra
M S P Eastham and H Kalf
66 Recent applications of generalized inverses
S L Campbell
67 Riesz and Fredholm theory in Banach algebra
B A Barnes, G J Murphy, M R F Smyth and T T West
68 Evolution equations and their applications
F Kappel and W Schappacher
69 Generalized solutions of Hamilton-Jacobi equations
P L Lions
70 Nonlinear partial differential equations and their applications: Collège de France Seminar. Volume III
H Brezis and J L Lions
71 Spectral theory and wave operators for the Schrödinger equation
A M Berthier
72 Approximation of Hilbert space operators I
D A Herrero
73 Vector valued Nevanlinna Theory
H J W Ziegler
74 Instability, nonexistence and weighted energy methods in fluid dynamics and related theories
B Straughan
75 Local bifurcation and symmetry
A Vanderbauwhede
76 Clifford analysis
F Brackx, R Delanghe and F Sommen
77 Nonlinear equivalence, reduction of PDEs to ODEs and fast convergent numerical methods
E E Rosinger
78 Free boundary problems, theory and applications. Volume I
A Fasano and M Primicerio
79 Free boundary problems, theory and applications. Volume II
A Fasano and M Primicerio
80 Symplectic geometry
A Crumeyrolle and J Grifone
81 An algorithmic analysis of a communication model with retransmission of flawed messages
D M Lucantoni
82 Geometric games and their applications
W H Ruckle
83 Additive groups of rings
S Feigelstock
84 Nonlinear partial differential equations and their applications: Collège de France Seminar. Volume IV
H Brezis and J L Lions
85 Multiplicative functionals on topological algebras
T Husain
86 Hamilton-Jacobi equations in Hilbert spaces
V Barbu and G Da Prato
87 Harmonic maps with symmetry, harmonic morphisms and deformations of metrics
P Baird
88 Similarity solutions of nonlinear partial differential equations
L Dresner
89 Contributions to nonlinear partial differential equations
C Bardos, A Damlamian, J I Díaz and J Hernández
90 Banach and Hilbert spaces of vector-valued functions
J Burbea and P Masani
91 Control and observation of neutral systems
D Salamon
92 Banach bundles, Banach modules and automorphisms of C*-algebras
M J Dupré and R M Gillette
93 Nonlinear partial differential equations and their applications: Collège de France Seminar. Volume V
H Brezis and J L Lions
94 Computer algebra in applied mathematics: an introduction to MACSYMA
R H Rand
95 Advances in nonlinear waves. Volume I
L Debnath
96 FC-groups
M J Tomkinson
97 Topics in relaxation and ellipsoidal methods
M Akgül
98 Analogue of the group algebra for topological semigroups
H Dzinotyiweyi
99 Stochastic functional differential equations
S E A Mohammed

100 Optimal control of variational inequalities
V Barbu
101 Partial differential equations and dynamical systems
W E Fitzgibbon III
102 Approximation of Hilbert space operators. Volume II
C Apostol, L A Fialkow, D A Herrero and D Voiculescu
103 Nondiscrete induction and iterative processes
V Ptak and F-A Potra
104 Analytic functions – growth aspects
O P Juneja and G P Kapoor
105 Theory of Tikhonov regularization for Fredholm equations of the first kind
C W Groetsch
106 Nonlinear partial differential equations and free boundaries. Volume I
J I Díaz
107 Tight and taut immersions of manifolds
T E Cecil and P J Ryan
108 A layering method for viscous, incompressible L_p flows occupying R^n
A Douglis and E B Fabes
109 Nonlinear partial differential equations and their applications: Collège de France Seminar. Volume VI
H Brezis and J L Lions
110 Finite generalized quadrangles
S E Payne and J A Thas
111 Advances in nonlinear waves. Volume II
L Debnath
112 Topics in several complex variables
E Ramírez de Arellano and D Sundararaman
113 Differential equations, flow invariance and applications
N H Pavel
114 Geometrical combinatorics
F C Holroyd and R J Wilson
115 Generators of strongly continuous semigroups
J A van Casteren
116 Growth of algebras and Gelfand–Kirillov dimension
G R Krause and T H Lenagan
117 Theory of bases and cones
P K Kamthan and M Gupta
118 Linear groups and permutations
A R Camina and E A Whelan
119 General Wiener–Hopf factorization methods
F-O Speck
120 Free boundary problems: applications and theory, Volume III
A Bossavit, A Damlamian and M Fremond
121 Free boundary problems: applications and theory, Volume IV
A Bossavit, A Damlamian and M Fremond
122 Nonlinear partial differential equations and their applications: Collège de France Seminar. Volume VII
H Brezis and J L Lions
123 Geometric methods in operator algebras
H Araki and E G Effros
124 Infinite dimensional analysis–stochastic processes
S Albeverio
125 Ennio de Giorgi Colloquium
P Krée
126 Almost-periodic functions in abstract spaces
S Zaidman
127 Nonlinear variational problems
A Marino, L Modica, S Spagnolo and M Degiovanni
128 Second-order systems of partial differential equations in the plane
L K Hua, W Lin and C-Q Wu
129 Asymptotics of high-order ordinary differential equations
R B Paris and A D Wood
130 Stochastic differential equations
R Wu
131 Differential geometry
L A Cordero
132 Nonlinear differential equations
J K Hale and P Martinez-Amores
133 Approximation theory and applications
S P Singh
134 Near-rings and their links with groups
J D P Meldrum
135 Estimating eigenvalues with *a posteriori/a priori* inequalities
J R Kuttler and V G Sigillito
136 Regular semigroups as extensions
F J Pastijn and M Petrich
137 Representations of rank one Lie groups
D H Collingwood
138 Fractional calculus
G F Roach and A C McBride
139 Hamilton's principle in continuum mechanics
A Bedford
140 Numerical analysis
D F Griffiths and G A Watson
141 Semigroups, theory and applications. Volume I
H Brezis, M G Crandall and F Kappel
142 Distribution theorems of L-functions
D Joyner
143 Recent developments in structured continua
D De Kee and P Kaloni
144 Functional analysis and two-point differential operators
J Locker
145 Numerical methods for partial differential equations
S I Hariharan and T H Moulden
146 Completely bounded maps and dilations
V I Paulsen
147 Harmonic analysis on the Heisenberg nilpotent Lie group
W Schempp
148 Contributions to modern calculus of variations
L Cesari
149 Nonlinear parabolic equations: qualitative properties of solutions
L Boccardo and A Tesei
150 From local times to global geometry, control and physics
K D Elworthy

151 A stochastic maximum principle for optimal control of diffusions
U G Haussmann
152 Semigroups, theory and applications. Volume II
H Brezis, M G Crandall and F Kappel
153 A general theory of integration in function spaces
P Muldowney
154 Oakland Conference on partial differential equations and applied mathematics
L R Bragg and J W Dettman
155 Contributions to nonlinear partial differential equations. Volume II
J I Díaz and P L Lions
156 Semigroups of linear operators: an introduction
A C McBride
157 Ordinary and partial differential equations
B D Sleeman and R J Jarvis
158 Hyperbolic equations
F Colombini and M K V Murthy
159 Linear topologies on a ring: an overview
J S Golan
160 Dynamical systems and bifurcation theory
M I Camacho, M J Pacifico and F Takens
161 Branched coverings and algebraic functions
M Namba
162 Perturbation bounds for matrix eigenvalues
R Bhatia
163 Defect minimization in operator equations: theory and applications
R Reemtsen
164 Multidimensional Brownian excursions and potential theory
K Burdzy
165 Viscosity solutions and optimal control
R J Elliott
166 Nonlinear partial differential equations and their applications. Collège de France Seminar. Volume VIII
H Brezis and J L Lions
167 Theory and applications of inverse problems
H Haario
168 Energy stability and convection
G P Galdi and B Straughan
169 Additive groups of rings. Volume II
S Feigelstock
170 Numerical analysis 1987
D F Griffiths and G A Watson
171 Surveys of some recent results in operator theory. Volume I
J B Conway and B B Morrel
172 Amenable Banach algebras
J-P Pier
173 Pseudo-orbits of contact forms
A Bahri
174 Poisson algebras and Poisson manifolds
K H Bhaskara and K Viswanath

John B Conway &
Bernard B Morrel (Editors)

Indiana University / Indiana University – Purdue University
at Indianapolis

Surveys of some recent results in operator theory VOLUME I

Copublished in the United States with
John Wiley & Sons, Inc., New York

Longman Scientific & Technical
Longman Group UK Limited
Longman House, Burnt Mill, Harlow
Essex CM20 2JE, England
and Associated Companies throughout the world.

Copublished in the United States with
John Wiley & Sons, Inc., 605 Third Avenue, New York, NY 10158

First published 1988

AMS Subject Classification : 47A99

ISSN 0269-3674

British Library Cataloguing in Publication Data
Surveys of some recent results in operator
theory.—(Pitman research notes in
mathematics series; ISSN 0269–3674; 171).
Vol. 1
1. Operator theory
I. Conway, John B. II. Morrel, B. B.
515.7′24 QA329
ISBN 0-582-00519-1

Library of Congress Cataloging-in-Publication Data
Surveys of some recent results in operator theory.
(Pitman research notes in mathematics series,
ISSN 0269-3674 ; 171-)
Bibliography: p.
1. Operator theory. I. Conway, Joseph B.
II. Morrel, Bernard B. III. Series: Pitman research
notes in mathematics series ; 171, etc.
QA329.S87 1988 515.7′24 87-29896
ISBN 0-470-21030-3 (USA only)

Printed and bound in Great Britain by
Biddles Ltd, Guildford and King's Lynn

Contents

Preface

This is the first of two volumes entitled *Surveys of Some Recent Results in Operator Theory*. The papers herein are based on series of lectures given at Indiana University as part of the Special Year in Operator Theory held there during the 1985-86 academic year. While both volumes fall under the general rubric of Operator Theory, this volume is primarily concerned with the study of single linear operators, while the second volume will be directed toward the study of algebras of bounded linear operators on Hilbert space.

As originally envisioned, each of the lecture series in the Special Year would develop a topic of current interest in Operator Theory, starting from a point accessible to advanced graduate students in the area, and proceeding rapidly to current and future researchers. We feel that the results have far exceeded our expectations.

In any enterprise of this sort, there are many who deserve thanks and recognition. First and foremost we thank the authors: Sheldon Axler, Joe Ball, Earl Berkson, Richard Carey, Carl Cowen, Lawrence A. Fialkow, Vern I. Paulsen, and Jim Thomson. Thanks also to Hari Bercovici, Ciprian Foiaş, Mort Lowengrub, Joe Stampfli, and Bill Ziemer of Indiana University their support and encouragement. Deep thanks to Jan Want of Texas A&M University for her careful preparation of the manuscript in TeX: thanks to her, its elegant style matches its substance.

Finally, thanks to Indiana University, the National Science Foundation, and the Argonne Universities Trust Fund, without whose financial support the Special Year would never have taken place.

John B. Conway
Bernard B. Morrel

Bergman Spaces and Their Operators

by

Sheldon Axler

I. Introduction

Let $\mathbb{D}$ denote the open unit disk in the complex plane $\mathbb{C}$, and let dA denote the usual two–dimensional area measure on $\mathbb{D}$. For $1 \leq p < \infty$, the Bergman space L^p_a (the subscript "a" stands for *analytic*) is the set of analytic functions $F : \mathbb{D} \to \mathbb{C}$ such that

$$\int_{\mathbb{D}} |f|^p dA < \infty.$$

We will discuss the following topics concerning Bergman spaces:

1. The dual of L^p_a for $1 < p < \infty$.
2. The Bloch space and the dual of L^1_a.
3. Carleson measures on L^p_a.
4. Toeplitz operators in L^2_a.
5. Hankel operators on L^2_a.

The subject of Bergman spaces and their operators is too vast to cover in a manuscript of this length. Of the topics selected, proofs of the main theorems are given only when I felt I could offer an improvement upon previously published proofs.

1. The Dual of L^p_a for $1 < p < \infty$

The main goal of this section is to prove that for $1 < p < \infty$, the dual of the Bergman space L^p_a can be identified with $L^{p'}_a$; here p' denotes the number such that $(1/p) + (1/p') = 1$. We will begin by finding the reproducing kernels and an explicit formula for the orthogonal projection of $L^2(\mathbb{D}, dA)$ onto L^2_a. The main work of the section lies in showing that this projection is also a bounded projection from $L^p(\mathbb{D}, dA)$ onto L^p_a for every p between 1 and ∞. This result will then be used to prove that the dual of L^p_a is $L^{p'}_a$. We will also see that harmonic conjugation is a bounded operator in the $L^p(\mathbb{D}, dA)$ norm for $1 < p < \infty$, and surprisingly, that harmonic conjugation is also continuous in the $L^1(\mathbb{D}, dA)$ norm.

A Bounded Projection onto the Bergman Space

For a Lebesgue measurable function $f : \mathbb{D} \to \mathbb{C}$, the $L^p(\mathbb{D}, dA)$ norm of f will always be denoted by $\|f\|_p$, so

$$\|f\|_p = \left(\int_{\mathbb{D}} |f|^p \, dA\right)^{1/p}.$$

Of course, $L^2(\mathbb{D}, dA)$ is a Hilbert space with inner product given by

1.1 $$< f, g >= \int_{\mathbb{D}} f\overline{g} \, dA.$$

Fix $w \in \mathbb{D}$. We want to find a function $k_w \in L^2_a$ such that

1.2 $$f(w) =< f, k_w >$$

for all $f \in L^2_a$. A function k_w with this property is called the reproducing kernel for the point w. To discover a formula for k_w, consider the power series expansion

1.3 $$k_w(z) = \sum_{n=0}^{\infty} b_n z^n.$$

Fixing m, and taking $f(z) = z^m$ in 1.2, we have

$$w^m =< z^m, k_w >= \int_{\mathbb{D}} z^m \left(\sum_{n=0}^{\infty} \overline{b_n z^n}\right) dA(z) = \overline{b_m}\pi/(m+1),$$

where the above integral is computed by replacing z by polar co-ordinates $re^{i\theta}$, replacing dA with $r \, d\theta \, dr$, and then interchanging the integral and summation (don't worry about the justification; this is only motivation). The above equation says that $b_m = \pi^{-1}(m+1)\overline{w^m}$, so by 1.3, k_w should be given by the formula

1.4 $$k_w(z) = \pi^{-1}\sum_{n=0}^{\infty}(n+1)(\overline{w}z)^n = \pi^{-1}(1-\overline{w}z)^{-2},$$

where the above infinite sum is evaluated by letting $\lambda = \overline{w}z$ and noting that

1.5 $$\sum_{n=0}^{\infty}(n+1)\lambda^n = \left[\sum_{n=0}^{\infty}\lambda^{n+1}\right]' = [\lambda/(1-\lambda)]' = 1/(1-\lambda)^2.$$

Let's extend the domain of the inner product given by 1.1 to include all pairs of functions f, g such that $fg \in L^1(\mathbb{D}, dA)$; for such a pair of functions $< f, g >$ is still defined by

1.6
$$< f, g >= \int_{\mathbb{D}} f\overline{g}\, dA.$$

Note that for fixed $w \in \mathbb{D}$, the function k_w given by 1.4 is bounded on $\mathbb{D}$. Thus it makes sense to ask whether 1.2 holds not only for $f \in L^2_a$, but also for $f \in L^1_a$ (which includes L^p_a for all $1 < p < \infty$). The following proposition says that this is the case and that the reproducing kernel k_w is indeed given by 1.4; the proof is shorter than the motivation that has preceded it.

Proposition 1.7. *Let $1 \le p < \infty$, let $f \in L^p_a$, and let $w \in \mathbb{D}$. Then*

$$f(w) = \pi^{-1} \int_{\mathbb{D}} f(z)(1 - w\overline{z})^{-2} dA(z).$$

Proof: Since f is analytic on $\mathbb{D}$, f has a power series expansion

$$f(z) = \sum_{n=0}^{\infty} a_n z^n.$$

We have not proved that the partial sums of this series converge to f in the $L^p(\mathbb{D}, dA)$ norm, but this does not matter because we can integrate over a disk of radius t and then take a limit as t tends to 1 (throughout these sections, all limits taken as $t \to 1$ means that t is increasing to 1). Thus

$$\begin{aligned}
\pi^{-1} \int_{\mathbb{D}} f(z)(1 - w\overline{z})^{-2} dA(z) &= \lim_{t\to 1} \pi^{-1} \int_{t\mathbb{D}} f(z)(1 - w\overline{z})^{-2} dA(z) \\
&= \lim_{t\to 1} \pi^{-1} \int_{t\mathbb{D}} f(z) \sum_{n=0}^{\infty} (n+1)(w\overline{z})^n dA(z) \qquad \text{[from 1.5]} \\
&= \lim_{t\to 1} \pi^{-1} \sum_{n=0}^{\infty} w^n (n+1) \int_{t\mathbb{D}} f(z)\overline{z^n} dA(z) = \sum_{n=0}^{\infty} a_n w^n \\
&= f(w),
\end{aligned}$$

where the last integral is computed by switching to polar co-ordinates and replacing f with its power series. ■

The symbol ■, used above, will always denote the end of a proof.

Note that the above proposition clearly implies that point evaluation (the map that sends f to $f(w)$) is bounded on L^p_a for each $p \in [1,\infty]$ and each $w \in \mathbb{D}$ and uniformly on compact subsets of $\mathbb{D}$). Thus if $\{f_n\}$ is a Cauchy sequence in L^p_a, then $\{f_n\}$ converges uniformly on compact subsets of $\mathbb{D}$ to some function f, which must be analytic on $\mathbb{D}$. Fatou's Lemma shows that f is in $L^p(\mathbb{D}, dA)$, so $f \in L^p_a$, and the usual argument shows that $\{f_n\}$ converges to f in the norm on L^p_a. Conclusion: If $1 \le p < \infty$, then L^p_a is a Banach space.

We will need to know the $L^2(\mathbb{D}, dA)$ norm of the reproducing kernel. This simple formula will be so useful throughout these sections that it deserves to be called a lemma.

Lemma 1.8. *Let $w \in \mathbb{D}$. Then*

$$\pi^{-2}\int_{\mathbb{D}} |1-\overline{w}z|^{-4}\,dA(z) = \|k_w\|_2^2 = \pi^{-1}(1-|w|^2)^{-2}.$$

Proof: The first equality follows immediately from the definition (1.4) of k_w. For the second equality, we use the reproducing property of k_w:

$$\|k_w\|_2^2 = \langle k_w, k_w \rangle = \pi^{-1}(1-|w|^2)^{-2}. \quad \blacksquare$$

Let P denote the orthogonal projection of the Hilbert space $L^2(\mathbb{D}, dA)$ onto L^2_a. For $f \in L^2(\mathbb{D}, dA)$ we can use the reproducing kernel k_w (see 1.4 and 1.7) to give an explicit formula for Pf:

1.9 $$(Pf)(w) = \langle Pf, k_w \rangle = \langle f, k_w \rangle = \pi^{-1}\int_{\mathbb{D}} f(z)(1-w\overline{z})^{-2}\,dA(z).$$

The right-hand side of the above equation makes sense whenever $f \in L^1(\mathbb{D}, dA)$, and so we extend the domain of P to $L^1(\mathbb{D}, dA)$; for $f \in L^1(\mathbb{D}, dA)$ and $w \in \mathbb{D}$, we define $(Pf)(w)$ by 1.9. Since we can differentiate under the integral sign, clearly Pf is analytic on $\mathbb{D}$ for each $f \in L^1(\mathbb{D}, dA)$. By 1.7, Pf equals f for all $f \in L^1_a$.

We already know that P is a bounded projection of $L^2(\mathbb{D}, dA)$ onto L^2_a. The following theorem says that this is also true for p between 1 and ∞. I believe that 1.10 was first proved by Zaharjuta and Judovič [27], who also noted that the boundedness of P immediately leads to the duality between L^p_a and $L^{p'}_a$ (see 1.16 of this section). The proof given by Zaharjuta and Judovič uses the theory of singular integral operators.

Since the integral 1.9 defining P is not a singular integral, it seems that singular integral theory should not be required for the proof that P is bounded. Forelli and

Rudin [10] gave a proof (with the unit ball in $\mathbb{C}^n$ replacing the unit disk) that does not use singular integral theory; they also noted that the boundedness of P immediately leads to the boundedness of harmonic conjugation (see 1.21 of this section, which uses their proof). The proof of 1.10 given below uses the ideas of the Forelli-Rudin proof, with some simplifications that eliminate the need to use the binomial theorem, the gamma function, and Stirling's formula.

Theorem 1.10. *Let $1 < p < \infty$. Then P is a bounded projection of $L^p(\mathbb{D}, dA)$ onto L^p_a.*

Proof: By 1.7, we only need to show that P is a bounded map of $L^p(\mathbb{D}, dA)$ into L^p_a. To do this, let $f \in L^p(\mathbb{D}, dA)$. Let h be a positive measurable function on $\mathbb{D}$ (we will specify h later). Then for $w \in \mathbb{D}$ we have

$$\begin{aligned} |(Pf)(w)| &\leq \pi^{-1} \int_{\mathbb{D}} |f(z)| |1 - w\bar{z}|^{-2} dA(z) \\ &= \pi^{-1} \int_{\mathbb{D}} h(z)|f(z)| \, |1 - w\bar{z}|^{-2} h(z)^{-1} dA(z) \\ &\leq \pi^{-1} \left(\int_{\mathbb{D}} h(z)^{p'} |1 - w\bar{z}|^{-2} dA(z) \right)^{1/p'} \\ &\quad \cdot \left(\int_{\mathbb{D}} |f(z)|^p h(z)^{-p} |1 - w\bar{z}|^{-2} dA(z) \right)^{1/p}. \end{aligned} \tag{1.11}$$

Suppose we could find a positive function h and a constant $c < \infty$ such that

$$\int_{\mathbb{D}} h(z)^{p'} |1 - w\bar{z}|^{-2} dA(z) \leq c\, h(w)^{p'} \qquad \text{for all } w \in \mathbb{D}, \tag{1.12}$$

and

$$\int_{\mathbb{D}} h(z)^{p} |1 - w\bar{z}|^{-2} dA(z) \leq c\, h(w)^{p} \qquad \text{for all } w \in \mathbb{D}. \tag{1.13}$$

Then from 1.11 and 1.12 we would have

$$|(Pf)(w)|^p \leq \pi^{-p} c^{p/p'} h(w)^p \left(\int_{\mathbb{D}} |f(z)|^p h(z)^{-p} |1 - w\bar{z}|^{-2} dA(z) \right),$$

so

$$\begin{aligned} &\int_{\mathbb{D}} |(Pf)(w)|^p dA(w) \\ &\quad \leq \pi^{-p} c^{p/p'} \int_{\mathbb{D}} |f(z)|^p \, h(z)^{-p} \left(\int_{\mathbb{D}} h(w)^p |1 - w\bar{z}|^{-2} dA(w) \right) dA(z) \\ &\quad \leq (c/\pi)^p \int_{\mathbb{D}} |f(z)|^p dA(z) \qquad \text{[from 1.13 with } w \text{ and } z \text{ interchanged].} \end{aligned}$$

The last inequality is precisely what we want, so the proof will be finished as soon as we can show that there is a positive measurable function h satisfying 1.12 and 1.13. One function that works is given by

1.14 $$h(z) = (1 - |z|^2)^{-1/(pp')}.$$

To verify that 1.12 and 1.13 are satisfied by this choice of h, we fix $w \in \mathbb{D}$ and begin with the left-hand side of 1.12:

$$\int_{\mathbb{D}} h(z)^{p'} |1 - w\bar{z}|^{-2} dA(z) = \int_{\mathbb{D}} (1 - |z|^2)^{1/p} |1 - \overline{w} z|^{-2} dA(z).$$

On the right-hand side of the above equation, we make a change of variables, replacing z by $(w - \lambda)/(1 - \overline{w}\lambda)$, so λ is the new variable. The map that takes λ to $(w - \lambda)/(1 - \overline{w}\lambda)$ is a one-to-one analytic map of $\mathbb{D}$ onto $\mathbb{D}$, so after the change of variables the region of integration is still $\mathbb{D}$. Computing the usual Jacobian, we must replace $dA(z)$ by $(1 - |w|^2)^2 |1 - \overline{w}\lambda|^{-4} dA(\lambda)$. Putting all this together (look at 3.1 and 3.9 if the manipulations are not clear) shows that

$$\begin{aligned}\int_{\mathbb{D}} h(z)^{p'} |1 - w\bar{z}|^{-2} dA(z) &= (1 - |w|^2)^{-1/p} \int_{\mathbb{D}} (1 - |\lambda|^2)^{-1/p} |1 - \overline{w}\lambda|^{(2/p)-2} dA(\lambda) \\ &= h(w)^{p'} \int_{\mathbb{D}} (1 - |\lambda|^2)^{-1/p} |1 - \overline{w}\lambda|^{(2/p)-2} dA(\lambda).\end{aligned}$$

The equation above shows that 1.12 will hold if

$$\int_{\mathbb{D}} (1 - |\lambda|^2)^{-1/p} |1 - \overline{w}\lambda|^{(2/p)-2} dA(\lambda)$$

is bounded independent of $w \in \mathbb{D}$. It is not hard to show that the above integral is bounded independent of $w \in \mathbb{D}$ if the integral is finite when we take $w = 1$. So now we must prove that

1.15 $$\int_{\mathbb{D}} (1 - |\lambda|^2)^{-1/p} |1 - \lambda|^{(2/p)-2} dA(\lambda) < \infty.$$

In the above integral, make the change of variables (polar coordinates plus a translation) $\lambda = 1 - re^{i\theta}$, getting

$$\begin{aligned}\int_{\mathbb{D}} (1 - |\lambda|^2)^{-1/p} |1 - \lambda|^{(2/p)-2} dA(\lambda) &= \int_{-\pi/2}^{\pi/2} \int_0^{2\cos\theta} (2\cos\theta - r)^{-1/p} r^{(1/p)-1} dr\, d\theta \\ &\leq \int_{-\pi/2}^{\pi/2} \int_0^{\cos\theta} (\cos\theta)^{-1/p} r^{(1/p)-1} dr\, d\theta \\ &+ \int_{-\pi/2}^{\pi/2} \int_{\cos\theta}^{2\cos\theta} (2\cos\theta - r)^{-1/p} (\cos\theta)^{(1/p)-1} dr\, d\theta \\ &< \infty.\end{aligned}$$

Thus 1.15 is true, and so 1.12 holds for the function h given by 1.14 and an appropriate choice of the constant c. Since h is symmetric in p and p', the same proof shows that 1.13 also holds, and so the proof of 1.10 is complete. ■

The above proof of 1.10 actually proves the following slightly stronger statement, which shows that the boundedness of P does not depend upon any cancellation properties of the kernel $(1 - w\overline{z})^{-2}$. If $1 < p < \infty$ and if $f \in L^p(\mathbb{D}, dA)$, then

$$\int_{\mathbb{D}} |f(z)|\ |1 - w\overline{z}|^{-2} dA(z)$$

is also in $L^p(\mathbb{D}, dA)$ (as a function of w) and has a norm bounded by a constant (independent of f) times $\|f\|_p$.

Duality

Now we will use the boundedness of the projection P to prove that the dual of L^p_a can be identified with $L^{p'}_a$. It is clear that if $g \in L^{p'}_a$, then the mapping that takes f to $< f, g >$ (defined by 1.6) is a bounded linear functional on L^p_a. The main content of the following theorem is that every bounded linear functional on L^p_a is of this form. Note that distinct functions in $L^{p'}_a$ induce distinct linear functionals on L^p_a (because if $< z^n, g_1 > = < z^n, g_2 >$ for all n, then the Taylor coefficients of g_1 equal the Taylor coefficients of g_2).

The identification given here between $L^{p'}_a$ and the dual of L^p_a is conjugate linear. A linear identification could be obtained by considering the pairing that takes $f \in L^p_a$ and $g \in L^{p'}_a$ to

$$\int_{\mathbb{D}} f(z)g(\overline{z})\ dA(z).$$

I have chosen to use the conjugate linear identification so that consistency can be maintained with the inner product notation.

As usual, to say that two norms $\|\ \|_a$ and $\|\ \|_b$ are equivalent means that there is a finite constant c such that $\|g\|_a \le c\|g\|_b$ and $\|g\|_b \le c\|g\|_a$ for all g in the space under consideration.

Theorem 1.16. *Let $1 < p < \infty$. Then the dual of L^p_a can be identified with $L^{p'}_a$. More precisely, every bounded linear functional on L^p_a is of the form*

$$f \mapsto \int_{\mathbb{D}} f\overline{g}\ dA$$

for some unique $g \in L^{p'}_a$. Furthermore, the norm of the linear functional on L^p_a induced by $g \in L^{p'}_a$ is equivalent to $\|g\|_{p'}$.

Proof: Suppose ψ is a bounded linear functional on L^p_a. By the Hahn–Banach Theorem and the Riesz Representation Theorem, there is a function $h \in L^{p'}(\mathbb{D}, dA)$ such that $\|h\|_{p'} = \|\psi\|$ and

$$\psi(f) = \int_{\mathbb{D}} f\overline{h}\, dA$$

for all $f \in L^p_a$. Let $g = Ph$. By 1.10, we know that $g \in L^{p'}_a$ and there is a constant c (independent of g) such that $\|g\|_{p'} \leq c\|h\|_{p'}$. If $f \in L^p_a$, then

$$\begin{aligned}
\int_{\mathbb{D}} f\overline{g} dA &= \int_{\mathbb{D}} f(z)\pi^{-1} \int_{\mathbb{D}} \overline{h}(w)(1 - w\overline{z})^{-2} dA(w) dA(z) \\
&= \int_{\mathbb{D}} \overline{h}(w)\pi^{-1} \int_{\mathbb{D}} f(z)(1 - w\overline{z})^{-2} dA(z) dA(w) \qquad \text{[by 1.7]} \\
&= \int_{\mathbb{D}} \overline{h}(w) f(w) dA(w) \\
&= \psi(f)
\end{aligned}$$

Thus ψ is the linear functional on L^p_a induced by g. It is clear that $\|\psi\| \leq \|g\|_{p'}$ and since $\|g\|_{p'} \leq c\|h\|_{p'} = c\|\psi\|$, the proof is complete. ■

A very careful reader will note that the comment following the proof of 1.10 seems to be needed to justify the use of Fubini's Theorem in the above proof.

Harmonic Conjugation

If u is a real valued harmonic function on the unit disk $\mathbb{D}$, then the harmonic conjugate of u, denoted $\widetilde{u}$, is the unique real valued harmonic function on $\mathbb{D}$ such that $\widetilde{u}(0) = 0$ and $u + i\widetilde{u}$ is analytic. We will see (1.21) that harmonic conjugation is a bounded operator in the $L^p(\mathbb{D}, dA)$ norm for $1 \leq p < \infty$.

Note that 1.21 asserts that harmonic conjugation is a bounded operator in the $L^1(\mathbb{D}, dA)$ norm. This result will be surprising to those who expect the theory of the Bergman spaces L^p_a to be completely analogous to the theory of the Hardy spaces $H^p(\partial\mathbb{D})$. (We will not need results from Hardy space theory for any proofs in these lectures. However, from time to time we will discuss the similarity, or lack of similarity, between Hardy space results and Bergman space results. Good references for basic material on the Hardy spaces can be found in the books of Hoffman [14], Duren [8], and Koosis [15].)

For $1 < p < \infty$, the boundedness of the projection P on $L^p(\mathbb{D}, dA)$ will be the key tool in the proof of 1.21. The projection P is not bounded on $L^1(\mathbb{D}, dA)$. However, the following lemma will allow us find another operator Q that is a bounded projection from $L^1(\mathbb{D}, dA)$ onto L^1_a.

Lemma 1.17. *Let f be an analytic function on $\mathbb{D}$ such that*

$$\int_{\mathbb{D}} |f(z)|(1-|z|^2)^2 dA(z) < \infty.$$

Let $w \in \mathbb{D}$. Then

$$f(w) = 3\pi^{-1} \int_{\mathbb{D}} f(z)(1-|z|^2)^2(1-w\overline{z})^{-4} dA(z).$$

Proof: By the Lebesgue Dominated Convergence Theorem,

$$\begin{aligned} &3\pi^{-1} \int_{\mathbb{D}} f(z)(1-|z|^2)^2(1-w\overline{z})^{-4} dA(z) \\ &\quad = \lim_{t \to 1} 3\pi^{-1} \int_{t\mathbb{D}} f(z)(t^2-|z|^2)^2(1-w\overline{z})^{-4} dA(z). \end{aligned} \tag{1.18}$$

Now in the integral on the right-hand side of the above equality, replace $(1-w\overline{z})^{-4}$ by its series

$$(1-w\overline{z})^{-4} = \sum_{n=0}^{\infty}(n+1)(n+2)(n+3)w^n(\overline{z})^n/6,$$

(this formula is obtained by starting with the Taylor series

$$(1-\lambda)^{-1} = \sum_{n=0}^{\infty} \lambda^n,$$

differentiating both sides three times, and then replacing λ by $w\overline{z}$), replace f with its Taylor series $\sum_{n=0}^{\infty} a_n z^n$, and then replace z by polar co-ordinates and compute, getting that the right-hand side of 1.18 equals

$$\lim_{t \to 1} t^6 f(t^2 w).$$

Since the above limit equals $f(w)$, the proof is complete. ■

Inspired by 1.17, we now define, for each $f \in L^1(\mathbb{D}, da)$, a function Qf on $\mathbb{D}$ by

$$(Qf)(w) = 3\pi^{-1} \int_{\mathbb{D}} f(z)(1-|z|^2)^2(1-w\overline{z})^{-4} dA(z). \tag{1.19}$$

It is clear that Qf is analytic on $\mathbb{D}$. The following proposition will allow us to use Q in place of P when dealing with $L^1(\mathbb{D}, dA)$.

Proposition 1.20. *Q is a bounded projection of $L^1(\mathbb{D}, dA)$ onto L^1_a.*

Proof: Lemma 1.17 says that $Qf = f$ for all $f \in L^1_a$, so we need only prove that Q is bounded on $L^1(\mathbb{D}, dA)$. To do this, let $f \in L^1(\mathbb{D}, dA)$. If $w \in \mathbb{D}$, then

$$|(Qf)(w)| \leq 3\pi^{-1} \int_{\mathbb{D}} |f(z)|(1-|z|^2)^2 |1-w\overline{z}|^{-4} dA(z),$$

so

$$\begin{aligned}\int_{\mathbb{D}} |(Qf)(w)| dA(w) &\leq 3 \int_{\mathbb{D}} |f(z)|(1-|z|^2)^2 \pi^{-1} \int_{\mathbb{D}} |1-w\overline{z}|^{-4} dA(w) dA(z) \\ &= 3 \int_{\mathbb{D}} |f(z)| dA(z) \quad \text{[from 1.8]},\end{aligned}$$

and we are done. ■

Now we are ready to prove that harmonic conjugation is a bounded operator on $L^p(\mathbb{D}, dA)$ for $1 \leq p < \infty$.

Theorem 1.21. *Let $1 \leq p < \infty$. Then there is a finite constant c such that*

$$\|\tilde{u}\|_p \leq c\|u\|_p$$

for every real valued harmonic function u on $\mathbb{D}$.

Proof: First we consider the case where $1 < p < \infty$.

By 1.10 there is a finite constant c such that

$$\|Pu\|_p \leq (c/2)\|u\|_p$$

for every $u \in L^p(\mathbb{D}, dA)$.

If $f \in L^p_a$ and $w \in \mathbb{D}$, then

1.22
$$\begin{aligned}(P\overline{f})(w) &= \pi^{-1} \int_{\mathbb{D}} \overline{f}(z)(1-w\overline{z})^{-2} dA(z) \\ &= \pi^{-1} \overline{\left(\int_{\mathbb{D}} f(z)(1-\overline{w}z)^{-2} dA(z)\right)} \\ &= \overline{f}(0),\end{aligned}$$

where the last integral is evaluated by using the mean value property of the function $f(z)(1-\overline{w}z)^{-2}$, which is analytic in $\mathbb{D}$ (as a function of z).

Thus if $f \in L^p_a$ and $\text{Im } f(0) = 0$ (Im denotes imaginary part, and Re denotes real part), then

$$\|\text{Im } f\|_p \le \|f + f(0)\|_p = \|P(f + \overline{f}\|_p = 2\|P(\text{Re } f)\|_p \le c\|\text{Re } f\|_p.$$

This is almost what we want, except that we have proved it only under the assumption that both $\text{Re } f$ and $\text{Im } f$ are in $L^p(\mathbb{D}, dA)$.

To finish the proof (for the case $1 < p < \infty$), let u be a real valued harmonic function defined on $\mathbb{D}$, and for $0 < t < 1$, let u_t be the function on $\mathbb{D}$ defined by $u_t(z) = u(tz)$. By the paragraph above, we have

$$\|(u_t)^{\sim}\|_p \le c\|u_t\|_p,$$

and now taking limits as t increases to 1 gives the desired result.

The proof for the case where $p = 1$ is very similar. If $f \in L^1_a$ and $w \in \mathbb{D}$, then

$$\begin{aligned}(Q\overline{f})(w) &= 3\pi^{-1} \int_0^1 (1 - r^2)^2 r \left(\int_0^{2\pi} f(\text{r}e^{i\theta})(1 - \overline{w}\text{r}e^{i\theta})^{-4} d\theta \right)^{-} dr \\ &= \overline{f}(0),\end{aligned}$$

where the inner integral, whose value is $2\pi f(0)$, is evaluated by using the mean value property of the function $f(z)(1 - \overline{w}z)^{-4}$, which is analytic on $\mathbb{D}$ (as a function of z). Now the proof for the case $p = 1$ can be finished by duplicating the proof for the case $1 < p < \infty$, except that Q is used in place of P, the above equation is used in place of 1.22, and 1.20 is used in place of 1.10. ■

2. The Bloch Space and the Dual of L^1_a

In the last section we saw that for $1 < p < \infty$, the dual of L^p_a can be identified with $L^{p'}_a$. What about the dual of L^1_a? In this section we will see that the dual of L^1_a can be identified with the Bloch space, which is defined below. We will also prove that P is a bounded map of $L^\infty(\mathbb{D}, dA)$ onto the Bloch space. Finally, we will define the little Bloch space, prove that it is the pre-dual of L^1_a, and prove that P maps $C(\overline{\mathbb{D}})$ onto the little Bloch space.

The Bloch Space

For f analytic on the unit disk $\mathbb{D}$, the Bloch norm $\|f\|_{\mathcal{B}}$ of f is defined by

$$\|f\|_{\mathcal{B}} = \sup\{(1-|z|^2)|f'(z)| : z \in \mathbb{D}\}.$$

The Bloch space $\mathcal{B}$ is the set of all analytic functions f on $\mathbb{D}$ such that $\|f\|_{\mathcal{B}}$ is finite. If f is a constant function, then $\|f\|_{\mathcal{B}} = 0$, so the Bloch norm $\| \ \|_{\mathcal{B}}$ is not actually a norm on the Bloch space. However, $\|f\|_{\mathcal{B}} + |f(0)|$ does define a norm on $\mathcal{B}$, and whenever we refer to properties that usually require a norm (boundedness of linear maps on $\mathcal{B}$, dual of $\mathcal{B}$, etc.), it will be this norm that we have in mind.

The Bloch space plays an important role in geometric function theory. Anderson's article [1] gives a nice survey of what is known about the Bloch space. Here we will concentrate on those aspects of the Bloch space that are relevant to Bergman spaces.

We begin with an alternate characterization of the Bloch space. For further generalizations of the following lemma, see Theorem 39 of Hardy and Littlewood's paper [12] or Theorem 5.5 of [8].

Lemma 2.1. *The following quantities are equivalent (for functions f analytic on the unit disk):*

(A) $\|f\|_{\mathcal{B}}$;

(B) $|f'(0)| + \sup\{(1-|z|^2)^2|f''(z)| : z \in \mathbb{D}\}$.

Proof: Let f be analytic on $\mathbb{D}$. To show that the Block norm of f is bounded by a multiple of quantity (B), let $w \in \mathbb{D}$. Then

$$f'(w) - f'(0) = w \int_0^1 f''(tw)dt,$$

so

$$\begin{aligned}
|f'(w)| &\le |f'(0)| + \sup\{(1-|z|^2)^2|f''(z)| : z \in \mathbb{D}\}|w| \int_0^1 (1-|w|^2t^2)^{-2}dt \\
&\le |f'(0)| + \sup\{(1-|z|^2)^2|f''(z)| : z \in \mathbb{D}\}|w| \int_0^1 (1-|w|^2t)^{-2}dt \\
&\le |f'(0)| + \sup\{(1-|z|^2)^2|f''(z)| : z \in \mathbb{D}\}(1-|w|^2)^{-1},
\end{aligned}$$

which implies that $\|f\|_{\mathcal{B}}$ is bounded by quantity (B).

To show that quantity (B) is bounded by a multiple of $\|f\|_{\mathcal{B}}$, fix $z \in \mathbb{D}$. Let $r = (1-|z|)/2$. Then the Cauchy integral formula for first derivatives, applied to f', shows that

$$f''(z) = (2\pi i)^{-1} \int_{|w-z|=r} f'(w)(w-z)^{-2}dw,$$

so

$$|f''(z)| \le r^{-1} \sup\{|f'(w)| : |w - z| = r\}$$
$$\le 2(1 - |z|)^{-1} \sup\{|f'(w)| : |w| = |z| + r\},$$

where the last inequality comes from the Maximum Modulus Theorem. Thus

$$(1 - |z|)^2 |f''(z)| \le 2(1 - |z|) \sup\{|f'(w)| : |w| = |z| + r\}$$

2.2
$$= 4[1 - (|z| + 1)/2] \sup\{|f'(w)| : |w| = (|z| + 1)/2\}$$
$$\le 4\|f\|_{\mathcal{B}},$$

so

$$(1 - |z|^2)^2 |f''(z)| \le 16\|f\|_{\mathcal{B}}.$$

The above inequality, combined with the obvious inequality $|f'(0)| \le \|f\|_{\mathcal{B}}$, shows that quantity (B) is bounded by a multiple of the Bloch norm. ■

Recall from Section 1 that P denotes the orthogonal projection of $L^2(\mathbb{D}, dA)$ onto L^2_a. An explicit formula for P was given by 1.9. The following lemma states that P maps $L^\infty(\mathbb{D}, dA)$ into the Bloch space. Later (2.7), we will improve this lemma by showing that P maps $L^\infty(\mathbb{D}, dA)$ onto the Bloch space.

Lemma 2.3. *P maps $L^\infty(\mathbb{D}, dA)$ boundedly into the Bloch space.*

Proof: Let $f \in L^\infty(\mathbb{D}, dA)$. The formula defining Pf shows that

$$|(Pf)(0)| \le \|f\|_\infty.$$

Let $w \in \mathbb{D}$. We can differentiate under the integral sign in the formula defining Pf, getting

$$(Pf)'(w) = 2\pi^{-1} \int_{\mathbb{D}} \overline{z} f(z)(1 - w\overline{z})^{-3} \, dA(z).$$

The above formula clearly implies that

$$|(Pf)'(0)| \le 2\|f\|_\infty.$$

Differentiating the formula for $(Pf)'$ in the above paragraph, we have

$$(Pf)''(w) = 6\pi^{-1} \int_{\mathbb{D}} (\overline{z})^2 f(z)(1 - w\overline{z})^{-4} \, dA(z),$$

so

$$\begin{aligned}|(Pf)''(w)| &\le 6\|f\|_\infty \pi^{-1} \int_{\mathbb{D}} |1 - w\bar{z}|^{-4} dA(z)\\ &= 6\|f\|_\infty (1 - |w|^2)^{-2} \qquad \text{[by 1.8]}.\end{aligned}$$

Using the equivalence of the Bloch norm and quantity (B) of 2.1, we now see that $Pf \in \mathcal{B}$ and that $|(Pf)(0)| + \|Pf\|_{\mathcal{B}}$ is bounded by a multiple of $\|f\|_\infty$, completing the proof. ■

We want to prove that the Bloch space is the dual of L^1_a with the pairing $< f, g >$ (for $f \in L^1_a$ and $g \in \mathcal{B}$) given by 1.6. However, we have a slight problem, because there exist $f \in L^1_a$ and $g \in \mathcal{B}$ such that $f\bar{g}$ is not in $L^1(\mathbb{D}, dA)$. So we once more extend the domain of $<,>$, and define $< f, g >$ by

$$< f, g > = \lim_{t \to 1} \int_{t\mathbb{D}} f\bar{g}\, dA,$$

so the domain of $<,>$ is now the set of all pairs of functions on $\mathbb{D}$ such that the above limits exists.

For f analytic on $\mathbb{D}$, let Vf denote the analytic function on $\mathbb{D}$ defined by

$$(Vf)(z) = [f(z) - f(0)]/z.$$

Lemma 2.4. *Let f and g be analytic on $\mathbb{D}$ with power series*

$$f(z) = \sum_{n=0}^{\infty} a_n z^n \text{ and } g(z) = \sum_{n=0}^{\infty} b_n z^n.$$

Let $0 < t < 1$. Then

$$\begin{aligned}\int_{t\mathbb{D}} f\bar{g}\, dA &= \int_{t\mathbb{D}} (Vf)(z)\overline{g'(z)}(t^2 - |z|^2)\, dA(z) + \pi t^2 f(0)\overline{g(0)}\\ &= (1/2) \int_{t\mathbb{D}} f''(z)\overline{[V(Vg)](z)}(t^2 - |z|^2)^2 dA(z) + \pi t^2 f(0)\overline{g(0)}\\ &\quad + (1/2)\pi t^4 f'(0)\overline{g'(0)}\\ &= \pi \sum_{n=0}^{\infty} (n+1)^{-1} a_n \overline{b_n} t^{2n+2}.\end{aligned}$$

Proof: Note that

$$(Vf)(z) = a_1 + a_2 z + a_3 z^2 + \ldots$$

and that

$$[V(Vg)](z) = b_2 + b_3 z + b_4 z^2 + \ldots.$$

Now evaluate each of the integrals in the statement of Lemma 2.4 by replacing f, g, Vf, $[V(Vg)], f''$, and g' by their power series, switching to polar co-ordinates, and computing. ■

Lemma 2.5. *Let $1 \leq p < \infty$. Then the polynomials are dense in L^p_a.*

Proof: For f a function on $\mathbb{D}$ and $0 < t < 1$, let f_t be the function on $\mathbb{D}$ defined by

$$f_t(z) = f(tz).$$

If f is uniformly continuous on $\mathbb{D}$ (or equivalently, if f is continuous on the closed unit disk), then clearly $\|f_t - f\|_p \to 0$ as $t \to 1$. Since the functions continuous on the closed unit disk are dense in $L^p(\mathbb{D}, dA)$, a standard $\epsilon/3$ argument shows that $\|f_t - f\|_p \to 0$ as $t \to 1$ for all $f \in L^p(\mathbb{D}, dA)$. Of course, if $f \in L^p_a$, then f_t is analytic on a disk strictly bigger than $\mathbb{D}$, so the partial sums of the Taylor series for f_t converge uniformly on $\mathbb{D}$ to f_t. Thus the polynomials are dense in L^p_a. ■

Now we are ready to prove that the dual of L^1_a can be identified with the Bloch space. Other proofs of this theorem can be found in [9] and [2]. Those proofs identify L^1_a with the set of analytic functions on $\mathbb{D}$ whose derivative is integrable, so the pairing of L^1_a with $\mathcal{B}$ is perhaps not as natural as the one given below.

Theorem 2.6. *The dual of L^1_a can be identified with the Bloch space. More precisely, if $g \in \mathcal{B}$, then $< f, g >$ is defined for every $f \in L^1_a$ and the mapping*

$$f \to < f, g >$$

is a bounded linear functional on L^1_a. Furthermore, every bounded linear functional on L^1_a is of the above form for some unique $g \in \mathcal{B}$, and the norm of the linear functional on L^1_a induced by $g \in \mathcal{B}$ is equivalent to $\|g\|_{\mathcal{B}} + |g(0)|$.

Proof: For $f \in L^1_a$, the analytic function Vf behaves no worse than f near $\partial\mathbb{D}$, so we can conclude that $Vf \in L^1_a$. Now the Closed Graph Theorem implies that there is a finite constant c such that

$$\|Vf\|_1 \leq c\|f\|_1 \qquad \text{for all } f \in L^1_a.$$

Thus 2.5, combined with the Lebesgue Dominated Convergence Theorem, shows that if $f \in L^1_a$ and $g \in \mathcal{B}$, then $< f, g >$ exists and is given by the formula

$$< f, g > = \int_{\mathbb{D}} (Vf)(z)\overline{g'(z)}(1 - |z|^2)dA(z) + \pi f(0)\overline{g(0)}.$$

The above equation and the previous inequality also show that

$$| < f, g > | \leq (c + \pi)\|f\|_1(\|g\|_{\mathcal{B}} + |g(0)|).$$

Thus the mapping $f \mapsto < f, g >$ is a bounded linear functional on L^1_a for every $g \in \mathcal{B}$, and the norm of the linear functional on L^1_a induced by g is bounded by a multiple of $\|g\|_{\mathcal{B}} + |g(0)|$.

Distinct functions in $\mathcal{B}$ induce distinct linear functionals on L^1_a (because if $< z^n, g_1 > = < z^n, g_2 >$ for all n, then the Taylor coefficients of g_1 equal the Taylor coefficients of g_2).

Now we will show that every bounded linear functional on L^1_a is induced by a function in $\mathcal{B}$. Let ψ be a bounded linear functional on L^1_a. By the Hahn-Banach Theorem and the Riesz Representation Theorem, there exists a function $h \in L^\infty(\mathbb{D}, dA)$ such that $\|h\|_\infty = \|\psi\|$ and

$$\psi(f) = \int_{\mathbb{D}} f\overline{h}dA \qquad \text{for every } f \in L^1_a.$$

Let $g = Ph$. Lemma 2.3 tells us that $g \in \mathcal{B}$ and that $\|g\|_{\mathcal{B}} + |g(0)|$ is bounded by a multiple of $\|\psi\|$. The first part of this proof tells us that g induces a bounded linear functional on L^1_a; to complete the proof we need only verify that this bounded linear functional equals ψ. To do this, we need only verify that $< f, g > = \psi(f)$ for all f in a dense subset of L^1_a. Lemma 2.4 says that the polynomials are dense in L^1_a, so let f be a polynomial. Then

$$\begin{aligned} < f, g > &= < f, Ph > = \pi^{-1} \int_{\mathbb{D}} f(z) \int_{\mathbb{D}} \overline{h}(w)(1 - \overline{z}w)^{-2} dA(w) dA(z) \\ &= \int_{\mathbb{D}} \overline{h}(w) \pi^{-1} \int_{\mathbb{D}} f(z)(1 - \overline{z}w)^{-2} dA(z) dA(w) \\ &= \int_{\mathbb{D}} \overline{h}(w) f(w) dA(w) \qquad \text{[by 1.7]} \\ &= \psi(f), \end{aligned}$$

where the use of Fubini's Theorem above is justified because f and h are bounded and $|1 - \overline{z}w|^{-2}$ is integrable on $\mathbb{D} \times \mathbb{D}$, which can be proved either by direct calculation or by using the comment after the proof of 1.10. The above equation now shows that the linear functional on L_a^1 induced by g is equal to ψ, and so the proof is complete. ∎

Now we will see that the image of $L^\infty(\mathbb{D}, dA)$ under P (the orthogonal projection of $L^2(\mathbb{D}, dA)$ onto L_a^2) equals the Bloch space. A completely different approach to this theorem can be found in [7], where the theorem first appeared.

Theorem 2.7. *P maps $L^\infty(\mathbb{D}, dA)$ boundedly onto the Bloch space $\mathcal{B}$.*

Proof: We have already proved that P is a bounded map of $L^\infty(\mathbb{D}, dA)$ into the Bloch space (see 2.3), so we need only prove that P maps $L^\infty(\mathbb{D}, dA)$ onto $\mathcal{B}$.

Let $g \in \mathcal{B}$. We need to find a function $h \in L^\infty(\mathbb{D}, dA)$ such that $g = Ph$. For $f \in L_a^1$, define $\psi(f)$ by

$$\psi(f) = < f, g > .$$

By 2.6, ψ is a bounded linear functional on L_a^1. The proof of 2.6 shows that there is a function $h \in L^\infty(\mathbb{D}, dA)$ such that

$$\psi(f) = < f, Ph >$$

for all $f \in L_a^1$. Since there is only one function in $\mathcal{B}$ that induces ψ, we must have $g = Ph$. ∎

The Little Bloch Space

The little Bloch space $\mathcal{B}_0$ is the set of functions f analytic on the unit disk such that

$$(1 - |z|^2)f'(z) \to 0 \text{ as } |z| \to 1.$$

We will see that the little Bloch space is the pre-dual of the Bergman space L_a^1. First we need some preliminary results.

The following lemma is the little o version of 2.1.

Lemma 2.8. *Let f be analytic on the unit disk. Then $f \in \mathcal{B}_0$ if and only if*

$$(1 - |z|^2)^2 f''(z) \to 0 \text{ as } |z| \to 1.$$

Proof: If $f \in \mathcal{B}_0$, then 2.2 shows that

$$(1 - |z|^2)^2 f''(z) \to 0 \text{ as } |z| \to 1.$$

To prove the implication in the other direction, suppose now that

$$(1-|z|^2)^2 f''(z) \to 0 \text{ as } |z| \to 1.$$

Let $\epsilon > 0$, and let $r \in (0,1)$ be such that

$$(1-|z|^2)^2 |f''(z)| < \epsilon \qquad \text{whenever } r < |z| < 1.$$

If $w \in \mathbb{D}$ and $r < |w| < 1$, then

$$f'(w) - f'(0) = (w/|w|) \int_0^{|w|} f''(tw/|w|)dt,$$

so

$$\begin{aligned} |f'(w)| &\leq |f'(0)| + \int_0^r |f''(tw/|w|)|dt + \int_r^{|w|} |f''(tw/|w|)|dt \\ &\leq |f'(0)| + \int_0^r |f''(tw/|w|)|dt + \epsilon \int_r^{|w|} (1-t^2)^{-2} dt \\ &\leq |f'(0)| + \int_0^r |f''(tw/|w|)|dt + 2\epsilon(1-|w|^2)^{-1}. \end{aligned}$$

Multiplying both sides of the above inequality by $(1-|w|^2)$ shows that $f \in \mathcal{B}_0$. ■

In Section 1 we defined a map Q (see 1.19) that turned out to be a bounded projection of $L^1(\mathbb{D}, dA)$ onto L^1_a (see 1.20). To prove that the dual of the little Bloch space is L^1_a, we will need to extend the domain of Q to the space $M(\overline{\mathbb{D}})$ of finite Borel measures on $\overline{\mathbb{D}}$. Here $\overline{\mathbb{D}}$ denotes the closed unit disk in the complex plane and measures are understood to be complex valued. If μ is a finite Borel measure on $\overline{\mathbb{D}}$, then the total variation measure of μ is denoted by $|\mu|$. The total variation norm of μ is denoted by $\|\mu\|$ and is defined to be $|\mu|(\overline{\mathbb{D}})$; the set $M(\overline{\mathbb{D}})$ becomes a Banach space under this norm.

We can identify a function $f \in L^1_a$ with the measure $f\, dA$, and using this identification, it is easy to see from 1.19 how to extend the domain of Q. Namely, for μ a finite Borel measure on $\overline{\mathbb{D}}$, define a function $Q\mu$ on $\mathbb{D}$ by

$$(Q\mu)(w) = 3\pi^{-1} \int_{\overline{\mathbb{D}}} (1-|z|^2)^2 (1-w\bar{z})^{-4} d\mu(z).$$

It is clear that $Q\mu$ is analytic on $\mathbb{D}$. The following proposition is an extension of 1.20; the proof is really the same.

Proposition 2.9. *Q maps $M(\overline{\mathbb{D}})$, the space of finite Borel measures on $\overline{\mathbb{D}}$, boundedly into L^1_a.*

Proof: Let μ be a finite Borel measure on $\overline{\mathbb{D}}$. If $w \in \mathbb{D}$, then

$$|(Q\mu)(w)| \leq 3\pi^{-1} \int_{\overline{\mathbb{D}}} (1-|z|^2)^2 |1 - w\overline{z}|^{-4} d|\mu|(z),$$

so

$$\begin{aligned}\int_{\mathbb{D}} |(Q\mu)(w)| dA(w) &\leq 3 \int_{\overline{\mathbb{D}}} (1-|z|^2)^2 \pi^{-1} \int_{\mathbb{D}} |1 - w\overline{z}|^{-4} dA(w) d|\mu|(z) \\ &= \int_{\overline{\mathbb{D}}} 3 \, d|\mu|(z) \qquad \text{[from 1.8]} \\ &= 3\|\mu\|,\end{aligned}$$

and we are done. ∎

Now we can prove that the dual of the little Bloch space can be identified with the Bergman space L^1_a. As with the identification of the dual of L^1_a, different proofs (which use a slightly different pairing) can be found in [2] and [9].

Theorem 2.10. *The dual of the little Bloch space can be identified with L^1_a. More precisely, every bounded linear functional on $\mathcal{B}_0$ is of the form*

$$f \mapsto < f, g >$$

for some unique $g \in L^1_a$, and the norm of the linear functional on $\mathcal{B}_0$ induced by $g \in L^1_a$ is equivalent to $\|g\|_1$.

Proof: From 2.6 we already know that $< f, g >$ is defined for every $f \in \mathcal{B}_0$ and every $g \in L^1_a$ and that there is a constant c independent of f and g such that

$$| < f, g > | \leq c(\|f\|_{\mathcal{B}} + |f(0)|)\|g\|_1.$$

Thus each $g \in L^1_a$ induces a bounded linear functional on $\mathcal{B}_0$ with a norm that is bounded by a multiple of $\|g\|_1$. Furthermore, distinct functions in L^1_a induce distinct bounded linear functions on $\mathcal{B}_0$ (because if $< z^n, g_1 > = < z^n, g_2 >$ for all n, then the Taylor coefficients of g_1 equal the Taylor coefficients of g_2).

So all that remains is to prove that every bounded linear functional on $\mathcal{B}_0$ is induced by some function in L^1_a. To do this, first note that if $f \in \mathcal{B}_0$, then by 2.8, the function

$(1-|z|^2)^2 f''(z)$ can be extended to be continuous on $\overline{\mathbb{D}}$. Thus we can define a map $U : \mathcal{B}_0 \to C(\overline{\mathbb{D}}) \oplus \mathbb{C} \oplus \mathbb{C}$ by

$$Uf = (1-|z|^2)^2 f''(z) \oplus f'(0) \oplus f(0).$$

Here $C(\overline{\mathbb{D}}) \oplus \mathbb{C} \oplus \mathbb{C}$ is given the usual norm:

$$\|h \oplus \alpha \oplus \mathcal{B}\| = \|\langle\|_\infty + |\alpha| + |\mathcal{B}|.$$

Lemma 2.1 says that $\|Uf\|$ is equivalent to $\|f\|_{\mathcal{B}} + |f(0)|$.

Now let ψ be a bounded linear functional on $\mathcal{B}_0$. By the above paragraph, we can think of ψ as a bounded linear functional on the range of U, so by the Hahn-Banach Theorem, there is a finite Borel measure μ on $\overline{\mathbb{D}}$ and complex constants λ_1 and λ_0 such that $\|\mu\|, |\lambda_1|, |\lambda_0|$ are all bounded by a multiple of $\|\psi\|$ and

$$\psi(f) = \int_{\overline{\mathbb{D}}} (1-|z|^2)^2 f''(z)\, d\overline{\mu}(z) + \overline{\lambda_1} f'(0) + \overline{\lambda_0} f(0) \quad \text{for all } f \in \mathcal{B}_0,$$

where $\overline{\mu}$ denotes the complex conjugate of the measure μ.

Define a function g on $\mathbb{D}$ by

$$g(z) = \pi^{-1}\lambda_0 + 2\pi^{-1}\lambda_1 z + 2z^2 (Q\mu)(z).$$

By 2.9 and the above paragraph, g is in L^1_a and $\|g\|_1$ is bounded by a multiple of $\|\psi\|$. We will complete the proof by showing that the bounded linear functional on $\mathcal{B}_0$ induced by g is ψ. Note that $[V(Vg)](z) = 2(Q\mu)(z)$, where V is the difference quotient defined before 2.4.

Let $f \in \mathcal{B}_0$. By 2.4, we have

$$\begin{aligned}
< f, g > &= \int_{\mathbb{D}} f''(z)\overline{(Q\mu)(z)}(1-|z|^2)^2\, dA(z) + \overline{\lambda_0} f(0) + \overline{\lambda_1} f'(0) \\
&= 3\pi^{-1} \int_{\overline{\mathbb{D}}} (1-|w|^2)^2 \int_{\mathbb{D}} f''(z)(1-|z|^2)^2 (1-w\overline{z})^{-4}\, dA(z)\, d\overline{\mu}(w) + \overline{\lambda_0} f(0) \\
&\quad + \overline{\lambda_1} f'(0) \\
&= \int_{\overline{\mathbb{D}}} (1-|w|^2)^2 f''(w)\, d\overline{\mu}(w) + \overline{\lambda_0} f(0) + \overline{\lambda_1} f'(0) \qquad [\text{by } 1.17] \\
&= \psi(f).
\end{aligned}$$

Thus the bounded linear functional on $\mathcal{B}_0$ induced by g equals ψ, and the proof is complete. ■

We saw earlier that P maps $L^\infty(\mathbb{D}, dA)$ onto the Bloch space. Now we will prove that P maps the uniformly continuous functions on $\mathbb{D}$ onto the little Bloch space. A different approach to this theorem can be found in [7], where the theorem first appeared.

Theorem 2.11. *P maps $C(\overline{\mathbb{D}})$ boundedly onto $\mathcal{B}_0$.*

Proof: First we prove that P maps $C(\overline{\mathbb{D}})$ boundedly onto $\mathcal{B}_0$. By the Stone-Weierstrass Theorem, linear combinations of

$$\{z^m\overline{z^n} : m \text{ and } n \text{ are nonnegative integers}\}$$

form a dense subset of $C(\overline{\mathbb{D}})$. By 2.4, P maps $C(\overline{\mathbb{D}})$ boundedly into $\mathcal{B}$. Since $\mathcal{B}_0$ is a closed subspace of $\mathcal{B}$, to prove that P maps $C(\overline{\mathbb{D}})$ into $\mathcal{B}_0$ we need only prove that $P(z^m\overline{z^n}) \in \mathcal{B}_0$ whenever m and n are nonnegative integers. To do this, let $w \in \mathbb{D}$. Then

$$\begin{aligned}[P(z^m\overline{z^n})](w) &= \pi^{-1}\int_{\mathbb{D}} z^m\overline{z}^n(1-w\overline{z})^{-2}dA(z)\\ &= \pi^{-1}\sum_{k=0}^{\infty}(k+1)w^k\int_{\mathbb{D}} z^m\overline{z^{n+k}}dA(z) \quad \text{[from 1.5]}.\end{aligned}$$

Evaluating the above integral by changing to polar co-ordinates, we see that

$$\begin{aligned}&P(z^m\overline{z^n})(w) = (m-n+1)(m+1)^{-1}w^{m-n} \qquad \text{if } m \geq n,\\ &P(z^m\overline{z^n})(w) = 0 \qquad \text{if } m \leq n.\end{aligned}$$

Thus $P(z^m\overline{z^n})$ is in $\mathcal{B}_0$, and so P maps $C(\overline{\mathbb{D}})$ boundedly into $\mathcal{B}_0$.

To show that P maps $C(\overline{\mathbb{D}})$ onto $\mathcal{B}_0$, we will need to extend the domain of our inner product once more. If μ is a finite Borel measure on $\overline{\mathbb{D}}$ and g is a continuous function on $\overline{\mathbb{D}}$, then we define

$$< g, \mu >= \int_{\overline{\mathbb{D}}} g \; d\overline{\mu},$$

where g is integrated against the complex conjugate of μ to maintain consistency with the conjugate linear nature of the inner product. This pairing identifies the dual of $C(\overline{\mathbb{D}})$ with $M(\overline{\mathbb{D}})$.

Since P is a bounded map if $C(\overline{\mathbb{D}})$ into $\mathcal{B}_0$, the adjoint of P, denoted P^*, maps the dual of $\mathcal{B}_0$ into the dual of $C(\overline{\mathbb{D}})$. Identifying the dual of $\mathcal{B}_0$ with L^1_a (by 2.10) and the dual of $C(\overline{\mathbb{D}})$ with $M(\overline{\mathbb{D}})$, P^* is defined to be the unique map from L^1_a to $\mathcal{B}_0$ such that

$$< g, P^*f >=< Pg, f >$$

for all $g \in C(\overline{\mathbb{D}})$ and all $f \in L^1_a$.

Now we will compute P^*. Let $g \in C(\overline{\mathbb{D}})$ and let f be a polynomial (so that Fubini's Theorem can be applied below). Then

$$\begin{aligned} < g, P^* f > =< Pg, f >&= \int_{\mathbb{D}} (Pg)(w)\overline{f}(w)dA(w) \\ &= \pi^{-1} \int_{\mathbb{D}} g(z) \left(\int_{\mathbb{D}} f(w)(1-\overline{w}z)^{-2} \right)^{-} dA(w)dA(z) \\ &= \int_{\mathbb{D}} g(z)\overline{f}(z)dA(z). \end{aligned}$$

The above equation shows that P^* is the map that takes a function $f \in L_a^1$ to the measure of $f\ dA$. In particular, this means that P^* is a one-to-one map with closed range, and this implies that P is onto (see for example [22], Theorem 4.5). ■

Stephens Semmes correctly suggested to me that it should be easy to prove the *onto* parts of 2.7 and 2.11 by an explicit formula. If $g \in \mathcal{B}$, define h by

$$h(z) = (1-|z|^2)z|z|^{-2}(g'(z) - g'(0)) + g'(0)z + g(0).$$

Then it is easy to verify that $h \in L^\infty(\mathbb{D}, dA)$ and that $Ph = g$. Similarly, if $g \in \mathcal{B}_0$, let

$$h(z) = (1-|z|^2)z|z|^{-2}(g'(z) - g''(0)z - g'(0)) + g''(0)z^2/2 + g'(0)z + g(0).$$

Then again, it is not hard to verify that $h \in C(\overline{\mathbb{D}})$ and $Ph = g$. I have retained the proofs presented earlier because they may be useful when attempting to replace the disk by more general domains.

3. Carleson Measures on L_a^p

Let $1 \le p < \infty$. For precisely which positive Borel measures μ on $\mathbb{D}$ is L_a^p contained in $L^p(\mathbb{D}, d\mu)$? This section is devoted to answering this question (see 3.6) and exploring some consequences of the answer. To obtain the desired results, we will need to know about the pseudo-hyperbolic metric.

The Pseudo-Hyperbolic Metric

For $w \in \mathbb{D}$, let φ_w be the function defined by

$$\varphi_w(z) = (w-z)/(1-\overline{w}z).$$

The function φ_w is a one-to-one analytic map of $\mathbb{D}$ onto $\mathbb{D}$. For w and z points in $\mathbb{D}$, the pseudo-hyperbolic distance $d(w, z)$ between w and z is defined by

$$d(w, z) = |\varphi_w(z)|.$$

It is not hard to prove that d is actually a metric on $\mathbb{D}$; see, for example, [11], page 4.

For $w \in \mathbb{D}$ and $0 < r < 1$, the pseudo-hyperbolic disk $D(w, r)$ with (pseudo-hyperbolic) center w and (pseudo-hyperbolic) radius r is defined by

$$D(w, r) = \{z \in \mathbb{D} : d(w, z) < r\}.$$

Since φ_w is a fractional linear transformation, the pseudo-hyperbolic disk $D(w, r)$ is also a Euclidean disk. Except for the special case $D(0, r) - r\mathbb{D}$, the Euclidean center and Euclidean radius of $D(w, r)$ do not coincide with the pseudo-hyperbolic center and pseudo-hyperbolic radius.

For $w \in \mathbb{D}$, it is easy to verify that φ_w is its own inverse under composition:

$$(\varphi_w \cdot \varphi)w)(z) = z \qquad \text{for all } z \in \mathbb{D}.$$

Another simple calculation shows that φ_w preserves pseudo-hyperbolic distances:

$$d(\lambda, z) = d(\varphi_w(\lambda),_{\varphi_w}(z)) \qquad \text{for all } \lambda, z \in \mathbb{D}.$$

Thus φ_w maps a pseudo-hyperbolic disk centered at point λ to a pseudo-hyperbolic disk centered at $\varphi_w(\lambda)$ (and the pseudo-hyperbolic radius of the image disk equals the pseudo-hyperbolic radius of the domain disk):

$$\varphi_w(D(\lambda, r)) = D(\varphi_w(\lambda), r) \qquad \text{for all } \lambda \in \mathbb{D} \text{ and all } r \in (0, 1).$$

For $\lambda \in \mathbb{D}$, the substitution $z = \varphi_\lambda(w)$ produces the usual Jacobian change in measure given by $dA(z) = |\varphi'_\lambda(w)|^2 dA(w)$, and of course the domain of integration must also be changed appropriately. Combining all this with the specific function obtained from differentiating φ_λ shows that for each integrable function h we have the formula

3.1 $$\int_{D(\lambda,r)} h(z)dA(z) = (1 - |\lambda|^2)^2 \int_{D(0,r)} (h \circ \varphi_\lambda)(w)|1 - \overline{\lambda}w|^{-4} dA(w).$$

The notation $|K|$ will denote the area of a set $K \subset \mathbb{C}$. We begin by finding the area of each pseudo-hyperbolic disk.

Lemma 3.2. *Let $w \in \mathbb{D}$ and let $0 < r < 1$. Then*

$$|D(w,r)| = \pi r^2(1-|w|^2)^2(1-r^2|w|^2)^{-2}.$$

Proof:

$$\begin{aligned}
|D(w,r)| &= \int_{D(w,r)} 1\, dA \\
&= (1-|w|^2)^2 \int_{D(0,r)} |1-\overline{w}z|^{-4} dA(z) \qquad \text{[from 3.1]} \\
&= (1-|w|^2)^2 r^2 \int_{\mathbb{D}} |1-w\overline{r}\lambda|^{-4} dA(\lambda) \\
&= \pi r^2(1-|w|^2)^2(1-r^2|w|^2)^{-2},
\end{aligned}$$

where the last integral is evaluated by using 1.8. ■

The above lemma implies that if $w \in \mathbb{D}$ and $r \in (0,1)$, then the Euclidean radius of $D(w,r)$ is equal to $r(1-|w|^2)(1-r^2|w|^2)^{-1}$. For fixed r, this shows that the Euclidean radius of $D(w,r)$ is comparable to the distance from w to $\partial\mathbb{D}$.

We will need three more lemmas to characterize the measures μ such that L_a^p is contained in $L^p(\mathbb{D}, d\mu)$. The following lemma describes the size of the reproducing kernels on pseudo-hyperbolic disks.

Lemma 3.3. *Let $w \in \mathbb{D}$ and let $0 < r < 1$. Then*

$$\inf\{|k_w(z)| : z \in D(w,r)\} = \pi^{-1}(1-r|w|)^2(1-|w|^2)^{-2}$$

and

$$\sup\{|k_w(z)| : z \in D(w,r)\} = \pi^{-1}(1+r|w|)^2(1-|w|^2)^{-2}.$$

Proof: This becomes easy when we recall that $D(w,r) = \varphi_w(D(0,r))$. We have

$$\begin{aligned}
\inf\{|k_w(z)| : z \in D(w,r)\} &= \inf\{|k_w(\varphi_w(\lambda))| : \lambda \in D(0,r)\} \\
&= \pi^{-1}(1-|w|^2)^{-2} \inf\{|1-\overline{w}\lambda|^2 : |\lambda| < r\} \\
&= \pi^{-1}(1-|w|^2)^{-2}(1-r|w|)^2.
\end{aligned}$$

The supremum is computed in the same fashion. ■

Note that the constant c in the following lemma depends only on r and t, and not upon w, z, f, or p.

Lemma 3.4. *Let $0 < r < t < 1$. Then there is a finite constant c such that*

$$|f(z)|^p \leq c|D(w,t)|^{-1} \int_{D(w,t)} |f|^p dA$$

for all $w \in \mathbb{D}$, all $z \in D(w,r)$, all f analytic on $\mathbb{D}$, and all $p \in [1,\infty)$.

Proof: Let $w \in \mathbb{D}$, let $z \in D(w,r)$, let f be analytic on $\mathbb{D}$, and let $p \in [1,\infty)$. Let E denote the Euclidean disk centered at $\varphi_w(z)$ with radius $t - r$. Since $|\varphi_w(z)| < r$, the disk E is contained in $D(0,t)$. The average of any analytic function on a disk is equal to the value at the center of the disk; applying this fact to the analytic function $f \circ \varphi_w$ and the disk E gives

$$(f \circ \varphi_w)(\varphi_w(z)) = \pi^{-1}(t-r)^{-2} \int_E (f \circ \varphi_w)(\lambda) dA(\lambda).$$

The left-hand side of the above equation equals $f(z)$; applying Hölder's inequality to the right-hand side, and then replacing E by the larger set $D(0,t)$ shows that

$$\begin{aligned}
|f(z)|^p &\leq \pi^{-1}(t-r)^{-2} \int_{D(0,t)} |f \circ \varphi_w(\lambda)|^p dA(\lambda) \\
&\leq 16\pi^{-1}(t-r)^{-2} \int_{D(0,t)} |f \circ \varphi_w(\lambda)|^p |1 - \overline{w}\lambda|^{-4} dA(\lambda) \\
&= 16\pi^{-1}(t-r)^{-2}(1-|w|^2)^{-2} \int_{D(w,t)} |f|^p dA \qquad \text{[by 3.1]} \\
&\leq 16(t-r)^{-2}(1-t^2)^{-2}|D(w,t)|^{-1} \int_{D(w,t)} |f|^p dA \qquad \text{[by 3.2]},
\end{aligned}$$

and so $c = 16(t-r)^{-2}(1-t^2)^{-2}$ works. ■

The following covering lemma says that for fixed $r \in (0,1)$, we can cover the unit disk with pseudo-hyperbolic disks of radius r that do not intersect too often, even when the pseudo-hyperbolic radius is increased.

Lemma 3.5. *Let $0 < r < 1$. Then there is a sequence $w_1, w_2, \ldots$ in $\mathbb{D}$ and a positive integer M such that*

$$\cup_{n=1}^{\infty} D(w_n, r) = \mathbb{D}$$

and each $z \in \mathbb{D}$ is in at most M of the pseudo-hyperbolic disks

$$D(w_1,(r+1)/2), D(w_2,(r+1)/2), D(w_3,(r+1)/2), \ldots\ .$$

Proof: Let $B_1, B_2, \ldots$ be a sequence of pseudo-hyperbolic disks in $\mathbb{D}$ of radius $r/3$ such that

$$\cup_{n=1}^{\infty} B_n = \mathbb{D}.$$

Choose a subsequence $D_1, D_2, \ldots$ of $B_1, B_2, \ldots$ as follows. Start by letting $D_1 = B_1$. Now let D_2 be the first element in the sequence $B_1, B_2, \ldots$ that is disjoint from D_1. Then let D_3 be the first element in the sequence $B_1, B_2, \ldots$ that is disjoint from $D_1 \cup D_2$. Continue in this fashion.

Let w_n denote the pseudo-hyperbolic center of D_n. Using the triangle inequality, it is easy to see that

$$\cup_{n=1}^{\infty} D(w_n, r) = \mathbb{D}.$$

Fix $z \in \mathbb{D}$ and let W be the set defined by

$$W = \{\varphi_z(w_j) : j \text{ is an integer and } z \in D(w_j, (r+1)/2)\}.$$

To finish the proof, it suffices to show that the cardinality of W is bounded by a constant independent of z. To do this, note that each element of W has absolute value less than $(r+1)/2$. Furthermore, if $\varphi_z(w_j)$ and $\varphi_z(w_k)$ are distinct elements of W, then

$$\begin{aligned} |\phi_z(w_j) - \phi_z(w_k)| &\geq [1-(r+1)^2/4]d(\phi_z(w_j), \phi_z(w_k)) \\ &= [1-(r+1)^2/4]d(w_j, w_k) \\ &\geq [1-(r+1)^2/4]r/3. \end{aligned}$$

Thus W is a set of complex numbers of absolute value less than $(r+1)/2$ such that the Euclidean distance between any two distinct elements of w is at least $[1-(r+1)^2/4]r/3$. This clearly means that there is an integer M, depending only on r, such that the cardinality of W cannot exceed M. ■

Carleson Measures on Bergman Spaces

Let $1 \leq p < \infty$ and let μ be a positive Borel measure on $\mathbb{D}$. The Closed Graph Theorem shows that L^p_a is contained in $L^p(\mathbb{D}, d\mu)$ if and only if the inclusion map from L^p_a to $L^p(\mathbb{D}, d\mu)$ is a bounded linear operator. The following theorem gives a necessary and sufficient condition on μ for this to happen.

A very similar result, due to Hastings [13], uses curvilinear Carleson squares rather than pseudo-hyperbolic disks. I believe that most proofs are easier using pseudo-hyperbolic disks instead of Carleson squares. Furthermore, in practice it seems that condition (B) of the following theorem is easier to verify than the corresponding (equivalent) condition involving Carleson squares.

A different proof of the following theorem, in a more general context, can be found in Luecking's paper [16].

Note that condition (B) in the following theorem does not involve p. Thus, if μ is a positive Borel measure on $\mathbb{D}$, the following theorem implies that L^p_a is contained in $L^p(\mathbb{D}, d\mu)$ for some $p \in [1, \infty)$ if and only if L^p_a is contained in $L^p(\mathbb{D}, d\mu)$ for every $p \in [1, \infty)$.

Note also that condition (A) of the following theorem does not involve r. Thus, if μ is a positive Borel measure on $\mathbb{D}$, then quantity (B) below is finite for some $r \in (0, 1)$ if and only if quantity (B) is finite for every $r \in (0, 1)$.

Theorem 3.6. *Let $0 < r < 1$. Let $1 \leq p < \infty$ and let μ be a positive Borel measure on $\mathbb{D}$. Then the following two quantities are equivalent:*

(A) $\sup\{\int_{\mathbb{D}} |f|^p d\mu / \int_{\mathbb{D}} |f|^p dA : f \in L^p_a, f \not\equiv 0\}$;

(B) $\sup\{\mu(D(w,r))/|D(w,r)| : w \in \mathbb{D}\}$.

Furthermore, the constants of equivalency depend only upon r, and not upon p or μ.

Proof: First we will show that quantity (B) is bounded by a multiple of quantity (A).

Fix $w \in \mathbb{D}$. If we could let f equal the characteristic function of $D(w,r)$, then we would immediately get that quantity (B) is bounded by quantity (A). Unfortunately, the characteristic function of $D(w,r)$ is not analytic, so we use the next best function, which is an appropriate power of the reproducing kernel. Thus let

$$f(z) = \pi^{-2/p}(1 - \overline{w}z)^{-4/p}.$$

From our formula for the $L^2(\mathbb{D}, dA)$ norm of the reproducing kernel (see 1.8), we know that

$$\int_{\mathbb{D}} |f|^p dA = \pi^{-1}(1 - |w|^2)^{-2}. \tag{3.7}$$

Now we need to estimate the integral of $|f|^p$ with respect to the measure μ. We have

$$\begin{aligned} \int_{\mathbb{D}} |f|^p d\mu \geq \int_{D(w,r)} |f|^p d\mu &\geq \inf\{|k_w|^2 : z \in D(w,r)\}\mu(D(w,r)) \\ &\geq \pi^{-2}(1 - r|w|)^4(1 - |w|^2)^{-4}\mu(D(w,r)) \qquad \text{[by 3.3]}. \end{aligned}$$

Combining the above inequality with 3.7 gives

$$\begin{aligned}\int_{\mathbb{D}} |f|^p d\mu / \int_{\mathbb{D}} |f|^p dA &\geq \pi^{-1}(1-r)^4(1-|w|^2)^{-2}\mu(D(w,r)) \\ &= (1-r)^4 r^2(1-r^2|w|^2)^{-2}\mu(D(w,r)/|D(w,r)| \qquad \text{[by 3.2]} \\ &\geq r^2(1-r)^4\mu(D(w,r))/|D(w,r)|.\end{aligned}$$

The above inequality shows that quantity (B) is bounded by $r^{-2}(1-r)^{-4}$ times quantity (A).

To prove that quantity (A) is bounded by a multiple of quantity (B), let $w_1, w_2, \dots$ and M be as in 3.5. Let $f \in L^p_a$, $f \not\equiv 0$. Then

$$\begin{aligned}\int_{\mathbb{D}} |f|^p d\mu &\leq \sum_{n=1}^{\infty} \int_{D(w_n,r)} |f|^p d\mu \\ &\leq \sum_{n=1}^{\infty} \sup\{|f(z)|^p : z \in D(w_n,r)\}\mu(D(w_n,r)) \\ &\leq c\sum_{n=0}^{\infty} \mu(D(w_n,r))|D(w_n,(r+1)/2)|^{-1} \int_{D(w_n,(r+1)/2)} |f|^p dA,\end{aligned}$$

where the constant c comes from 3.4 with $t = (r+1)/2$.

Since $D(w_n,r)$ is a subset of $D(w_n,(r+1)/2)$, the above inequality becomes

$$\begin{aligned}\int_{\mathbb{D}} |f|^p d\mu &\leq c\sum_{n=1}^{\infty} \mu(D(w_n,r)|D(w,r)|^{-1} \int_{D(w_n,(r+1)/2} |f|^p dA \\ &\leq c\big[\sup\{\mu(D(w,r))/|D(w,r)| : w \in \mathbb{D}\big] \sum_{n=1}^{\infty} \int_{D(w_n,(r+1)/2)} |f|^p dA \\ &\leq cM\big[\sup\{\mu(D(w,r))/|D(w,r)| : w \in \mathbb{D}\}\big] \int_{\mathbb{D}} |f|^p dA,\end{aligned}$$

and so quantity (A) is bounded by a multiple of quantity (B). ■

Atomic Decomposition of Bergman Spaces

The following proposition is an important consequence of the characterization of Carleson type measures given above. It states that if a sequence in $\mathbb{D}$ is separated in the pseudo-hyperbolic metric, then the values of L^p_a functions on the sequence cannot grow too fast. This proposition is a key ingredient of the atomic decomposition, which will be discussed later in this section.

Proposition 3.8. *Let $\lambda_1, \lambda_2, \ldots$ be a sequence in $\mathbb{D}$ such that*

$$\inf\{d(\lambda_m, \lambda_n) : m \neq n\} > 0.$$

Then there is a finite constant c such that

$$\sum_{n=1}^{\infty} |f(\lambda_n)|^p (1 - |\lambda_n|^2)^2 \leq c\|f\|_p^p$$

for all $p \in [1, \infty)$ and all $f \in L_a^p$.

Proof: Let δ_n denote the point mass at λ_n, and let μ be the measure defined by

$$\mu = \sum_{n=1}^{\infty} (1 - |\lambda_n|^2)^2 \delta_n,$$

so

$$\int_{\mathbb{D}} |f|^p \, d\mu = \sum_{n=1}^{\infty} |f(\lambda_n)|^p (1 - |\lambda_n|^2)^2.$$

Thus, by 3.6, the proof will be finished if we can find $r \in (0, 1)$ such that

$$\sup\{\mu(D(w, r))/|D(w, r)| : w \in \mathbb{D}\} < \infty.$$

Let

$$r = \inf\{d(\lambda_m, \lambda_n) : n \neq m\}/2.$$

To verify that this value of r satisfies the above condition, fix $w \in \mathbb{D}$. If $\lambda_m \in D(w, r)$ and $\lambda_n \in D(w, r)$, then

$$\begin{aligned} d(\lambda_m, \lambda_n) &\leq d(\lambda_m, w) + d(w, \lambda_n) \\ &< 2r. \end{aligned}$$

The above inequality, combined with the definition of r, implies that $m = n$. Thus, $D(w, r)$ contains at most one point in the sequence $\lambda_1, \lambda_2, \ldots$.

If $D(w, r)$ contains no points in the sequence $\lambda_1, \lambda_2, \ldots$, then clearly $\mu(D(w, r)) = 0$. By the above paragraph, the only other possibility is that $D(w, r)$ contains only one point, say λ_n, from the sequence $\lambda_1, \lambda_2, \ldots$. In this case,

$$\mu(D(w, r)) = (1 - |\lambda|_n^2)^2.$$

Since $\lambda_n \in D(w,r) = \phi_w(D(0,r))$, there is a point $z \in D(0,r)$ such that $\lambda_n = \phi_w(z)$. The formula

3.9 $$1 - |\phi_w(z)|^2 = (1-|w|^2)(1-|z|^2)|1-\overline{w}z|^{-2}$$

is easily verified by replacing $|\phi_w(z)|^2$ by $\phi_w(z)\overline{\phi_w(z)}$ and using simple algebra. Combining the above two equations gives

$$\begin{aligned}\mu(D(w,r)) &= (1-|w|^2)^2(1-|z|^2)^2|1-\overline{w}z|^{-4} \\ &\leq (1-|w|^2)^2(1-r)^{-4}.\end{aligned}$$

Now combine the above inequality with the formula for $|D(w,r)|$ (see 3.2) to get

$$\mu(D(w,r))/|D(w,r)| \leq \pi^{-1}r^{-2}(1-r)^{-4}.$$

Thus

$$\sup\{\mu(D(w,r))/|D(w,r)| : w \in \mathbb{D}\} < \infty,$$

and we are done. ■

The atomic decomposition says, roughly, that every element of L^p_a is an appropriate sum of normalized reproducing kernels. So we will need to know, at least asymptotically, the $L^p(\mathbb{D}, dA)$ norms of the reproducing kernels. The following lemma provides this information.

Lemma 3.10. *Let $1 < p < \infty$. Then the following quantities are equivalent (for $w \in \mathbb{D}$):*

(A) $\|k_w\|_p$;

(B) $(1-|w|^2)^{-2/p'}$.

Proof: Recall the following formulas:

$$\begin{aligned}k_w(z) &= \pi^{-1}(1-\overline{w}z)^{-2}; \\ \varphi_w(z) &= (w-z)(1-\overline{w}z)^{-1}; \\ \varphi'_w(z) &= (|w|^2-1)(1-\overline{w}z)^{-2}.\end{aligned}$$

Thus for $w \in \mathbb{D}$, we have

$$\begin{aligned}\|k_w\|_p^p &= \pi^{-p}\int_{\mathbb{D}} |1-\overline{w}z|^{-2p}\,dA(z) \\ &= \pi^{-p}(1-|w|^2)^{-p}\int_{\mathbb{D}} |\phi'_w(z)|^{p-2}|\phi'_w(z)|^2\,dA(z) \\ &= \pi^{-p}(1-|w|^2)^{-p}\int_{\mathbb{D}} |\phi'_w(\phi_w(z))|^{p-2}\,dA(z),\end{aligned}$$

where the last equality uses the change of variables formula and the formula $\phi_w(\phi_w(z)) = z$. But $\phi_w'(\phi_w(z)) = (1 - \overline{w}z)^2(|w|^2 - 1)^{-1}$, so the above equality becomes

$$\begin{aligned}\|k_w\|_p^p &= \pi^{-p}(1 - |w|^2)^{2-2p} \int_{\mathbb{D}} |1 - \overline{w}z|^{2p-4} dA(z) \\ &= \pi^{-p}(1 - |w|^2)^{-2p/p'} \int_{\mathbb{D}} |1 - \overline{w}z|^{2p-4} dA(z).\end{aligned}$$

The above equality shows that the proof will be complete if we prove that

$$3.11 \qquad \int_{\mathbb{D}} |1 - \overline{w}z|^{2p-4} dA(z)$$

is bounded and bounded away from 0 (independently of $w \in \mathbb{D}$).

To prove that the above integral is bounded away from 0 is easy: The integeral is made smaller if we integrate over $(1/2)\mathbb{D}$ rather than $\mathbb{D}$. But on $(1/2)\mathbb{D}$, the integral is always between $(1/2)^{2p-4}$ and $(3/2)^{2p-4}$.

To show that 3.11 is bounded, we consider separately the cases $2 \le p < \infty$ and $1 < p < 2$. If $2 \le p < \infty$, then clearly 3.11 is less than $\pi 2^{2p-4}$.

Now suppose that $1 < p < 2$ (note that $2p - 4$ is negative, so the idea used in the previous paragraph does not work). Fix $w \in \mathbb{D}$ with $|w| > 1/2$ (clearly 3.11 is bounded for $|w| \le 1/2$). In 3.11, make a change *a* variables $\lambda = \overline{w}z$, so $dA(z) = |w|^{-2} dA(\lambda) \le 4dA(\lambda)$. The new domain of integration should be $|w|\mathbb{D}$, but we can make the integral bigger by integrating over $\mathbb{D}$. So

$$\int_{\mathbb{D}} |1 - \overline{w}z|^{2p-4} dA(z) \le 4 \int_{\mathbb{D}} |1 - \lambda|^{2p-4} dA(\lambda).$$

The right-hand side of the above inequality does not depend upon w, so the proof is finished by verifying that

$$\int_{\mathbb{D}} |1 - \lambda|^{2p-4} dA(\lambda) < \infty.$$

This is easily done by changing to polar coordinates based at 1 rather than the origin.
■

The following theorem is called an atomic decomposition because every function in a Bergman space is written as a sum of very simple elements (normalized reproducing kernels), which we think of as atoms of the space. The theorem as stated here is a special case of results of Luecking [17], whose proof we outline. Another approach to atomic decomposition is given in Rochberg's paper [21], which also gives applications of the atomic decomposition.

Note that given any positive number ε, it is always possible to find a sequence $\{\lambda_n\}$ in $\mathbb{D}$ satisfying the condition of the theorem; see the first paragraph of the proof of 3.5.

Theorem 3.12. *Let $1 < p < \infty$. Then there exists a positive number ε with the property that whenever $\lambda_1, \lambda_2, \ldots$ is a sequence in $\mathbb{D}$ such that $\inf\{d(\lambda_m, \lambda_n) : m \neq n\} > \varepsilon/3$ and*

$$\cup_{n=1}^{\infty} D(\lambda_n, \varepsilon) = \mathbb{D},$$

then every $f \in L_a^p$ can be written in the form

$$f = \sum_{n=1}^{\infty} a_n k_{\lambda_n} / \|k_{\lambda_n}\|_p,$$

where $\{a_n\} \in \ell^p$ and the above infinite sum converges in the norm on L_a^p. Furthermore, the $\{a_n\}$ can be chosen so that $\|f\|_p$ is equivalent to $(\sum_{n=1}^{\infty} |a_n|^p)^{1/p}$.

Proof: We will just indicate the framework of the proof.

Let $\lambda_1, \lambda_2, \ldots$ be a sequence in $\mathbb{D}$ such that

$$\inf\{d(\lambda_m, \lambda_n) : m \neq n\} > 0.$$

Define a map R from $L_a^{p'}$ to $\ell^{p'}$ by

$$(Rf)_n = f(\lambda_n)/\|k_{\lambda_n}\|_p.$$

By 3.8 and 3.10, R is a bounded map from $L_a^{p'}$ into $\ell^{p'}$.

The dual of $\ell^{p'}$ can be identified with ℓ^p, and by 1.16 the dual of $L_a^{p'}$ can be identified with L_a^p. Thus the adjoint of R, denoted R^*, is the bounded map from ℓ^p into L_a^p defined by

$$< f, R^* x > = < Rf, x > \qquad \text{for } f \in L_a^{p'} \text{ and } x \in \ell^p.$$

Here the inner product on the left is the pairing between $L_a^{p'}$ and L_a^p given in Section 1: multiply the first function by the complex conjugate of the second function and integrate. The inner product on the right of the above equation represents the conjugate linear pairing between $\ell^{p'}$ and ℓ^p: multiply each coordinate of the first vector by the complex conjugate of the corresponding coordinate of the second vector and sum.

To compute R^*, let e_n denote the vector that equals 1 in the nth coordinate and equals 0 in all the other coordinates. Let $f \in L_a^{p'}$. Then

$$\begin{aligned} < f, R^* e_n > &= < Rf, e_n > = (Rf)_n = f(\lambda_n)/\|k_{\lambda_n}\|_p \\ &= < f, k_{\lambda_n} > /\|k_{\lambda_n}\|_p \\ &= < f, k_{\lambda_n}/\|k_{\lambda_n}\|_p > . \end{aligned}$$

The above equation shows that

$$R^* e_n = k_{\lambda_n} / \|k_{\lambda_n}\|_p.$$

Thus

$$R^*(a_1, a_2, \ldots) = \sum_{n=1}^{\infty} a_n k_{\lambda_n} / \|k_{\lambda_n}\|_p$$

for all $(a_1, a_2, \ldots) \in \ell^p$, where the infinite sum converges in the L^p_a norm because R^* is continuous and $(a_1, a_2, \ldots, a_N, 0, 0, \ldots)$ converges to $(a_1, a_2, \ldots)$ in ℓ^p as $N \to \infty$.

Thus to complete the proof of the theorem, we need only show that R^* is onto (the statement that we can choose $\{a_n\}$ so that $\|f\|_p$ is equivalent to the ℓ^p norm of $\{a_n\}$ then follows from the Open Mapping Theorem). However, for any bounded linear operator R, the adjoint R^* is onto if and only if R is bounded below (see [22], Theorem 4.14). We will not repeat the details here, but simply state that in [17], Luecking shows that for ε sufficiently small, the condition

$$\cup_{n=1}^{\infty} D(\lambda_n, \varepsilon) = \mathbb{D}$$

implies that R is bounded below. ■

4. Toeplitz Operators on L^2_a

Recall that P denotes the orthogonal projection of $L^2(\mathbb{D}, dA)$ onto L^2_a. For $f \in L^\infty(\mathbb{D}, dA)$, the Toeplitz operator with symbol f is the operator T_f from L^2_a to L^2_a defined by

$$T_f(g) = P(fg).$$

It is clear that T_f is a bounded operator and that

$$\|T_f\| \leq \|f\|_\infty.$$

This section focuses on compact Toeplitz operators and on algebras generated by Toeplitz operators.

Compact Toeplitz Operators

The only compact Hardy space Toeplitz operator is the Hardy space Toeplitz operator with symbol 0. The following two lemmas show that Bergman space Toeplitz operators behave differently than their Hardy space counterparts.

Lemma 4.1. *Let K be a compact subset of $\mathbb{D}$, and let $f \in L^\infty(\mathbb{D}, dA)$ be such that f is identically 0 on $\mathbb{D} \sim K$. Then T_f is a compact operator.*

Proof: Let $\{g_n\}$ be a sequence that is norm bounded in L^2_a. Then $\{g_n\}$ is uniformly bounded on each compact subset of $\mathbb{D}$ (by 1.7), and so $\{g_n\}$ is a normal family. Thus there exists a function g analytic on $\mathbb{D}$ such that some subsequence $\{g_{n_i}\}$ of $\{g_n\}$ converges uniformly on K to g. Thus $\{gf_{n_i}\}$ converges in $L^2(\mathbb{D}, dA)$ norm to gf, and so $\{T_f(g_{n_i})\}$ converges in L^2_a norm to $P(gf)$.

We have shown that if $\{g_n\}$ is a norm bounded sequence in L^2_a, then some subsequence of $\{T_f g_n\}$ converges in L^2_a. This means that T_f is compact. ■

Lemma 4.2. *Let $f \in C(\overline{\mathbb{D}})$ be such that f is identically zero on $\partial\mathbb{D}$. Then T_f is a compact operator.*

Proof: The function f can be uniformly approximated on $\overline{\mathbb{D}}$ by functions that are identically zero in a neighborhood of $\partial\mathbb{D}$. Now use 4.1. ■

Lemmas 4.1 and 4.2 suggest that we ask about a possible converse: If $f \in L^\infty(\mathbb{D}, dA)$ and T_f is a compact operator, must $f(w) \to 0$ as w tends to $\partial\mathbb{D}$ (at least in some almost everywhere sense)? Unfortunately this question has a negative answer, as we will soon see. First we need the following lemma, which states that the normalized reproducing kernels tend weakly (in L^2_a) to 0 as we approach the boundary of $\mathbb{D}$.

Lemma 4.3. *$k_w/\|k_w\|_2$ tends weakly to 0 in L^2_a as $|w| \to 1$.*

Proof: If $f \in L^2_a$, then by the reproducing property of k_w and the formula for $\|k_w\|_2$ (see 1.8), we have

$$< f, k_w/\|k_w\|_2 >= \pi^{1/2}(1 - |w|^2)f(w).$$

The above formula shows that if f is a bounded function in L^2_a, then $< f, k_w/\|k_w\|_2 >\to 0$ as $|w| \to 1$. Since the polynomials are dense in L^2_a, this implies that

$$< f, k_w/\|k_w\|_2 >\to 0 \text{ as } |w| \to 1$$

for all $f \in L^2_a$, which means that $k_w/\|k_w\|_2$ tends weakly to 0 in L^2_a as $|w| \to 1$. ■

For nonnegative functions $f \in L^\infty(\mathbb{D}, dA)$, the following proposition can be generalized to give necessary and sufficient conditions for T_f to belong to a Schatten p-ideal; see [18].

Proposition 4.4. *Let f be a nonnegative function in $L^\infty(\mathbb{D}, dA)$. Then the following are equivalent statements:*

(A) T_f is a compact operator.

(B) There exists a number $r \in (0,1)$ such that

$$|D(w,r)|^{-1} \int_{D(w,r)} f \, dA \to 0 \text{ as } |w| \to 1.$$

(C) For every $r \in (0,1)$, we have

$$|D(w,r)|^{-1} \int_{D(w,r)} f \, dA \to 0 \text{ as } |w| \to 1.$$

Proof: First we will prove that (B) implies (A). So suppose that (B) is true. For $s \in (0,1)$, let T_s denote the Toeplitz operator whose symbol is the function that equals f on $s\mathbb{D}$ and 0 on $\mathbb{D} \sim s\mathbb{D}$. By 4.1, T_s is a compact operator for every $s \in (0,1)$. Now

$$\begin{aligned} \|T_f - T_s\|^2 &\le \sup\{\int_{\mathbb{D}\sim s\mathbb{D}} f^2|g|^2 \, dA : g \in L^2_a \text{ and } \|g\|_2 = 1\} \\ &\le c \sup\{|D(w,r)|^{-1} \int_{(\mathbb{D}\sim s\mathbb{D})\cap D(w,r)} f^2 dA : w \in \mathbb{D}\} \\ &\le c\|f\|_\infty \sup\{|D(w,r)|^{-1} \int_{(\mathbb{D}\sim s\mathbb{D})\cap D(w,r)} f dA : w \in \mathbb{D}\} \end{aligned}$$

where the middle inequality and the constant c come from applying the Carleson measure condition for Bergman spaces (see 3.6) to the measure (characteristic function of $\mathbb{D} \sim s\mathbb{D}) \cdot f dA$.

The above inequality, combined with the hypothesis on r from (B), shows that $\|T_f - T_s\|$ tends to 0 as s increases to 1. Since each T_s is compact, this implies that T_f is compact, so (A) holds.

Now we will prove that (A) implies (C). So suppose that T_f is a compact operator. Let $f \in (1,0)$. Then

$$|D(w,r)|^{-1} \int_{D(w,r)} f \, dA \le c \int_{D(w,r)} f(z)|k_w(z)|^2/\|k_w\|_2^2 dA,$$

where the constant c depends only on r (not on w), and the inequality comes from estimating $|D(w,r)|$ from 3.2, estimating $|k_w(z)|$ from 3.3, and estimating $\|k_w\|_2$ from

1.8. Integrating over the unit disk in place of $D(w,r)$, the above inequality becomes

$$\begin{aligned}|D(w,r)|^{-1}\int_{D(w,r)} f\,dA &\le c\int_{\mathbb{D}} f(z)|k_w(z)|^2/\|k_w\|_2^2\,dA\\ &= c < T_f k_w(z)/\|k_w\|_2, k_w(z)/\|k_w\|_2 >\\ &\le c\|T_f k_w\|_2/\|k_w\|_2.\end{aligned}$$

By the compactness of T_f and 4.3, the last quantity above tends to 0 as $|w| \to 1$. Thus (C) holds.

It is obvious that (C) implies (B). ■

To use the above proposition to construct a compact Toeplitz operator whose symbol does not have 0 as a limit on $\partial\mathbb{D}$, fix $r \in (0,1)$ (say $r = 1/2$) and let $t_1, t_2, \ldots,$ be a sequence of numbers strictly increasing to 1, with $t_1 = 0$. Define $f \in L^\infty(\mathbb{D}, dA)$ by letting

$$f(z) = 0 \text{ if } t_{2n-1} \le |z| < t_{2n} \text{ for some integer } n, \text{ and}$$

$$f(z) = 1 \text{ if } t_{2n} \le |z| < t_{2n+1} \text{ for some integer } n.$$

If t_{2n} is close to t_{2n+1} (compared to the distance from t_{2n-1} to t_{2n}), then it is easy to verify that condition (B) of 4.4 is satisfied, so that T_f is a compact operator.

So unlike Hardy space Toeplitz operators, on the Bergman space there are compact Toeplitz operators with symbol in $L^\infty(\mathbb{D}, dA)$ that are not 0 on $\partial\mathbb{D}$. However, the analogy with Hardy spaces can be recovered by restricting our symbols to a smaller algebra than $L^\infty(\mathbb{D}, dA)$.

Let H^∞ denote the space of all bounded analytic functions on the open unit disk $\mathbb{D}$. Recall that $L^\infty(\partial\mathbb{D}, d\theta)$ is the closed algebra generated by $H^\infty(\partial\mathbb{D})$ and the complex conjugates of all the functions in $H^\infty(\partial\mathbb{D})$; here $H^\infty(\partial\mathbb{D})$ denotes the set of functions on $\partial\mathbb{D}$ obtained by taking radial limits of functions in H^∞.

Motivated by the above analogy with $L^\infty(\partial\mathbb{D}, d\theta)$, let $\mathcal{U}$ denote the norm closed subalgebra of $L^\infty(\mathbb{D}, dA)$ generated by H^∞ and the complex conjugates of all the functions in H^∞. We will see that Toeplitz operators with symbols in $\mathcal{U}$ are considerably more manageable than arbitrary Toeplitz operators.

Since $L^\infty(\partial\mathbb{D}, dA)$ can be identified, via the Poisson integral, with the set of bounded harmonic functions on $\mathbb{D}$, the following proposition strengthens the analogy between $L^\infty(\partial\mathbb{D}, dA)$ (for Hardy space Toeplitz operators) and $\mathcal{U}$ (for Bergman space Toeplitz operators).

Proposition 4.5. *$\mathcal{U}$ is the norm closed subalgebra of $L^\infty(\mathbb{D}, dA)$ generated by the bounded harmonic functions.*

Proof: Since the real and imaginary parts of every analytic function are harmonic functions, it is clear that $\mathcal{U}$ is contained in the norm closed subalgebra of $L^\infty(\mathbb{D}, dA)$ generated by the bounded harmonic functions.

To prove the inclusion in the other direction, let u be a real-valued bounded harmonic function on $\mathbb{D}$. Then $e^{u+i\tilde{u}} \in H^\infty$, so e^{2u}, which equals $e^{u+i\tilde{u}}$ times $e^{u-i\tilde{u}}$, is in $\mathcal{U}$. Now the closure of $\{e^{2u(z)} : z \in \mathbb{D}\}$ is a closed subinterval of $\mathbb{R}$ not containing 0, so the logarithm can be uniformly approximated on this set by polynomials. Thus $2u \in \mathcal{U}$, which of course implies that $u \in \mathcal{U}$. ■

We will need three lemmas to classify the compact Toeplitz operators whose symbol is in $\mathcal{U}$. The first lemma states that every bounded analytic function on $\mathbb{D}$ is uniformly continuous when we put the pseudo-hyperbolic metric on $\mathbb{D}$ (but retain the usual Euclidean metric on the range).

Lemma 4.6. *Let $f \in H^\infty$. Then*

$$|f(w) - f(z)| \leq 2\|f\|_\infty d(w, z)$$

for all w and z in $\mathbb{D}$.

Proof: First suppose that f is analytic in a neighborhood of $\overline{\mathbb{D}}$ and that $\|f\|_\infty = 1$. Fix $w \in \mathbb{D}$, and consider the function whose value at z equals

$$\left([f(w) - f(z)][1 - \overline{f(w)}f(z)]^{-1}\right) / \left([w - z][1 - \overline{w}z]^{-1}\right).$$

This function is analytic in a neighborhood of $\overline{\mathbb{D}}$ and has modulus at most 1 on $\partial\mathbb{D}$. By the maximum modulus theorem, the function is less than or equal to 1 on $\mathbb{D}$. Thus we have shown that

$$|f(w) - f(z)||1 - \overline{f(w)}f(z)|^{-1} \leq d(w, z)$$

for all w in z in $\mathbb{D}$. The above inequality implies that

$$|f(w) - f(z)| \leq 2d(w, z).$$

The assumption that f is analytic on a neighborhood of $\overline{\mathbb{D}}$ can be eliminated by replacing f by the dilate f_t and taking the limit as t increases to 1.

Now if f is an arbitrary function in H^∞, apply the above inequality to $f/\|f\|_\infty$ to obtain the desired result. ■

The following lemma states that for each $w \in \mathbb{D}$ and each $r \in (0,1)$, the fraction of the mass (in the L^2_a sense) of the reproducing kernel k_w that lies in the pseudo-hyperbolic disk centered at w with pseudo-hyperbolic radius r is precisely r^2.

Lemma 4.7. *Let $w \in \mathbb{D}$ and let $r \in (0,1)$. Then*

$$\int_{D(w,r)} |k_w|^2 dA = r^2 \|k_w\|_2^2.$$

Proof: Simple calculus shows that $|k_w(z)| = \pi^{-1}(1-|w|^2)^{-1}|\phi_w'(z)|$. Thus

$$\begin{aligned}\int_{D(w,r)} |k_w|^2 dA &= \pi^{-2}(1-|w|^2)^{-2}\int_{D(w,r)} |\phi_w'(z)|^2 dA(z)\\ &= \pi^{-2}(1-|w|^2)^{-2}|\phi_w(D(w,r))|\\ &= \pi^{-2}(1-|w|^2)^{-2}|D(0,r)| = \pi^{-1}r^2(1-|w|^2)^{-2}\\ &= r^2\|k\|_2^2,\end{aligned}$$

where the last equality comes from 3.3. ■

To prove the next lemma we will need to know some simple facts about the algebraic properties of map $f \to T_f$. It is easy to verify that

$$T_f^* = T_{\overline{f}} \qquad \text{for all } f \in L^\infty(\mathbb{D}, dA),$$

and

$$T_{f+g} = T_f + T_g \qquad \text{for all } f, g \in L^\infty(\mathbb{D}, dA).$$

It is not true that T_{fg} is equal to T_fT_g for all $f, g \in L^\infty(\mathbb{D}, dA)$. However, if $g \in H^\infty$, then $T_g h = gh$ for all $h \in L^2_a$; from this it easily follows that

$$T_{gf} = T_fT_g \qquad \text{for all } f \in L^\infty(\mathbb{D}, dA) \text{ and all } g \in H^\infty.$$

Taking adjoints of both sides of the above equation and replacing f with $\overline{f}$ shows that

$$T_{\overline{g}f} = T_{\overline{g}}T_f \qquad \text{for all } f \in L^\infty(\mathbb{D}, dA) \text{ and all } g \in H^\infty.$$

The following lemma tells us that there is some algebraic structure to the set of functions f in $\mathcal{U}$ such that T_f is compact.

Lemma 4.8. *The set*

$$\{f \in \mathcal{U} : T_f \text{ is a compact operator}\}$$

is a closed ideal of $\mathcal{U}$.

Proof: Let $J = \{f \in \mathcal{U} : T_f$ is a compact operator$\}$. It is clear that J is a closed subspace of $\mathcal{U}$.

Suppose $f \in J$, so T_f is a compact operator. Let $u \in \mathcal{U}$ be a function of the form

4.9 $$u = \overline{g_1}h_1 + \ldots + \overline{g_n}h_n,$$

where $g_1, \ldots, g_n$ and $h_1, \ldots, h_n$ are all in H^∞. Then

$$T_{fu} = T^*_{g_1} T_f T_{h_1} + \ldots + T^*_{g_n} T_f T_{h_n},$$

and so T_{fu} is a compact operator and $fu \in J$. Functions of the form 4.9 are dense in $\mathcal{U}$, so J is an ideal of $\mathcal{U}$. ∎

The following theorem completely classifies the compact Toeplitz operators with symbol in the closed algebra generated by the bounded harmonic functions. This theorem (along with 4.8) was first proved by McDonald and Sundberg as Proposition 5 of [19]. The proof given here partly follows the ideas of McDonald and Sundberg, but by using 4.7 we can avoid having to use Hastings's result about Carleson measures on Bergman spaces [13].

Theorem 4.10. *Let* $f \in \mathcal{U}$. *Then* T_f *is a compact operator if and only if* $f \in C(\overline{\mathbb{D}})$ *and* f *is identically* 0 *on* $\partial\mathbb{D}$.

Proof: We have already proved that if $f \in C(\overline{\mathbb{D}})$ and f is identically 0 on $\partial\mathbb{D}$, then T_f is a compact operator (see 4.2). To prove the implication in the other direction, suppose that $f \in \mathcal{U}$ and that T_f is a compact operator. We will assume that f cannot be continuously extended to be identically 0 on $\partial\mathbb{D}$, and complete the proof by finding a contradiction.

Our assumption about f means that there is a positive number δ and a sequence $w_1, w_2, \ldots$ in $\mathbb{D}$ such that $|w_n| \to 1$ as $n \to \infty$ and

$$|f(w_n)|^2 \geq \delta$$

for all n. The uniform continuity of f in the pseudo-hyperbolic metric (see 4.6) implies that there is a sufficiently small positive number r such that

4.11 $$|f(z)|^2 \geq \delta/2 \qquad \text{for all } n \text{ and all } z \in D(w_n, r).$$

Since $|f|^2 = f\overline{f}$, Lemma 4.8 tells us that $T_{|f|^2}$ is also a compact operator. Now

$$\begin{aligned} < T_{|f|^2} k_{w_n}/\|k_{w_n}\|_2, k_{w_n}/\ \|k_{w_n}\|_2 > &= \int_{\mathbb{D}} |f|^2 |k_{w_n}|^2 dA/\|k_{w_n}\|_2^2 \\ &\geq \int_{D(w_n,r)} |f|^2 |k_{w_n}|^2 dA/\|k_{w_n}\|_2^2 \\ &\geq (\delta/2) \int_{D(w_n,r)} |k_{w_n}|^2 dA/\|k_{w_n}\|_2^2 \quad \text{[by 4.11]} \\ &= r^2 \delta/2 \quad \text{[by 4.7]}. \end{aligned}$$

However, $k_{w_n}/\|k_{w_n}\|_2$ tends weakly to 0 as $n \to \infty$ (by 4.3), and since $T_{|f|^2}$ is a compact operator, this implies that

$$< T_{|f|^2} k_{w_n}/\|k_{w_n}\|_2, k_{w_n}/\|k_{w_n}\|_2 >$$

should tend to 0 as $n \to \infty$. Thus the above inequality shows that we have reached a contradiction. ■

An open question is to find a necessary and sufficient condition on $f \in L^\infty(\mathbb{D}, dA)$ for T_f to be a compact operator. The proof of 4.10 does not work if $\mathcal{U}$ is replaced by $L^\infty(\mathbb{D}, dA)$ because the set

$$\{f \in L^\infty(\mathbb{D}, dA) : T_f \text{ is a compact operator}\}$$

is not an ideal of $L^\infty(\mathbb{D}, dA)$ (compare to 4.8). In fact, Donald Sarason has constructed a function $f \in L^\infty(\mathbb{D}, dA)$ such that T_f is a compact operator but f^2 is identically 1; thus the above set is not even closed under multiplication.

We have seen that $\mathcal{U}$ in some sense plays the role that $L^\infty(\partial\mathbb{D}, d\theta)$ plays for Hardy spaces. This leads to the following conjecture: P maps $\mathcal{U}$ onto the Bloch space. As motivation for this conjecture, recall that P maps $L^\infty(\mathbb{D}, dA)$ onto the Bloch space (see 2.7), and that P maps the space of bounded harmonic functions on $\mathbb{D}$ (which can be identified with the algebra $L^\infty(\partial\mathbb{D}, d\theta)$) onto BMOA.

Toeplitz Algebras

For B a subset of $L^\infty(\mathbb{D}, dA)$, let $\mathcal{T}(B)$ denote the norm closed algebra of operators generated by $\{T_f : f \in B\}$. The commutator ideal of $\mathcal{T}(B)$, denoted $\mathcal{C}(B)$, is the smallest norm closed two-sided ideal of $\mathcal{T}(B)$ containing

$$\{SR - RS : R, S \in \mathcal{T}(B)\}.$$

It is easy to see that $\mathcal{C}(B)$ equals the smallest norm closed two-sided ideal of $\mathcal{T}(B)$ containing

$$\{T_f T_g - T_g T_f : f, g \in B\}.$$

For B a subset of $L^\infty(\mathbb{D}, dA)$, the quotient $\mathcal{T}(B)/\mathcal{C}(B)$ is a commutative Banach algebra, and we can hope to learn something about the Toeplitz algebra $\mathcal{T}(B)$ by applying commutative Banach algebra techniques to $\mathcal{T}(B)/\mathcal{C}(B)$. For example, we can try to identify the maximal ideal space of $\mathcal{T}(B)/\mathcal{C}(B)$ with some familiar space, and then try to find the Gelfand transform.

If B is a subset of $L^\infty(\mathbb{D}, dA)$ that is closed under complex conjugation, then clearly $\mathcal{T}(B)$ is a C^*-algebra, so that $\mathcal{T}(B)/\mathcal{C}(B)$ is a commutataive C^*-algebra. Thus in this case $\mathcal{T}(B)/\mathcal{C}(B)$ will be isometrically isomorphic to the space of continuous complex valued functions on the maximal ideal space of $\mathcal{T}(B)/\mathcal{C}(B)$.

The following theorem, due to Coburn [6], gives a complete description of the algebra $\mathcal{T}(C(\overline{\mathbb{D}}))$.

Theorem 4.12. *$\mathcal{C}(C(\overline{\mathbb{D}}))$ is equal to the set of compact operators on L^2_a. $\mathcal{T}(C(\overline{\mathbb{D}}))$ is equal to the set*

$$\{T_f + K : f \in C(\overline{\mathbb{D}}) \text{ and } K \text{ is a compact operator on } L^2_a\}.$$

The map that takes $T_f + \mathcal{C}(C(\overline{\mathbb{D}}))$ to $f|\partial\mathbb{D}$ is well defined and extends to an isometric isomorphism between $\mathcal{T}(C(\overline{\mathbb{D}}))/\mathcal{C}(C(\overline{\mathbb{D}}))$ and $C(\partial\mathbb{D})$. The maximal ideal space of $\mathcal{T}(C(\overline{\mathbb{D}}))/\mathcal{C}(C(\overline{\mathbb{D}}))$ can be identified with $\partial\mathbb{D}$.

There is a remarkable description of $\mathcal{T}(\mathcal{U})/\mathcal{C}(\mathcal{U})$ due to McDonald and Sundberg [19]. They proved that $\mathcal{T}(\mathcal{U})/\mathcal{C}(\mathcal{U})$ can be identified with $C(M_1)$, where M_1 denotes the set of one-point parts of the maximal ideal space of H^∞. Carl Sundberg (pre-print) has recently found some major simplifications of the hardest part of the original proof. Another part of the original proof can be simplified by using pseudo-hyperbolic disks rather than Carleson squares, and using 4.7 instead of some more difficult estimates used in the original proof.

5. Hankel Operators on L^2_a

Closely related to Toeplitz operators are the Hankel operators. Recall that P denotes the orthogonal projection of $L^2(\mathbb{D}, dA)$ onto L^2_a, so that $1 - P$ is the orthogonal projection of $L^2(\mathbb{D}, dA)$ onto $(L^2_a)^\perp$. For $f \in L^\infty(\mathbb{D}, dA)$, the Hankel operator with symbol f is the operator $H_f : L^2_a \to (L^2_a)^\perp$ defined by

$$H_f(g) = (1 - P)(fg).$$

It is clear that if $f \in L^\infty(\mathbb{D}, dA)$, then H_f is a bounded operator and $\|H_f\| \le \|f\|_\infty$. This section will focus on the characterization of compact Hankel operators with conjugate analytic symbol and the alternate descriptions of the Bloch space that are needed for this characterization.

Compact Hankel Operators

The connection between Hankel and Toeplitz operators is provided by the following formula, which can be verified by an easy calculation:

5.1
$$T_{fg} - T_f T_g = H^*_{\overline{f}} H_g \qquad \text{for all } f, g \in L^\infty(\mathbb{D}, dA).$$

An important special case of the above formula is obtained by letting one of the functions be analytic and letting the other be its complex conjugate:

5.2
$$T^*_f T_f - T_f T^*_f = H^*_{\overline{f}} H_{\overline{f}} \qquad \text{for all } f \in H^\infty.$$

If S is a bounded operator on a Hilbert space, then $S^*S - SS^*$ is called the self-commutator of S. Of course, if the self-commutator of S equals 0, then S is called a normal operator. If the self-commutator of S is a compact operator, then S is called an essentially normal operator. It is important to know which operators are essentially normal, because the theory developed by Brown, Douglas, and Fillmore [5] can be applied to such operators.

For $f \in H^\infty$, the operator T_f is simply the operator of multiplication by f on L^2_a. These multiplication operators furnish basic examples of subnormal operators, so it is desirable to know as much as possible about them.

For $f \in H^\infty$, equation 5.2 tells us that the Toeplitz operator T_f is essentially normal if and only if the Hankel operator $H_{\overline{f}}$ is compact. This raises the following question: For which $f \in H^\infty$ is $H_{\overline{f}}$ compact?

To answer this question, it is useful to consider Hankel operators whose symbol may be unbounded. For $f \in L^2(\mathbb{D}, dA)$, we slightly modify the definition of H_f given above by restricting the domain of H_f to H^∞. So now H_f maps H^∞ into $(L_a^2)^\perp$ by the formula $H_f(g) = (1 - P)(fg)$. The set H^∞ is dense in L_a^2. Thus if H_f is a bounded operator (when we put the L_a^2 norm on H^∞), then H_f extends to a bounded operator from L_a^2 to $(L_a^2)^\perp$, also denoted by H_f. Of course, if $f \in L^\infty(\mathbb{D}, dA)$, then our new definition of H_f agrees with the earlier definition. However, it is possible that even for an unbounded function f, the operator H_f can be bounded.

Now before asking for which $f \in H^\infty$ is the Hankel operator $H_{\overline{f}}$ a compact operator, it is more natural to ask first for which $f \in L_a^2$ is $H_{\overline{f}}$ a bounded operator. The answer, given by the following theorem, turns out to be the familiar Bloch space, which we discussed in Section 2 when we saw that the Bloch space $\mathcal{B}$ can be identified with the dual of the Bergman space L_a^1. The norm $\|H_{\overline{f}}\|$ below refers to the norm of the operator $H_{\overline{f}}$ when both the domain and the range are given the $L^2(\mathbb{D}, dA)$ norm.

Theorem 5.3. *For functions $f \in L_a^2$, the Bloch norm $\|f\|_{\mathcal{B}}$ and the operator norm $\|H_{\overline{f}}\|$ are equivalent. In particular, $H_{\overline{f}}$ is bounded if and only if f is in the Bloch space.*

Knowing now that for $f \in L_a^2$ the Hankel operator $H_{\overline{f}}$ is bounded if and only if f is in the Bloch space, it should not be hard to guess that $H_{\overline{f}}$ is compact if and only if f is in the little Bloch space $\mathcal{B}_0$. The following theorem says that this is indeed correct. Recall from Section 2 that the little Bloch space can be identified with the pre-dual of L_a^1.

Theorem 5.4. *Let $f \in L_a^2$. Then $H_{\overline{f}}$ is compact if and only if f is in the little Bloch space $\mathcal{B}_0$.*

The above two theorems were recently proved by me in [3], so the details of the proofs will not be given here.

Recently J. Arazy, S. Fisher, and J. Peetre found a necessary and sufficient condition for a Hankel operator with a conjugate analytic symbol to be in a Schatten p-class. Their result states that if $f \in L_a^2$ and $1 < p < \infty$, then $H_{\overline{f}}$ is in the Schatten p-class if and only if

$$\int_{\mathbb{D}} |f'(z)|^p (1 - |z|^2)^{p-2} dA(z) < \infty.$$

This is the same condition that is necessary and sufficient for a Hardy space Hankel operator with conjugate analytic symbol to be in the Schatten p-class (see [20]). For

$p = 1$, Arazy, Fisher, and Peetre proved that if $f \in L^2_a$ and $H_{\overline{f}}$ is a trace class operator, then f is constant on $\mathbb{D}$. In contrast, there are many Hardy space Hankel operators with nonconstant conjugate analytic symbol that are in the trace class.

Analogies Between the Bloch Space and BMOA

Recall that a function $f \in L^1(\partial\mathbb{D}, d\theta/2\pi)$ is said to be in the space BMO (bounded mean oscillation) if the supremum, taken over all intervals, of the average amount by which f differs from its average is finite. If in addition f is of analytic type (meaning that the harmonic extension of f to $\mathbb{D}$, via the Poisson kernel, is analytic), then f is said to be in BMOA. The space VMO (vanishing mean oscillation) is the set of functions f in BMO such that the average amount by which f differs from its average on intervals tends to 0 as the length of the intervals tends to 0. The space VMOA is the set of functions in VMO that are of analytic type.

The dual of the Hardy space H^1 can be identified with BMOA, and the pre-dual of H^1 can be identified with VMOA. Furthermore, a Hardy space Hankel operator with symbol $\overline{f}$, where $f \in H^2$, is bounded if and only if $f \in$ BMOA (and is compact if and only if $f \in$ VMOA).

Thus 5.3 and 5.4, combined with the duality results in Section 2, show that the Bloch space and the little Bloch space play the same role for Bergman spaces as BMOA and VMOA play for the Hardy spaces.

The proofs for the Hardy space Hankel operator results use the fact that the orthogonal complement of H^2 in $L^2(\partial\mathbb{D}, d\theta/2\pi)$ differs by only one dimension from the complex conjugate of H^2. In the Bergman space setting, the orthogonal complement of L^2_a in $L^2(\mathbb{D}, dA)$ is much bigger than the complex conjugate of any set of analytic functions, and this is one of the major stumbling blocks that prevents the Hardy space proofs from working for the Bergman space.

The proofs of 5.3 and 5.4 use alternate descriptions of the Bloch space and the little Bloch space, as given by the following theorem and 5.6.

For f analytic on $\mathbb{D}$ and $\lambda \in \mathbb{D}$ and $r \in (0,1)$, let $f_{D(\lambda,r)}$ denote the average of f on the pseudo-hyperbolic disk $D(\lambda, r)$. Thus

$$f_{D(\lambda,r)} = |D(\lambda,r)|^{-1} \int_{D(\lambda,r)} f \, dA.$$

Theorem 5.5. *Let $1 \le p < \infty$ and let $0 < r < 1$. Then the following quantities are equivalent (for functions f analytic on the unit disk $\mathbb{D}$):*

(A) $\|f\|_{\mathcal{B}}$;
(B) $\sup\{[|D(\lambda,r)|^{-1}\int_{D(\lambda,r)}|f(z)-f_{D(\lambda,r)}|^p dA(z)]^{1/p}:\lambda\in\mathbb{D}\}$;
(C) $\sup\{[|D(\lambda,r)|^{-1}\int_{D(\lambda,r)}|f(z)-f(\lambda)|^p dA(z)]^{1/p}:\lambda\in\mathbb{D}\}$;
(D) $\sup\{\|f\circ\phi_\lambda-f(\lambda)\|_p:\lambda\in\mathbb{D}\}$;
(E) $\sup\{\text{distance }[\overline{f}|D(\lambda,r),H^\infty(D(\lambda,r))]:\lambda\in\mathbb{D}\}$;
(F) $\sup\{[\text{area } f(D(\lambda,r))]^{1/2}:\lambda\in\mathbb{D}\}$;
(G) $\sup\{[\int_{D(\lambda,r)}|f'(z)|^2 dA(z)]^{1/2}:\lambda\in\mathbb{D}\}$.

A proof of the above theorem is given in [3]. Although I have been unable to find a proof elsewhere in print, the equivalence of quantities (A) through (D) probably has the status of a folk theorem. The equivalence of quantities (F) and (G) with the Bloch norm is due to Yamashita [26].

Some of the quantities in 5.5 depend upon p or r or both, but since quantity (A) does not depend upon either p or r, the theorem implies that changing the values of p and r produces equivalent quantities. Since quantity (A) of 5.5 is just the Bloch norm, a function f analytic on $\mathbb{D}$ is in the Bloch space if and only if one (or equivalently, all) of the quantities (B) through (G) (for either some fixed values of p and r or all values of p and r) is finite.

The equivalence of quantities (A) and (B) shows that the Bloch space $\mathcal{B}$ is analogous to BMOA, where pseudo-hyperbolic disks of a fixed pseudo-hyperbolic radius have replaced intervals and area measure has replaced arc length measure. We will see that other parts of 5.5 deepen this analogy.

Since $D(\lambda,r)$ is a Euclidean disk as well as a pseudo-hyperbolic disk, $f_{D(\lambda,r)}$, which is used in computing quantity (B), is just the value of the analytic function f at the Euclidean center of $D(\lambda,r)$. Quantity (C) computes the average amount by which f differs from its value at the pseudo-hyperbolic center of the disk in question. The equivalence of quantities (B) and (C) shows that we can use either the pseudo-hyperbolic center or the Euclidean center in computing this BMO-type quantity.

I learned that the Bloch space can be regarded as an area version of BMOA from [7], Section 5, where it is shown that the Bloch norm is equivalent to a modified version of quantity (B), using the sets

$$\{B\cap\mathbb{D}: B \text{ is a Euclidean disk with center in } \mathbb{D}\}$$

rather than the pseudo-hyperbolic disks of a fixed pseudo-hyperbolic radius.

Quantity (D) is analogous to the Möbius invariant version of the usual BMO norm on the circle $\partial\mathbb{D}$ (see [11], Chapter VI, Corollary 1.4). The only part of 5.5 that is used in proving 5.3 and 5.4 is the equivalence of quantities (A) and (D).

The distance in quantity (E) is measured in the sup-norm. Perhaps the equivalence of quantity (E) with the Bloch norm is less surprising when we consider the analogy between $\mathcal{B}$ and BMOA, and recall that for a function $f \in H^\infty$, the BMO norm of (the boundary value function of) f is equivalent to distance $[\overline{f}, H^\infty]$. Thus one way to interpret the equivalence of quantities (A) and (E) is that the Bloch space $\mathcal{B}$ is simply the set of analytic functions on $\mathbb{D}$ whose restrictions to pseudo-hyperbolic disks (of fixed pseudo-hyperbolic radius) are uniformly in BMO.

As mentioned earlier, the equivalence of quantities (F) and (G) with the Bloch norm is due to Yamashita [26]. A different proof of the equivalence of quantities (F) and (G) can be found in [3].

In quantity (G), the integral $\int_{D(\lambda,r)} |f'(z)|^2 dA(z)$, familiar from the theory of Dirichlet spaces, is just the area of $f(D(\lambda, r))$ counting multiplicities (so, for example, if a region is covered twice, then its area is counted twice), whereas quantity (F) uses the area of the set $f(D(\lambda, r))$, disregarding multiplicities. The equivalence of quantities (A) and (G) says that the Bloch space $\mathcal{B}$ is the set of analytic functions on $\mathbb{D}$ whose restrictions to pseudo-hyperbolic disks (of fixed pseudo-hyperbolic radius) are uniformly in the Dirichlet space. As mentioned above, the equivalence of quantities (A) and (E) has a similar interpretation, with BMO replacing the Dirichlet space. The equivalence of quantities (E) and (G) was surprising to me, because the BMO norm and the Dirichlet norm are not equivalent.

The following theorem gives a description of the little Bloch space that corresponds to the description of the Bloch space given by 5.5. See [3] for the proof.

Theorem 5.6. *Let $1 \le p < \infty$ and let $0 < r < 1$. Then the following are equivalent (for functions f analytic on the unit disk $\mathbb{D}$):*

(a) $f \in \mathcal{B}_0$;
(b) $|D(\lambda, r)|^{-1} \int_{D(\lambda,r)} |f(z) - f_{D(\lambda,r)}|^p dA(z) \to 0$ *as* $|\lambda| \to 1$;
(c) $|D(\lambda, r)|^{-1} \int_{D(\lambda,r)} |f(z) - f(\lambda)|^p dA(z) \to 0$ *as* $|\lambda| \to 1$;
(d) $\|f \circ \phi_\lambda - f(\lambda)\|_p \to 0$ *as* $|\lambda| \to 1$;
(e) distance $[\overline{f}|D(\lambda, r), H^\infty(D(\lambda, r))] \to 0$ *as* $|\lambda| \to 1$;
(f) area $f(D(\lambda, r)) \to 0$ *as* $|\lambda| \to 1$;

(g) $\int_{D(\lambda,r)} |f'(z)|^2 dA(z) \to 0$ as $|\lambda| \to 1$.

The space $\mathcal{B}_0 \cap H^\infty$ (which, by 5.4, consists precisely of the bounded analytic functions whose complex conjugates give compact Hankel operators) has an interesting alternate description. It turns out that a function $f \in H^\infty$ is in $\mathcal{B}_0 \cap H^\infty$ if and only if f is constant on each Gleason part (except for the usual disk $\mathbb{D}$) of the maximal ideal space of H^∞; see [11], Chapter X, Exercise 11(b). For $f \in H^\infty$, the Hardy space Hankel operator with symbol $\overline{f}$ is compact if and only if f is constant on the support of each measure on the maximal ideal space of $L^\infty(\mathbb{D}, dA)$ that represents an element of the maximal ideal space of H^∞ (except, again, the point evaluations at points of the usual disk $\mathbb{D}$). Thus we see that when we move from the Hardy space context to the Bergman space context, support sets get replaced by Gleason parts. In my opinion, this analogy can be further exploited.

The following lemma is one of the main tools used in the characterization of bounded (and compact) Hankel operators with conjugate analytic symbols. The proof of this curious estimate uses the equivalence of (A) and (D) of 5.5 for $p = 6$. Again, see [3] for a proof.

Lemma 5.7. *There is a finite constant c such that*

$$\int_{\mathbb{D}} |f(w) - f(z)||1 - \overline{w}z|^{-2}(1 - |w|^2)^{-1/2} dA(w) \leq c(1 - |z|^2)^{-1/2} \|f\|_{\mathcal{B}}$$

for all $f \in \mathcal{B}$ and all $z \in \mathbb{D}$.

Questions

We now discuss some open questions and directions for further research.

The proofs of 5.3 and 5.4 show that for $f \in L^2_a$, the Hankel operator $H_{\overline{f}}$ is bounded if and only if $\sup\{\|H_{\overline{f}} k_\lambda\|_2 / \|k_\lambda\|_2 : \lambda \in \mathbb{D}\} < \infty$, and that $H_{\overline{f}}$ is compact if and only if $\|H_{\overline{f}} k_\lambda\|_2 / \|k_\lambda\|_2 \to 0$ as $|\lambda| \to 1$. Thus the boundedness and compactness of $H_{\overline{f}}$ is determined only by the action on the normalized reproducing kernels. This same behavior can be shown to hold also for Hankel operators on the Hardy space $H^2(\partial\mathbb{D})$. Is there a theorem that would show that the reproducing kernels alone determine boundedness and compactness of Hankel operators on a large class of function spaces?

The dual of the Hardy space $H^1(\partial\mathbb{D})$ can be identified with BMOA (and the pre-dual of $H^1(\partial\mathbb{D})$ can be identified with VMOA). Recall from Section 2 that the dual of the Bergman space L^1_a can be identified with the Bloch space $\mathcal{B}$ (and the pre-dual of L^1_a

can be identified with $\mathcal{B}_{\prime}$). Thus 5.3 and 5.4 (and the analogous theorems for the Hardy space context) lead to the following question: Is there a theorem that would show that boundedness and compactness of Hankel operators on a large class of function spaces depends only upon duality?

There exist Blaschke products in $\mathcal{B}_0$ with infinitely many zeros; see [24]. The following problem, raised by Sarason [24], is repeated here because of the clear connection with 5.4: Characterize the Blaschke products in $\mathcal{B}_0$ by means of the distribution of their zeros.

For functions f and g in $H^\infty(\mathbb{D})$, find necessary and sufficient conditions for the commutator $T_f^* T_g - T_g T_f^*$ to be compact. Equation 5.1 shows that

$$T_f^* T_g - T_g T_f^* = H_{\bar{g}}^* H_{\bar{f}},$$

so this problem asks when the product of a Hankel operator and the adjoint of a Hankel operator is compact. Based upon 5.4, the discussion of $H^\infty \cap \mathcal{B}_0$ following the statement of 5.6, and on the solution of this problem for the Hardy space (see [4], Theorem 1, Lemma 3, and Theorem 3), my guess is that for f and g in $H^\infty(\mathbb{D})$, the commutator $T_f^* T_g - T_g T_f^*$ is compact if and only if for each Gleason part G of the maximal ideal space of $H^\infty(\mathbb{D})$ (except the usual disk $\mathbb{D}$), either f or g is constant on G. This condition is equivalent to the condition that $(1-|z|^2)\min\{|f'(z)|, |g'(z)|\} \to 0$ as $z \to 1$. (Added in 1987: This conjecture has now been proved. Dechao Zheng has proved both directions of the conjecture. Simultaneously and independently, Pamela Gorkin and I proved one direction of the conjecture.)

Another possible area of research would be to characterize the compact Hankel operators on L_a^2 with symbol in $L^\infty(\mathbb{D}, dA)$, removing the restriction of 5.4 that the symbol must be the complex conjugate of an analytic function. Kehe Zhu, in a forthcoming paper, has made an important step in this direction by characterizing the functions f in $L^\infty(\mathbb{D}, dA)$ such that both H_f and $H_{\bar{f}}$ are compact.

Finally, one could ask whether appropriate generalizations of 5.3 through 5.6 hold when the unit disk $\mathbb{D}$ is replaced by an arbitrary region in the plane or in $\mathbb{C}^N$.

References

1. Anderson, J. M., Bloch functions: The basic theory, *Operators and Function Theory*, edited by S. C. Power, D. Reidel, Dordrecht, 1985, 1–17.

2. Anderson, J. M., J. Clunie, and Ch. Pommerenke, On Bloch functions and normal functions, Journal für die reine und angewandte Mathematik **270** (1974), 12–37.
3. Axler, Sheldon, The Bergman space, the Bloch space, and commutators of multiplication perators, Duke Mathematical Journal **53** (1986), 315–332.
4. ———, Sun-Yang A. Chang, and Donald Sarason, Products of Toeplitz operators, *Integral Equations and Operator Theory* **1** (1978), 285–309.
5. Brown, L. G., R. G. Douglas, and P. A. Fillmore, Unitary equivalence modulo the compact operators and extensions of C^*–algebras, in Proceedings of a Conference on Operator Theory, Halifax, Nova Scotia, 1973, edited by P. A. Fillmore, 58–128. Lecture Notes in Mathematics, vol. 345, Springer-Verlag, Berlin, 1973.
6. Coburn, L. A., Singular integral operators and Toeplitz operators on odd spheres, Indiana University Mathematics Journal **23** (1973), 433–439.
7. Coifman, R. R., R. Rochberg, and Guido Weiss, Factorization theorems for Hardy spaces in several variables, Annals of Mathematics **103** (1976), 611–635.
8. Duren, Peter L., *Theory of H^p Spaces*, Academic Press, New York, 1970.
9. ———, B. W. Rumberg, and A. L. Shields, Linear functionals on H^p spaces with $0 < p < 1$, Journal für die reine und angewandte Mathematik **238** (1969), 32–60.
10. Forelli, Frank and Walter Rudin, Projections on spaces of holomorphic functions in balls, Indiana University Mathematics Journal **24** (1974), 593–602.
11. Garnett, John B., *Bounded Analytic Functions*, Academic Press, New York, 1981.
12. Hardy, G. H. and J. E. Littlewood, Some properties of functional integrals. II, Mathematische Zeitschrift **34** (1932), 403–439.
13. Hastings, William W., A Carleson measure theorem for Bergman spaces, Proceedings of the American Mathematical Society **52** (1975), 237–241.
14. Hoffman, Kenneth, *Banach Spaces of Analytic Functions*, Prentice-Hall, Englewood Cliffs, 1962.
15. Koosis, Paul, *Introduction to H_p Spaces*, Cambridge University Press, Cambridge, 1980.
16. Luecking, Daniel, A technique for characterizing Carleson measures on Bergman spaces, Proceedings of the American Mathematical Society **87** (1983), 656–660.
17. ———, Representation and duality in weighted spaces of analytic functions, Indiana University Mathematics Journal **34** (1985), 319–336.
18. ———, Trace ideal criteria for Toeplitz operators, preprint.

19. McDonald, G. and C. Sundberg, Toeplitz operators on the disc, Indiana University Mathematics Journal **28** (1979), 595–611.
20. Peller, V. V., Hankel operators of class $\mathcal{C}_p$ and their applications (rational approximation, Gaussian processes, the problem of majorizing operators), Mathematics of the USSR-Sbornik **41** (1982), 443–479.
21. Rochberg, Richard, Decomposition theorems for Bergman spaces and their applications, in *Operators and Function Theory*, 225–277, edited by S. C. Power, D. Reidel, Dordrecht, 1985.
22. Rudin, Walter, *Functional Analysis*, McGraw–Hill, New York, 1973.
23. ———, *Function Theory in the Unit Ball of* $\mathbf{C}^n$, Springer-Verlag, New York, 1980.
24. Sarason, Donald, Blaschke products in $\mathcal{B}_0$, in *Linear and Complex Analysis Problem Book*, 337–338, edited by V. P. Havin, S. V. Hruscěv, and N. K. Nikol'skii, Lecture Notes in Mathematics, vol. 1043, Springer-Verlag, Berlin, 1984.
25. Timoney, Richard M., Bloch functions in several complex variables, I, Bulletin of the London Mathematical Society **12** (1980), 241–267.
26. Yamashita, Shinji, Criteria for functions to be Bloch, Bulletin of the Australian Mathematical Society **21** (1980), 223–227.
27. Zaharjuta, V. P. and V. I. Judovič, The general form of a linear functional in H_p', (Russian), Uspekhi Matematicheskikh Nauk **19** (1964), 139–142.

Sheldon Axler
Department of Mathematics
Michigan State University
East Lansing, Michigan 48824-1027 U.S.A.

Nevanlinna-Pick Interpolation: Generalizations and Applications

by

Joseph A. Ball

The classical Nevanlinna-Pick interpolation problem is the construction of a function analytic on the unit disk which assumes prescribed values at prescribed points and which satisfies a prescribed uniform bound on its modulus. We first review the original algorithmic approach of Nevalinna and the existence criterion of Pick. We then develop the recent operator-theoretic approach of Ball-Helton. This leads to a solution of an abstract interpolation problem which contains several types of recently studied generalizations as special cases. We indicate how the method can be used to obtain explicit formulas for the solution. Finally we discuss other types of interpolation problems also of interest in engineering which as of yet do not have as clean a solution.

1. Introduction to Nevanlinna-Pick Interpolation

Let $\{z_1, \ldots, z_n\}$ be an n-tuple of distinct points in the open unit disk $\mathcal{D}$. Let $\{w_1, \ldots, w_n\}$ be an n-tuple of complex numbers. The classical problem of Nevanlinna and Pick is to formulate necessary and sufficient conditions for the existence of a function f, analytic in $\mathcal{D}$, bounded in modulus by 1 and such that $f(z_i) = w_i (1 \leq i \leq n)$. We then say that f interpolates the sequences $\{z_i\}$ and $\{w_i\}$. When solutions exist, we would moreover like to have a characterization of the set of all solutions.

Nevanlinna's idea (see Nevanlinna [1929]) was as follows. First note that the maps of the form

$$z \to L_w z = \frac{z - w}{1 - z\overline{w}}$$

for a complex number w with $|w| < 1$ map the unit disk $\mathcal{D}$ biholomorphically onto itself and the boundary of the disk T onto itself. Let $\mathcal{B}$ denote the set of analytic functions in $\mathcal{D}$ whose modulus is bounded by 1. Suppose f is in $\mathcal{B}$, and z_1 is in $\mathcal{D}$. If $|f(z_1)| = 1$, then by the maximum modulus theorem $f = f(z_1)$ identically. If $|f(z_1)| < 1$, then

$$L_{f(z_1)} \circ f(z) = \frac{f(z) - f(z_1)}{1 - f(z)\overline{f(z_1)}}$$

is in $\mathcal{B}$, and is zero at z_1. But then

$$f_1(z) = \frac{f(z) - f(z_1)}{1 - f(z)\overline{f(z_1)}} / \frac{z - z_1}{1 - z\overline{z}_1}$$

is analytic on $\mathcal{D}$. Since

$$\frac{z - z_1}{1 - z\overline{z}_1}$$

has modulus 1 on the boundary T of the disk, we see that $f(z_1)$ has modulus at most 1 on T. Then by the maximum modulus theorem, we get $f_1 \in \mathcal{B}$. We have established the following invariant form of Schwarz's lemma: *If $z_1 \in \mathcal{D}$, then the function f is in $\mathcal{B}$ if and only if $|f(z_1)| = 1$ and $f = f(z_1)$ identically, or $|f(z_1| < 1$ and the function*

$$f(z_1) = \frac{f(z) - f(z_1)}{1 - f(z)\overline{f(z_1)}} / \frac{z - z_1}{1 - z\overline{z}_1} \tag{1.1}$$

is in $\mathcal{B}$.

Moreover, f interpolates the n-tuples $\{z_i\}$ and $\{w_i\}(1 \leq i \leq n)$ if and only if $|w_1| = 1$ and $w_1 = w_2 = \ldots = w_n$ in which case $f(z) = w_1$ is the unique solution, or $|w_1| < 1$ and f_1 interpolates the $(n-1)$-tuples $\{z_i\}$ and $\{w_i^1\}(2 \leq i \leq n)$ where

$$w_i^1 = \frac{w_i - w_1}{1 - w_i\overline{w}_1} / \frac{z_i - z_1}{1 - z_i\overline{z}_1}.$$

Of course if $|w_1| > 1$, no interpolating function in $\mathcal{B}$ can exist. For the case $|w_1| < 1$, we can solve (1.1) for f in terms of f_1 to get

$$f(z) = \frac{A_1(z)f_1(z) + B_1(z)}{C_1(z)f_1(z) + D_1(z)} \tag{1.2}$$

where

$$\begin{aligned} A_1(z) &= z - z_1 \\ B_1(z) &= w_1(1 - z\overline{z}_1) \\ C_1(z) &= \overline{w}_1(z - z_1) \\ D_1(z) &= 1 - \overline{z}_1 z. \end{aligned} \tag{1.3}$$

We apply now the same reasoning to f_1. If $|w_2^1| = 1$, then we must have $w_2^1 = w_3^1 = \cdots = w_n^1$ and $f_1(z) = w_2^1$ for interpolating functions in $\mathcal{B}$ to exist, and formula (1.2) with w_2^1 inserted for $f_1(z)$ gives the unique solution of the original interpolation problem. If

$|w_2^1| < 1$, then f_1 is in $\mathcal{B}$ and interpolates the $(n-1)$-tuples $\{z_i\}$ and $\{w_i^1\}(2 \le i \le n)$ if and only if

1.4
$$f_2(z) = \frac{f_1(z) - w_2^1}{1 - f_1(z)\overline{w_2^1}} / \frac{z - z_2}{1 - z\overline{z}_2}$$

is in $\mathcal{B}$ and interpolates the $(n-2)$-tuples $\{z_i\}$ and $\{w_i^2\}(3 \le i \le n)$, where

$$w_i^2 = \frac{w_i^1 - w_2^1}{1 - w_i^1\overline{w_2^1}} / \frac{z_i - z_2}{1 - z_i\overline{z}_2}.$$

Using (1.4) and (1.2) we may solve for f in terms of f_2 to get a formula of the form

$$f(z) = \frac{A_2(z)f_2(z) + B_2(z)}{C_2(z)f_2(z) + D_2(z)}.$$

We may continue the process by induction. If the process has continued to step k, we have a formula for the general solution f of the form

1.5
$$f(z) = \frac{A_k(z)f_k(z) + B_k(z)}{C_k(z)f_k(z) + D_k(z)}$$

where $f_k \in \mathcal{B}$ interpolates the $(n-k)$-tuples $\{z_i\}$ and $\{w_i^k\}(k+1 \le i \le n)$. If $|w_{k+1}^k| > 1$, no solutions exist. If $|w_{k+1}^k| = 1$, then solutions exist if and only if $w_{k+1}^k = w_{k+2}^k = \ldots = w_n^k$ and then $f_k(z) = w_{k+1}^k$; plugging this into (1.5) then gives the unique solution to the original interpolaion problem, and the inductive process stops here. Otherwise, $|w_{k+1}^k| < 1$ and $f_k \in \mathcal{B}$ interpolates $\{z_i\}$ and $\{w_i^k\}(k+1 \le i \le n)$ if and only if

1.6
$$f_{k+1}(z) = \frac{f_k(z) - w_{k+1}^k}{1 - f_k(z)\overline{w_{k+1}^k}} / \frac{z_i - z_{k+1}}{1 - z\overline{z}_{k+1}}.$$

is in $\mathcal{B}$ and interpolates the $(n-k-1)$-tuples $\{z_i\}$ and $\{w_i^{k+1}\}(k+2 \le i \le n)$, where

1.7
$$w_i^{k+1} = \frac{w_i^k - w_{k+1}^k}{1 - w_i^k\overline{w_{k+1}^k}} / \frac{z_i - z_{k+1}}{1 - z_i\overline{z}_{k+1}}.$$

Moreover, we may solve (1.6) for f_k in terms of f_{k+1} to get

1.8
$$f_k(z) = \frac{\alpha_{k+1}(z)f_{k+1}(z) + \beta_{k+1}(z)}{\gamma_{k+1}(z)f_{k+1}(z) + \delta_{k+1}(z)}$$

where

1.9
$$\begin{aligned} \alpha_{k+1}(z) &= z - z_{k+1} \\ \beta_{k+1}(z) &= w_{k+1}^k(1 - z\overline{z}_{k+1}) \\ \gamma_{k+1}(z) &= \overline{w_{k+1}^k}(z - z_{k+1}) \\ \delta_{k+1}(z) &= 1 - \overline{z}_{k+1}z. \end{aligned}$$

If we combine this with (1.5) we get

$$f(z) = \frac{A_{k+1}(z)f_{k+1}(z) + B_{k+1}(z)}{C_{k+1}(z)f_{k+1}(z) + D_{k+1}(z)} \tag{1.10}$$

where

$$\begin{aligned} A_{k+1} &= A_k\alpha_{k+1} + B_k\gamma_{k+1} \\ B_{k+1} &= A_k\beta_{k+1} + B_k\delta_{k+1} \\ C_{k+1} &= C_k\alpha_{k+1} + D_k\gamma_{k+1} \\ D_{k+1} &= C_k\beta_{k+1} + D_k\delta_{k+1} \end{aligned} \tag{1.11}$$

are polynomials of degree at most $k+1$ in z.

Nevalinna's solution criterion then is somewhat implicit. If the inductive process given above continues through n steps, then there are infinitely many solutions given by the linear fractional formula

$$f(z) = \frac{A_n(z)f_n(z) + B_n(z)}{C_n(z)f_n(z) + D_n(z)}, \tag{1.12}$$

where f_n is an arbitrary function in $\mathcal{B}$ (satisfying no additional interpolation constraints), and A_n, B_n, C_n, D_n are determined inductively by (1.3), (1.7), (1.9), and (1.11). If the process stops at step $k < n$, then solutions exist if and only if $|w_{k+1}^k| = 1$ and $w_{k+1}^k = w_{k+2}^k = \ldots = w_n^k$, in which case the solution is unique and is given by (1.5) with $f_k(z) = w_{k+1}^k$.

Prior to Nevanlinna's work Pick (see Pick [1916]) had obtained an existence criterion which is much more explicit. Given the n-tuple $\{z_i\}$ of distinct points in $\mathcal{D}$ and the n-tuple $\{w_i\}$ of complex numbers, form the $n \times n$ Hermitian matrix (often called the "Pick matrix")

$$\Lambda = \left[\frac{1 - w_j\overline{w}_i}{1 - z_j\overline{z}_i}\right] 1 \leq i, j \leq n \tag{1.12}$$

Then the result of Pick is that there exists an $f \in \mathcal{B}$ which interpolates the n-tuples $\{z_i\}$ and $\{w_i\}(1 \leq i \leq n)$ if and only if the matrix Λ is positive semidefinite; moreover, the solution is unique if and only if Λ is singular. For the case $n = 2$, it is easy to see that Pick's condition is equivalent to Nevanlinna's. Indeed, in this case Λ has the form

$$\Lambda = \begin{bmatrix} \dfrac{1 - |w_1|^2}{1 - |z_1|^2} & \dfrac{1 - w_2\overline{w}_1}{1 - z_2\overline{z}_1} \\ \dfrac{1 - w_1\overline{w}_2}{1 - z_1\overline{z}_2} & \dfrac{1 - |w_2|^2}{1 - |z_2|^2} \end{bmatrix}.$$

If $|w_1| = 1$, then the upper left corner of Λ is zero and the positive semidefiniteness of Λ forces the off-diagonal entries to be zero. This then forces $w_2 = w_1$ and $\Lambda = 0$, and both Pick's and Nevanlinna's conditions imply that we are in the situation of uniqueness. If $|w_1| < 1$, then the upper left corner of Λ is positive and Λ is positive semidefinite if and only if det $\Lambda \geq 0$, i.e.

1.13
$$\left|\frac{1 - w_2\overline{w}_1}{1 - z_2\overline{z}_1}\right|^2 \leq \frac{1 - |w_1|^2}{1 - |z_1|^2}\frac{1 - |w_2|^2}{1 - |z_2|^2}.$$

Uniqueness holds (according to Pick) when equality holds in (1.13). On the other hand, by Nevanlinna's analysis, solutions exist if and only if $|w_2^1| \leq 1$, which we write as

$$\left|\frac{w_2 - w_1}{1 - w_2\overline{w}_1}\right|^2 \leq \left|\frac{z_2 - z_1}{1 - z_2\overline{z}_1}\right|^2$$

and the case of uniqueness corresponds to equality. Rewrite this as

1.14
$$1 - \left|\frac{z_2 - z_1}{1 - z_2\overline{z}_1}\right|^2 \leq 1 - \left|\frac{w_2 - w_1}{1 - w_2\overline{w}_1}\right|^2.$$

It is now a simple matter of algebra to show that (1.13) and (1.14) are equivalent. Recall that Nevanlinna's condition for 2 points just amounts to the invariant form of Schwarz's lemma. We conclude that the positive semidefiniteness condition on the Pick matrix amounts to a generalization of the Schwarz lemma to more than 2 points. An elementary linear algebra-type proof of the equivalence of the Pick and Nevanlinna criteria for the general case of n points can be found in Marshall [1974].

Notes: The inductive procedure of Nevanlinna described above is often called the Schur algorithm. Schur [1918] developed it in connection with the problem where one prescribes the first few Taylor coefficients at the origin of an unknown function in $\mathcal{B}$. A variant of this problem was also considered in Caratheodory [1907] (see also the references in Sarason [1967]). Expository accounts of the Schur algorithm can also be found in Garnett [1981] and Sarason [1985]. The limitation that the number of interpolating points be finite is not essential; indeed Nevanlinna studied the limiting behavior of his algorithm for the situation of infinitely many interpolating points.

Some Facts About Krein Spaces

To discuss the approach to Nevanlinna-Pick interpolation and its generalizations which Bill Helton and I have been developing in the past few years, we need some machinery about Krein spaces. A Krein space $\mathcal{K}$ by definition is the direct sum $\mathcal{K} = \mathcal{K}_+ \oplus \mathcal{K}_-$ of two Hilbert spaces $\mathcal{K}_+$ and $\mathcal{K}_-$ with an indefinite inner product $[\cdot,\cdot]_\mathcal{K}$ given by

$$1.15 \qquad [k_+ \oplus k_-, k'_+ \oplus k'_-]_\mathcal{K} = < k_+, k'_+ >_{\mathcal{K}_+} - < k_-, k'_- >_{\mathcal{K}_-} .$$

We emphasize that the decomposition $\mathcal{K} = \mathcal{K}_+ \oplus \mathcal{K}_-$ is often not given and indeed is never unique except in trivial cases. The Krein space a priori is just a complex linear space $\mathcal{K}$ with indefinite inner product $[\cdot,\cdot]_\mathcal{K}$ for which there exists some decomposition $\mathcal{K} = \mathcal{K}_+ \oplus \mathcal{K}_-$ into Hilbert spaces so that $[\cdot,\cdot]_\mathcal{K}$ splits as the difference of Hilbert space inner products as in (1.15). The geometry of Krein spaces differs in many respects from the geometry of Hilbert spaces (see Bognar [1974] for the general case and Gohberg-Lancaster-Rodman [1983] for the finite dimensional case). As a direct sum of Hilbert spaces, $\mathcal{K}$ can be made a Hilbert space with norm

$$\|k_+ \oplus k_-\|^2 = \|k_+\|^2_{\mathcal{K}_+} + \|k_-\|^2_{\mathcal{K}_-} .$$

While the norm itself depends on the choice of fundamental decomposition $\mathcal{K} = \mathcal{K}_+ \oplus \mathcal{K}_-$, the topology induced by the norm does not.

For $\mathcal{M}$ a subspace of $\mathcal{K}$, let $\mathcal{M}'$ denote the $[\cdot,\cdot]_\mathcal{K}$-orthogonal complement of $\mathcal{M}$

$$\mathcal{M}' = \{x \in \mathcal{K} : [x,y]_\mathcal{K} = 0 \text{ for all } y \in \mathcal{M}\}.$$

The subspace $\mathcal{M}$ is said to be *regular* (or *ortho-complemented*) if $\mathcal{K} = \mathcal{M} + \mathcal{M}'$. When this happens, then also $\mathcal{M} \cap \mathcal{M}' = (0)$ since no nonzero vector of $\mathcal{K}$ can be orthogonal to all of $\mathcal{K}$, so the sum $\mathcal{K} = \mathcal{M} + \mathcal{M}'$ is a $[\cdot,\cdot]_\mathcal{K}$-orthogonal direct sum. Equivalently, $\mathcal{M}$ is a regular subspace if and only if the restriction of $[\cdot,\cdot]_\mathcal{K}$ to $\mathcal{M}$ makes $\mathcal{M}$ a Krein space (see p. 104 of Bognar [1974]).

Let us say that a subspace $\mathcal{P} \subset \mathcal{K}$ is *positive* if $[x,x]_\mathcal{K} \geq 0$ for all $x \in \mathcal{P}$; we say $\mathcal{P}$ is *strictly positive* if $[x,x]_\mathcal{K} > 0$ for $0 \neq x \in \mathcal{P}$. The subspace $\mathcal{P}$ is *maximal positive* if $\mathcal{P}$ itself is positive and is not contained in any larger positive subspace. For $\mathcal{P}$ a positive subspace, we define the *positive cosignature* of $\mathcal{P}$ to be the codimension of $\mathcal{P}$ in some maximal positive subspace. As we shall see, such a maximal positive superspace always exists and the codimension of $\mathcal{P}$ is independent of the particular superspace chosen. The *positive signature* of $\mathcal{M}$ we define to be the dimension of a maximal positive subspace of $\mathcal{M}$. When we apply all these notions to the inner product $-[\cdot,\cdot]_\mathcal{K}$ in place of $[\cdot,\cdot]_\mathcal{K}$, we have defined *negative subspace, strictly negative subspace, maximal negative subspace, negative cosignature* and *negative signature.*

The basic lemma which will be needed in the next section is the following.

Lemma 1.1. (see Ball-Helton [1986]) *Suppose $\mathcal{K}$ is a Krein space, $\mathcal{P}$ is a regular positive subspace and $\mathcal{N}$ is a maximal negative subspace.*

a. *Then $\mathcal{N} + \mathcal{P}$ is closed and the positive cosignature of $\mathcal{P}$ coincides with the codimension (in $\mathcal{K}$) of $\mathcal{N} + \mathcal{P}$.*
b. *Equivalently, the positive cosignature of $\mathcal{P}$ is equal to the positive signature of $\mathcal{P}'$.*

Proof: Choose and fix some fundamental decomposition $\mathcal{K} = \mathcal{K}_+ \oplus \mathcal{K}_-$. For ease of matrix manipulation we use column vector notation

$$k = \begin{bmatrix} k_+ \\ k_- \end{bmatrix} \in \begin{bmatrix} \mathcal{K}_+ \\ \mathcal{K}_- \end{bmatrix} = \mathcal{K}.$$

One can check that a positive subspace $\mathcal{P}$ must have the form

1.16 $$\mathcal{P} = \begin{bmatrix} I \\ X \end{bmatrix} \mathcal{D} = \left\{ \begin{bmatrix} d \\ Xd \end{bmatrix} : d \in \mathcal{D} \right\}$$

where $\mathcal{D}$ is a closed subspace of $\mathcal{K}_+$ and $X : \mathcal{D} \to \mathcal{K}_-$ is a contraction operator ($\|X\| \leq 1$) in the operator norm induced by the Hilbert space norms on $\mathcal{D} \subset \mathcal{K}_+$ and $\mathcal{K}_-$. The positive subspace is in addition regular if and only if $\|X\| < 1$. In general, a subspace is maximal positive if and only if its $[\cdot, \cdot]$-orthogonal complement is negative. Using this criterion, one gets that the form of a maximal positive subspace is as in (1.16), but with $\mathcal{D} = \mathcal{K}_+$. If $\mathcal{P}$ is a positive subspace as in (1.16), then any maximal positive subspace $\widetilde{\mathcal{P}}$ containing $\mathcal{P}$ is a subspace of the form

$$\widetilde{\mathcal{P}} = \begin{bmatrix} I \\ Z \end{bmatrix} \mathcal{K}_+$$

where $Z : \mathcal{K}_+ \to \mathcal{K}_-$ is a contraction and $Z|\mathcal{D} = X$. From this we see that the positive cosignature of $\mathcal{P}$ is $\dim(\mathcal{K}_+ \ominus \mathcal{D})$.

Now suppose that $\mathcal{N}$ is a maximal negative subspace of $\mathcal{K}$. By the same type of analysis as above, we see that $\mathcal{N}$ has the form

$$\mathcal{N} = \begin{bmatrix} Y \\ I \end{bmatrix} \mathcal{K}_-$$

where $Y : \mathcal{K}_- \to \mathcal{K}_+$ is a contraction operator. Thus

1.17 $$\mathcal{P} + \mathcal{N} = \begin{bmatrix} I & Y \\ X & I \end{bmatrix} \begin{bmatrix} \mathcal{D} \\ \mathcal{K}_- \end{bmatrix}.$$

Let $\widetilde{X}$ be the linear extension of X to all of $\mathcal{K}_+$ defined by setting $\widetilde{X}x = 0$ if $x \in \mathcal{K}_+ \ominus \mathcal{D}$. Assume $\mathcal{P}$ is a regular positive subspace, so $\|X\| < 1$; then also $\|\widetilde{X}\| < 1$, and $I - \widetilde{X}Y$ is invertible on $\mathcal{K}_+$. The factorization

$$\begin{bmatrix} I & Y \\ \widetilde{X} & I \end{bmatrix} = \begin{bmatrix} I & 0 \\ \widetilde{X} & I \end{bmatrix} \begin{bmatrix} I & 0 \\ 0 & I - \widetilde{X}Y \end{bmatrix} \begin{bmatrix} I & Y \\ 0 & I \end{bmatrix}$$

shows that

$$\begin{bmatrix} I & Y \\ \widetilde{X} & I \end{bmatrix}$$

is an invertible operator on

$$\begin{bmatrix} \mathcal{K}_+ \\ \mathcal{K}_- \end{bmatrix}.$$

Since $\mathcal{D}$ is a closed subspace of $\mathcal{K}_+$, we see from (1.17) that $\mathcal{P} + \mathcal{N}$ is closed. Also from (1.17) we get that $\mathcal{P} + \mathcal{N}$ had codimension in

$$\begin{bmatrix} \mathcal{K}_+ \\ \mathcal{K}_- \end{bmatrix}$$

equal to the codimension of

$$\begin{bmatrix} \mathcal{D} \\ \mathcal{K}_- \end{bmatrix} \text{ in } \begin{bmatrix} \mathcal{K}_+ \\ \mathcal{K}_- \end{bmatrix},$$

i.e. $\dim(\mathcal{K}_+ \ominus \mathcal{D})$. As mentioned above, this in turn is the negative cosignature of P. This establishes part (a) in Lemma 1.1.

If $\mathcal{P}$ is given by (1.16), one can compute that

$$\mathcal{P}' = \begin{bmatrix} X^* \\ I \end{bmatrix} \mathcal{K}_- + \begin{bmatrix} \mathcal{K}_+ \ominus \mathcal{D} \\ 0 \end{bmatrix}. \tag{1.18}$$

Since $\|X\| < 1$, the subspace

$$\begin{bmatrix} X^* \\ I \end{bmatrix} \mathcal{K}_-$$

is a strictly negative subspace. Clearly

$$\begin{bmatrix} \mathcal{K}_+ \ominus \mathcal{D} \\ 0 \end{bmatrix}$$

is positive. Since the range of X^* is orthogonal to $\mathcal{K}_+ \ominus \mathcal{D}$, these two spaces are $[\cdot,\cdot]_{\mathcal{K}}$-orthogonal. We conclude that

$$\begin{bmatrix} \mathcal{K}_+ \ominus \mathcal{D} \\ 0 \end{bmatrix}$$

is a maximal positive subspace of $\mathcal{P}'$ and thus the positive signature of $\mathcal{P}'$ is also $\dim \mathcal{K}_+ \ominus \mathcal{D}$. This verifies part (b).

2. Nevanlinna-Pick Interpolation: The Grassmannian Formulation

In this section we review the approach to Nevanlinna-Pick interpolation from Ball-Helton [1983]; in the next section we shall put this in an abstract axiomatic framework which picks up many generalizations and variations as special cases.

We return to the simplest Nevanlinna-Pick interpolation problem as discussed in S1. We are given an n-tuple $\{z_1, \ldots, z_n\}$ of distinct points in the unit disk $\mathcal{D}$ and an n-tuple $\{w_1, \ldots, w_n\}$ of complex numbers. The problem is to find a function F in the class $\mathcal{B}$ (functions analytic on the disk with modulus bounded by 1) which interpolates the $\{z_i\}$ and $\{w_i\}$ $(1 \leq i \leq n)$. The problem as stands is highly nonlinear (unless all the w_i's are 0). We get a more linear problem by considering the graph of the unknown function rather than the function itself. We denote by H^∞ the algebra of bounded analytic functions on the unit disk $\mathcal{D}$, by H^2 the usual Hardy space of analytic functions on $\mathcal{D}$, and by L^2 the Hilbert space of square integrable functions on the unit circle T. When convenient, we identify H^2 as a subspace of L^2 in the usual way by taking nontangential limits to the boundary (see Hoffman [1965]). We can identify H^∞ with a subalgebra of the bounded linear operators $\mathcal{L}(H^2)$ on H^2 by considering the multiplication operator

$$M_F : f \in H^2 \to Ff \in H^2$$

induced by an F in H^∞. These operators are characterized by the fact that they commute with the shift operator M_z

$$(M_z f)(z) = zf(z)$$

on H^2.

The direct sum Hilbert space $H^2 \oplus H^2$ we write in column form

$$\begin{bmatrix} H^2 \\ H^2 \end{bmatrix}.$$

If X is any operator on H^2, its graph space $\mathcal{G}_X$ is the subspace of

$$\begin{bmatrix} H^2 \\ H^2 \end{bmatrix}$$

given by

$$\mathcal{G}_X = \begin{bmatrix} X \\ I \end{bmatrix} H^2 = \left\{ \begin{bmatrix} X(f) \\ f \end{bmatrix} : f \in H^2 \right\}.$$

Graph spaces

$$\mathcal{G} \subset \begin{bmatrix} H^2 \\ H^2 \end{bmatrix}$$

are characterized by the conditions that

$$P_{\begin{bmatrix} 0 \\ H^2 \end{bmatrix}} \mathcal{G} = \begin{bmatrix} 0 \\ H^2 \end{bmatrix}$$

and that

$$\mathcal{G} \cap \begin{bmatrix} H^2 \\ 0 \end{bmatrix} = (0),$$

i.e. by the direct sum condition

$$(i) \qquad \mathcal{G} \dot{+} \begin{bmatrix} H^2 \\ 0 \end{bmatrix} = \begin{bmatrix} H^2 \\ H^2 \end{bmatrix}.$$

Graph spaces $\mathcal{G}$ of the form $\mathcal{G}_{M_F}$ for an $\mathcal{H}^\infty$ function F are characterized among graph spaces by the additional condition

$$(ii) \qquad M_z \mathcal{G} \subset \mathcal{G}$$

i.e., invariance under M_z. This is an immediate consequence of the fact mentioned above that $X \in \mathcal{L}(H^2)$ has the form $X = M_F$ for an $F \in H^\infty$ if and only if $XM_z = M_z X$. For brevity we shall write $\mathcal{G}_F$ rather than $\mathcal{G}_{M_F}$; the meaning will be clear from the context.

For the interpolation problem, we are interested in $F \in \mathcal{B}$, i.e., $F \in H^\infty$ and $\|F\| \leq 1$. An easy fact about an operator M_F is that its operator norm $\|M_F\|$ coincides with the infinity norm $\|F\|_\infty$ of its symbol. Thus our next task is to characterize for which graph spaces $\mathcal{G} = \mathcal{G}_X$ is the associated *angle-operator* X a contraction ($\|X\| \leq 1$). To deal with this problem we are led to introduce the indefinite inner product $[\cdot,\cdot]_J$ on

$$\begin{bmatrix} H^2 \\ H^2 \end{bmatrix}$$

induced by the matrix

$$J = \begin{bmatrix} 1 & 0 \\ 0 & -1 \end{bmatrix}$$

$$\left[\begin{bmatrix} f \\ g \end{bmatrix}, \begin{bmatrix} f' \\ g' \end{bmatrix} \right]_J = \left\langle J \begin{bmatrix} f \\ g \end{bmatrix}, \begin{bmatrix} f' \\ g' \end{bmatrix} \right\rangle = \langle f, f' \rangle_{H^2} - \langle g, g' \rangle_{H^2}.$$

Clearly

$$\begin{bmatrix} H^2 \\ H^2 \end{bmatrix}$$

is a Krein space in this inner product. Given that a subspace $\mathcal{G}$ has the form $\mathcal{G}_X$, one easily sees that $\|X\| \leq 1$ if and only if

(*iii*) $\quad \mathcal{G}$ is a negative subspace with respect to $[\cdot,\cdot]_J$, where $J = \begin{bmatrix} 1 & 0 \\ 0 & -1 \end{bmatrix}$.

So far we have established that *a subspace*

$$\mathcal{G} \subset \begin{bmatrix} H^2 \\ H^2 \end{bmatrix}$$

has the form $\mathcal{G}_F$ *for an* $F \in \mathcal{B}^\infty$ *if and only if* $\mathcal{G}$ *satisfies* (i), (ii), *and* (iii).

It remains to impose the condition that F interpolates $\{z_i\}$ and $\{w_i\}$ $(1 \leq i \leq n)$. Thus suppose

$$G = \begin{bmatrix} Ff \\ f \end{bmatrix} \in \mathcal{G}_F$$

for an interpolating F. Then we see that for $1 \leq i \leq n$,

$$(\star) \qquad \left\langle G(z_i), \begin{bmatrix} 1 \\ -\overline{w}_i \end{bmatrix} \right\rangle_{\mathbb{C}^2} = F(z_i)f(z_i) - f(z_i)w_i = 0 \textit{ for all } f \in H^2.$$

Conversely, if $(\star)$ holds for all $f \in H^2$, then F interpolates. This suggests that we introduce the subspace $\mathcal{M} = \mathcal{M}(\{z_i\},\{w_i\})$ given by

$$2.1 \qquad \mathcal{M} = \left\{ G \in \begin{bmatrix} H^2 \\ H^2 \end{bmatrix} : \left\langle G(z_i), \begin{bmatrix} 1 \\ -\overline{w}_i \end{bmatrix} \right\rangle_{\mathbb{C}^2} = 0 \text{ for } 1 \leq i \leq n. \right\}$$

We have established that a subspace $\mathcal{G}$ known to be of the form $\mathcal{G}_F$ for an $F \in H^\infty$ is such that its angle operator F interpolates $\{z_i\}$ and $\{w_i\}$ $(1 \leq i \leq n)$ if and only if

(*iv*) $\qquad \mathcal{G} \subset \mathcal{M}$ where $\mathcal{M}$ is given by (2.1).

If we combine (i)-(iv), we see that the original Nevanlinna-Pick interpolation problem for a function $f \in \mathcal{B}$ translates to finding subspaces $\mathcal{G}$ of

$$\begin{bmatrix} H^2 \\ H^2 \end{bmatrix}$$

which satisfy all of conditions (i)-(iv).

Let us ignore the shift invariance condition (ii) for the moment in order to understand a more tractable problem. For simplicity we also assume that $\mathcal{M}$ is regular as a subspace of the Krein space

$$\mathcal{K} = \begin{bmatrix} H^2 \\ H^2 \end{bmatrix};$$

this restriction is not essential. Then $\mathcal{M}$ is a Krein space itself in the $[\cdot,\cdot]_J$-inner product. By the discussion in Section 1b), $\mathcal{K}$-maximal negative subspaces $\mathcal{G}$ (i.e., subspaces satisfying (i) and (iii)) exist in abundance. If such a subspace is contained in $\mathcal{M}$ (i.e. satisfies (iv)), necessarily it must be $\mathcal{M}$-maximal negative (i.e. maximal negative as a subspace of the Krein space $\mathcal{M}$). As was the case for $\mathcal{K}$, we know that $\mathcal{M}$-maximal negative subspaces exist in abundance. The delicate part is: when is an $\mathcal{M}$-maximal negative subspace also $\mathcal{K}$-maximal negative? Note that in general a negative subspace $\mathcal{N}$ of $\mathcal{K}$ is $\mathcal{K}$-maximal negative if and only if it is a graph space. By the characterization (i) of a graph space we see that this happens if and only if

$$\mathcal{N} \dot{+} \begin{bmatrix} H^2 \\ 0 \end{bmatrix} = \begin{bmatrix} H^2 \\ H^2 \end{bmatrix}. \tag{2.2}$$

Assuming $\mathcal{N} \subset \mathcal{M}$, when we intersect each side of (2.2) with $\mathcal{M}$, we get

$$\mathcal{N} \dot{+} \mathcal{P} = \mathcal{M}, \tag{2.3}$$

where $\mathcal{P}$ is the regular positive subspace of $\mathcal{M}$

$$\begin{aligned} \mathcal{P} &= \begin{bmatrix} H^2 \\ 0 \end{bmatrix} \cap \mathcal{M} \\ &= \left\{ \begin{bmatrix} f \\ 0 \end{bmatrix} : f \in H^2 \text{ and } f(z_i) = 0 \text{ for } 1 \le i \le n \right\}. \end{aligned}$$

Conversely, if $\mathcal{N}$ satisfies (2.3) and $\mathcal{N}$ is negative, then $\mathcal{N}$ must be $\mathcal{K}$-maximal negative; note that

$$P_{\begin{bmatrix} 0 \\ H^2 \end{bmatrix}} \mathcal{M} = \begin{bmatrix} 0 \\ H^2 \end{bmatrix}.$$

By Lemma 1.1, (2.3) occurs for any $\mathcal{M}$-maximal negative subspace $\mathcal{N}$ if and only if $\mathcal{P}$ is $\mathcal{M}$-maximal positive. We have established so far the following.

Proposition 2.1. *There exist subspaces $\mathcal{G}$ satisfying (i), (iii), and (iv) if and only if the subspace*

$$\mathcal{P} = \begin{bmatrix} H^2 \\ 0 \end{bmatrix} \cap \mathcal{M}$$

is $\mathcal{M}$-maximal positive. When this is the case, any $\mathcal{M}$-maximal negative subspace $\mathcal{G}$ satisfies (i), (iii), and (iv).

In case $\mathcal{M}$ is not regular, the result still holds; precisely one must demand that $\mathcal{P}$ is maximal as a strictly positive subspace of $\mathcal{M}$.

In view of Proposition 2.1 it is of interest to know exactly when is $\mathcal{P}$ $\mathcal{M}$-maximal positive. The answer is:

Proposition 2.2. *If $\mathcal{M} = \mathcal{M}(\{z_i\}, \{w_i\})$ is as in (2.1) and*

$$\mathcal{P} = \begin{bmatrix} H^2 \\ 0 \end{bmatrix} \cap \mathcal{M},$$

then $\mathcal{P}$ is $\mathcal{M}$-maximal strictly positive if and only if the Pick matrix

$$\Lambda = \left[\frac{1 - w_j \overline{w}_i}{1 - z_j \overline{z}_i} \right]_{1 \leq i,j \leq n}$$

is positive semidefinite. Also $\mathcal{M}$ is regular if and only if Λ is invertible.

Proof: The $[\cdot,\cdot]_J$-orthogonal complement of a subspace $\mathcal{R}$ we denote by $\mathcal{R}^{\perp J}$. To show that $\mathcal{P}$ is $\mathcal{M}$-maximal strictly positive amounts to showing that $\mathcal{M} \cap \mathcal{P}^{\perp J}$ is negative. To compute $\mathcal{M} \cap \mathcal{P}^{\perp J}$ it is convenient to write $\mathcal{M}$ in a different form. Let $b(z)$ be the Blaschke product

$$b(z) = \prod_{j=1}^{n} \frac{z - z_j}{1 - z\overline{z}_j}$$

and let K be any convenient H^∞ function such that $K(z_i) = w_i$; by the theory of Lagrange interpolation one can in fact find such a K which is a polynomial. Then $\mathcal{M}$ can alternatively be written in the form

$$\mathcal{M} = \begin{bmatrix} K \\ 1 \end{bmatrix} H^2 + \begin{bmatrix} bH^2 \\ 0 \end{bmatrix}.$$

Note that

$$\mathcal{P} = \begin{bmatrix} bH^2 \\ 0 \end{bmatrix}.$$

The decomposition becomes a $[\cdot,\cdot]_J$-orthogonal direct sum if we write

$$\mathcal{M} = \begin{bmatrix} \Gamma \\ 1 \end{bmatrix} H^2 + \begin{bmatrix} bH^2 \\ 0 \end{bmatrix},$$

where Γ: $H^2 \to \mathcal{S} = H^2 \cap bH^{2\perp}$ is the operator $\Gamma : f \to P_{\mathcal{S}}(Kf)$. We have thus identified $\mathcal{M} \cap \mathcal{P}'$ as the space

$$\begin{bmatrix} \Gamma \\ 1 \end{bmatrix} H^2.$$

This space in turn is negative if and only if $\|\Gamma\| \leq 1$. It turns out that it is easier to work with $\Gamma^\star : \mathcal{S} \to H^2$ ($\Gamma^\star f = P_{H^2}(\bar{K} f)$). The space $\mathcal{S}$ has as a basis the kernel functions $k_{z_i}(z) = (1 - z\bar{z}_i)^{-1}$ ($i = 1, \ldots, n$); each of these basis vectors is an eigenfunction for $\Gamma^\star$

$$\Gamma^\star k_{z_i} = \overline{K(z_i)} k_{z_i} = \bar{w}_i k_{z_i}.$$

The general element of $\mathcal{S}$ is $\sum_{i=1}^n c_i k_{z_i}$. To check whether $\Gamma^\star$ is a contraction, we must verify

$$\left\| \Gamma^\star \left(\sum_{i=1}^n c_i k_{z_i} \right) \right\|_{H^2}^2 \leq \left\| \sum_{i=1}^n c_i k_{z_i} \right\|_{H^2}^2$$

i.e.

$$\left\| \sum_{i=1}^n c_i \bar{w}_i k_{z_i} \right\|^2 \leq \left\| \sum_{i=1}^n c_i k_{z_i} \right\|^2 . \tag{2.4}$$

Using the reproducing property of the kernel functions

$$\langle f, k_{z_i} \rangle_{H^2} = f(z_i) \quad \text{for } f \in H^2,$$

it is easy to check that (2.4) is equivalent to the positive semidefiniteness of Λ. Moreover, Λ is singular if and only if $\|\Gamma f\| = \|f\|$ for some $f \in H^2$; we leave it to the reader to check that this is equivalent to $\mathcal{M}$ not being regular.

So far we have neglected (ii) and have used only ideas about the geometry of Krein spaces. To bring in condition (ii) we need a new idea, namely, a generalization of the Beurling-Lax theorem to a Krein space setting.

Theorem 2.3. *Let $\mathcal{M}$ be as in (2.1) and suppose that $\mathcal{M}$ is regular. Then there is a rational 2×2 matrix function*

$$\Theta(z) = \begin{bmatrix} \Theta_{11}(z) & \Theta_{12}(z) \\ \Theta_{21}(z) & \Theta_{22}(z) \end{bmatrix}$$

such that

$$\mathcal{M} = \Theta \begin{bmatrix} H^2 \\ H^2 \end{bmatrix} \tag{2.5}$$

and

$$\Theta(z)^\star J\Theta(z) = J \qquad \text{for } |z| = 1, \tag{2.6}$$

where

$$J = \begin{bmatrix} 1 & 0 \\ 0 & -1 \end{bmatrix}.$$

Sketch of Proof. The idea is to adapt the Hilbert space proof due to P. R. Halmos involving the notion of a *wandering subspace* to the Krein space geometry. We let $\mathcal{L}$ be the subspace $\mathcal{M} \cap z\mathcal{M}^{\perp J}$. The space $\mathcal{L}$ is said to be wandering because $\mathcal{L}$ is J-orthogonal to all shifts $z^k\mathcal{L}(k \neq 0)$ of itself. One can show that $\mathcal{M}$, $\mathcal{M}^{\perp J}$ and L_2^2 have J-orthogonal decompositions

(a) $$\mathcal{M} = \oplus_{k=0}^{\infty}{}_J \; z^k\mathcal{L}$$

(b) $$\mathcal{M}^{\perp J} = \oplus_{k=-1}^{-\infty}{}_J \; z^k\mathcal{L}$$

(c) $$L_2^2 = \oplus_{k=-\infty}^{\infty}{}_J \; z^k\mathcal{L} = \mathcal{M}^{\perp \mathcal{J}} \oplus_{\mathcal{J}} \mathcal{M}.$$

It turns out that $\dim \mathcal{L} = 2$ for this case. If Θ is any map from the constants $\mathbb{C}^2$ into $\mathcal{L}$, one can extend Θ to a map from L_2^2 into L_2^2 by demanding that it commute with the shift operator M_z of multiplication by z. The extended map is then nothing but multiplication by the matrix-valued function $\Theta(z)$ defined by

$$\Theta(z)c = (\Theta c)(z), \qquad c \in \mathbb{C}^2.$$

Using the decompositions (a), (b), and (c), one sees that

$$\Theta H_2^2 = \mathcal{M}$$
$$\Theta H_2^{2\perp} = \mathcal{M}^{\perp J}$$
$$\Theta L_2^2 = L_2^2$$

and that any Θ having these properties must arise from this construction. If we further arrange that the original map $\Theta : \mathbb{C}^2 \to \mathcal{L}$ is a J-isometry, then the additional property $\Theta^\star(z)J\Theta(z) = J$ follows as well. For complete details see Ball-Helton [1983] or Ball [1981].

With this result we can complete Proposition 2.2 and show that the positive semidefiniteness of Λ is necessary and sufficient for the existence of subspaces satisfying all the conditions (i)-(iv); we will then have recovered the results of Pick and Nevanlinna through a completely different route. By Proposition 2.2, with the assumption that Λ is positive definite, our problem is reduced to looking for $\mathcal{M}$-maximal negative subspaces which are also invariant. We now have $\mathcal{M}$ represented as ΘH_2^2, or more precisely, as $M_\Theta H_2^2$, where $M_\Theta\colon H_2^2 \to H_2^2$ is the operator of multiplication by Θ. As M_Θ commutes with M_z, any M_z-invariant subspace $\mathcal{G} \subset \mathcal{M}$ necessarily has the form $\mathcal{G} = \Theta\mathcal{G}_1$, where $\mathcal{G}_1$ is an M_z-invariant subspace of H_2^2. Moreover, since $\Theta(z)^\star J\Theta(z) = J$ for $|z| = 1$, M_Θ is an isometry with respect to the inner product $[\cdot,\cdot]_J$, and thus preserves all the Krein space geometry. Therefore a subspace $\mathcal{G}$ is $\mathcal{M}$-maximal negative if and only if $\mathcal{G} = \Theta\mathcal{G}_1$, where $\mathcal{G}_1$ is maximal negative in H_2^2. The invariant maximal negative subspaces of H_2^2 have already been characterized as those of the form

$$\mathcal{G}_1 = \begin{bmatrix} G \\ I \end{bmatrix} H^2,$$

where $G \in \mathcal{B}$; in particular, they exist. We have thus established that the full problem (i)-(iv) always has solutions if Λ is positive definite. The case where Λ is merely positive semidefinite can be handled by an approximation argument.

We can go on to obtain the linear fractional parametrization of the set of all solutions for the case where Λ is positive definite. Indeed, if F is a solution, we know that for some $G \in \mathcal{B}$,

$$\begin{bmatrix} F \\ I \end{bmatrix} H^2 = \Theta \begin{bmatrix} G \\ I \end{bmatrix} H^2$$
$$\begin{bmatrix} \Theta_{11}G + \Theta_{12} \\ \Theta_{21}G + \Theta_{22} \end{bmatrix} H^2 .$$

From the equality of second components we see that $H^2 = (\Theta_{21}G + \Theta_{22})H^2$ and thus $(\Theta_{21}G + \Theta_{22})^{-1} \in H^\infty$. We can therefore write

$$\begin{aligned} \begin{bmatrix} F \\ I \end{bmatrix} H^2 &= \begin{bmatrix} (\Theta_{11}G + \Theta_{12})(\Theta_{21}G + \Theta_{22})^{-1} \\ I \end{bmatrix} (\Theta_{21}G + \Theta_{22})H^2 \\ &= \begin{bmatrix} (\Theta_{11}G + \Theta_{12})(\Theta_{21}G + \Theta_{22})^{-1} \\ I \end{bmatrix} H^2 \end{aligned}$$

and thus

$$F = (\Theta_{11}G + \Theta_{12})(\Theta_{21}G + \Theta_{22})^{-1} .$$

This establishes the following.

Theorem 2.4. *Suppose $\mathcal{M}$ is as in (2.1), the Pick matrix Λ as in Proposition 2.2 is positive definite and that $\Theta(z)$ is any matrix function satisfying (2.5) and (2.6) as in*

Theorem 2.3. Then $F \in \mathcal{B}$ and interpolates $\{z_i\}$ and $\{w_i\}$ $(1 \le i \le n)$ if and only if

2.7
$$F = (\Theta_{11}G + \Theta_{12})(\Theta_{21}G + \Theta_{22})^{-1}$$

for some $G \in \mathcal{B}$.

We mention here that a somewhat more general problem (first considered by Takagi [1924] and then by Adamjan-Arov-Krein [1971]) can be handled by the same methods. For ℓ a nonnegative integer, define $H^\infty(\ell)$ to be functions of the form $F = F_1\psi^{-1}$, where $F_1 \in H^\infty$ and ψ is a Blaschke product of degree at most ℓ; thus F is an L^∞ function with meromorphic continuation to the unit disk having at most ℓ poles there. We say that such a function interpolates $\{z_i\}$ and $\{w_i\}$ if $F(z_i) = w_i$ at each z_i at which F is analytic $(1 \le i \le n)$. The more general Nevanlinna-Pick interpolation problem is to ask what is the smallest ℓ for which there exists an $F \in \mathcal{B}(\ell)$ (where $\mathcal{B}(\ell)$ is the class of $H^\infty(\ell)$ functions with L^∞ norm on the unit circle at most 1) which interpolates $\{z_i\}$ and $\{w_i\}$. With any $F \in H^\infty(\ell)$, say with representation $F = F_1\psi^{-1}$, we associate its *graph space*

$$\mathcal{G}_F = \begin{bmatrix} F \\ I \end{bmatrix} \psi H^2.$$

We modify condition (i) to read

$$(i)_\ell P_{\begin{bmatrix} 0 \\ H^2 \end{bmatrix}} \mathcal{G} \text{ has codimension at most } \ell \text{ in } \begin{bmatrix} 0 \\ H^2 \end{bmatrix} \text{ and } \mathcal{G} \cap \begin{bmatrix} H^2 \\ 0 \end{bmatrix} = (0).$$

Conditions $(i)_\ell, (ii), (iii)$, and (iv) then characterize which subspaces $\mathcal{G} \subset H_2^2$ are of the form $\mathcal{G}_F$ for an $F \in \mathcal{B}(\ell)$ which interpolates. When one follows all the analysis through, one gets that the smallest ℓ for which there exist interpolating functions in $\mathcal{B}(\ell)$ is equal to the number of negative eigenvalues of the Pick matrix Λ. Moreover, if Λ is invertible (so $\mathcal{M}$ is regular), the formula (2.7) as in Theorem 2.4 (with $G \in \mathcal{B}$) parametrizes the class of all interpolating F in $\mathcal{B}(\ell)$ for this smallest ℓ.

Of course we have not completely recovered the results of Nevanlinna since we have not described how to compute the symbol Θ of the linear fractional map (2.7) explicitly from the interpolation data. We shall return to this topic in Section 4.

Notes. The application of Krein space geometry and generalized Beurling-Lax theorems to interpolation problems was first obtained in Ball-Helton [1983] (see also Ball [1981]). The details here follow Ball-Helton [1986c] rather than the original 1983 paper; in this way we shall be able to handle the boundary interpolation problem to be discussed in Section 5.

There exist several other operator-theoretic approaches to interpolation. One of the earliest is that of Sz.-Nagy-Koranyi [1956, 1958]; this was later refined in Ball [1983a] to handle more general interpolation problems. The commutant lifting framework of Sarason [1967] has proved to be quite powerful (see e.g. Helton [1978] and Rosenblum-Rovnyak [1980, 1985]). Soviet authors have also developed again different operator-theoretic approaches to interpolation (e.g. Krein-Nudelman [1977], Fedchina [1975a, 1975b], Kovalishina-Potapov [1982]).

It is also interesting to note that the generalized Beurling-Lax theorems have diverse applications besides interpolation, e.g. J-inner-outer and Wiener-Hopf factorization, computation of orbits of groups of linear fractional maps acting on matrix L^∞, lifting theory and applications to physical situations. This is covered in the series of papers Ball-Helton [1982a-b, 1983, 1984, 1986a-c]. For a survey, see Helton [1987].

3. An Axiomatic Framework and Generalizations

In this section we consider a more abstract form of the interpolation problem. Let us suppose that we are given some abstract algebra $\mathcal{A}$; usually $\mathcal{A}$ will also have some topology but this is not so important. By a Hilbert $\mathcal{A}$-module (or Hilbert module if $\mathcal{A}$ is understood), we mean a Hilbert space $\mathcal{H}$ on which is defined a multiplication by elements of $\mathcal{A}$:

$$(a, h) \leftarrow \mathcal{A} \times \mathcal{H} \rightarrow a \cdot h \in H$$

which satisfies all the usual axioms for a module. We also demand that multiplication by an element of the module is continuous as a transformation on the Hilbert space. If $\mathcal{K}_+$ and $\mathcal{K}_-$ are two Hilbert modules, we denote by $\text{Hom}(\mathcal{K}_-, \mathcal{K}_+)$ the set of bounded linear transformations T from the Hilbert space $\mathcal{K}_-$ to the Hilbert space $\mathcal{K}_+$ which also commute with the module multiplication

$$T(a \cdot k) = a \cdot T(k) \qquad \text{for } a \in \mathcal{A}, k \in \mathcal{K}_-.$$

Each $T \in \text{Hom}(\mathcal{K}_-, \mathcal{K}_+)$ has its usual operator norm $\|T\|$ associated with it.

Let us now suppose that $\mathcal{K}_-$ and $\mathcal{K}_+$ are two given Hilbert modules and $\mathcal{J}$ is a submodule of $\mathcal{K}_+$ (i.e., a subspace of $\mathcal{K}_+$ which is also invariant under the module action). By the *error class* $\mathcal{E}$ we mean the set of all $T \in \text{Hom}(\mathcal{K}_-, K_+)$ with image contained in $\mathcal{J}(T\mathcal{K}_- \subset \mathcal{J})$. Now suppose that K is some fixed module homomorphism in $\text{Hom}(\mathcal{K}_-, \mathcal{K}_+)$. The abstract interpolation problem which we wish to pose is:

ABINT: When is there an F in the coset $K + \mathcal{E}$ such that $\|F\| \leq 1$? When such F's exist, describe the collection of all such F's.

As a motivating example, let $\mathcal{A}$ be the algebra H^∞, let $\mathcal{K}_- = \mathcal{K}_+ = H^2$, let $\mathcal{J} = bH^2$, where b is the Blaschke product with zeros at the points $\{z_i\}$ and let K be any H^∞ function with $K(z_i) = w_i$ $(1 \leq i \leq n)$. Note that H^2 is an H^∞-module with multiplication given by pointwise multiplication of functions, that $\mathrm{Hom}(\mathcal{K}_-, \mathcal{K}_+)$ in this case consists of multiplication operators M_F induced by H^∞ functions F and that $\|M_F\| = \|F\|$. It is now easy to see that ABINT for this case reduces to the Nevanlinna-Pick interpolation problem studied in the earlier sections. We shall later give many more examples.

To solve ABINT, we must assume that a certain generalized Beurling-Lax theorem holds. We consider the direct sum Hilbert module

$$\mathcal{K} = \begin{bmatrix} \mathcal{K}_+ \\ \mathcal{K}_- \end{bmatrix}$$

and let J be the operator on $\mathcal{K}$ with block matrix form

$$J = \begin{bmatrix} I_{\mathcal{K}_+} & 0 \\ 0 & -I_{\mathcal{K}_-} \end{bmatrix}.$$

Then $\mathcal{K}$ becomes a Krein space in the inner product

$$[k, k']_J = \langle Jk, k' \rangle_{\mathcal{K}}.$$

The Beurling-Lax axiom we need is as follows.

BL Axiom: Suppose $\mathcal{M}$ is a submodule of $\mathcal{K} = \mathcal{K}_+ \oplus \mathcal{K}_-$ which is regular with respect to $[\cdot,\cdot]_J$. Then there exist Hilbert modules $\mathcal{K}'_+$ and $\mathcal{K}'_-$ and a module homomorphism $\Theta \in \mathrm{Hom}(\mathcal{K}'_+ \oplus \mathcal{K}'_-, \mathcal{K}_+ \oplus \mathcal{K}_-)$ such that

$$\mathcal{M} = \Theta(\mathcal{K}'_+ \oplus \mathcal{K}'_-)$$

and

$$\Theta^\star \begin{bmatrix} I_{\mathcal{K}_+} & 0 \\ 0 & -I_{\mathcal{K}_-} \end{bmatrix} \Theta = \begin{bmatrix} I_{\mathcal{K}'_+} & 0 \\ 0 & -I_{\mathcal{K}'_-} \end{bmatrix}.$$

If the BL axiom holds, the solution of ABINT goes through in the same way as was done for the concrete special case of Nevanlinna-Pick interpolation in Section 2. We next sketch the main ideas of this analysis.

The first step is to formulate the problem in terms of the graph space

$$\mathcal{G} = \mathcal{G}_F = \begin{bmatrix} F \\ I \end{bmatrix} \mathcal{K}_-$$

of a desired solution F. By following the development in Section 2, we see that a subspace $\mathcal{G}$ of $\mathcal{K} = \mathcal{K}_+ \oplus \mathcal{K}_-$ is of the form $\mathcal{G} = \mathcal{G}_F$, where F is a solution of ABINT if and only if

(i) $\mathcal{G} \dot{+} \begin{bmatrix} \mathcal{K}_+ \\ 0 \end{bmatrix} = \begin{bmatrix} \mathcal{K}_+ \\ \mathcal{K}_- \end{bmatrix}$.

(ii) $\mathcal{G}$ is a submodule of $\mathcal{K}$.

(iii) $\mathcal{G}$ is a negative subspace with respect to $[\cdot, \cdot]_J$, where

$$J = \begin{bmatrix} I_{\mathcal{K}_+} & 0 \\ 0 & -I_{\mathcal{K}_-} \end{bmatrix},$$

and

(iv) $\mathcal{G} \subset \mathcal{M}$, where $\mathcal{M} = \begin{bmatrix} K \\ I \end{bmatrix} \mathcal{K}_- + \begin{bmatrix} \mathcal{J} \\ 0 \end{bmatrix}$.

Proposition 2.1 takes the form: There exist subspaces $\mathcal{G}$ satisfying (i), (iii), and (iv) if and only if the subspace

$$\mathcal{P} = \begin{bmatrix} \mathcal{K}_+ \\ 0 \end{bmatrix} \cap \mathcal{M} = \begin{bmatrix} \mathcal{J} \\ 0 \end{bmatrix},$$

is $\mathcal{M}$-maximal positive. Moreover, if $\mathcal{P}$ is $\mathcal{M}$-maximal positive, then the subspaces $\mathcal{G}$ which satisfy (i), (iii), and (iv) are exactly the $\mathcal{M}$-maximal negative subspaces. Also, it is easy to check that $\mathcal{P}$ is $\mathcal{M}$-maximal positive if and only if the operator $\Gamma = P_{\mathcal{J}^\perp} K$ has $\|\Gamma\| \leq 1$, and that $\mathcal{M}$ is regular if and only if $I_{\mathcal{K}_-} - \Gamma^*\Gamma$ is invertible. To bring in the submodule condition (ii), we assume $\|\Gamma\| < 1$ and apply the BL Axiom to $\mathcal{M}$. Thus

$$\mathcal{M} = \Theta \begin{bmatrix} \mathcal{K}'_+ \\ \mathcal{K}'_- \end{bmatrix},$$

where

$$\Theta^* J \Theta = J' \left(J' = \begin{bmatrix} I_{\mathcal{K}'_+} & 0 \\ 0 & I_{\mathcal{K}'_-} \end{bmatrix} \right)$$

for a module homomorphism $\Theta \in \mathrm{Hom}(\mathcal{K}', \mathcal{K})$. Since Θ is a module homomorphism, a submodule $\mathcal{G} \subset \mathcal{M}$ is of the form $\mathcal{G} = \Theta \mathcal{G}_1$, where $\mathcal{G}_1$ is a submodule of $\mathcal{K}'$, and conversely. Since $[\Theta k', \Theta k']_J = [k', k']_{J'}$, a subspace $\mathcal{G}$ is $\mathcal{M}$-maximal negative with respect to $[\cdot, \cdot]_J$ if and only if $\mathcal{G} = \Theta \mathcal{G}_1$, where $\mathcal{G}_1$ is $\mathcal{K}'$-maximal negative with respect to $[\cdot, \cdot]_{J'}$. But $\mathcal{K}'$-maximal negative submodules $\mathcal{G}_1$ are easily characterized as subspaces of the form

$$\mathcal{G}_1 = \begin{bmatrix} G \\ I \end{bmatrix} \mathcal{K}'_-$$

where $G \in \mathrm{Hom}(\mathcal{K}'_-, \mathcal{K}'_+)$ has $\|G\| \leq 1$.

From the identity

$$\begin{bmatrix} F \\ I \end{bmatrix} \mathcal{K}_- = \begin{bmatrix} \Theta_{11} & \Theta_{12} \\ \Theta_{21} & \Theta_{22} \end{bmatrix} \begin{bmatrix} G \\ I \end{bmatrix} \mathcal{K}'_-$$

for a solution F of ABINT, we get $F = (\Theta_{11}G + \Theta_{12})(\Theta_{21}G + \Theta_{22})^{-1}$. We thus have the following result.

Theorem 3.1. *Suppose $\mathcal{K}_-, \mathcal{K}_+$ and $\mathcal{J} \subset \mathcal{K}_+$ are given Hilbert modules, K is a given module homomorphism from $\mathcal{K}_-$ into $\mathcal{K}_+$, and that the BL Axiom holds for the pair $(\mathcal{K}_-, \mathcal{K}_+)$. Then solutions of ABINT exist if and only if the operator $\Gamma = P_{\mathcal{J}^\perp}K$ has $\|\Gamma\| \leq 1$. Moreover, if $\|\Gamma\| < 1$, then the class of all solutions F of ABINT coincides with the class of all F having a representation*

$$F = (\Theta_{11}G + \Theta_{12})(\Theta_{21}G + \Theta_{22})^{-1}$$

for some $G \in \mathrm{Hom}(\mathcal{K}'_-, \mathcal{K}'_+)$ with $\|G\| \leq 1$. Here

$$\Theta = \begin{bmatrix} \Theta_{11} & \Theta_{12} \\ \Theta_{21} & \Theta_{22} \end{bmatrix}$$

is the representer for

$$\mathcal{M} = \begin{bmatrix} K \\ I \end{bmatrix} \mathcal{K}_- + \begin{bmatrix} \mathcal{J} \\ 0 \end{bmatrix}$$

as in the BL Axiom.

I expect some form of Theorem 3.1 holds for an abstract form of the generalized Nevanlinna-Pick interpolation problem ($\ell > 0$) mentioned at the end of Section 2, but this remains to be worked out.

We now turn to examples. We have already seen how the Nevanlinna-Pick interpolation problem is one example. Some others are:

Example 1. As before $\mathcal{A} = H^\infty$, but now we choose $\mathcal{K}_- = H^2_n$($\mathbb{C}^n$-valued H^2-functions), $\mathcal{K}_+ = L^2_m$(= $\mathbb{C}^m$-valued L^2-functions on the circle), and $\mathcal{J} = H^2_m$. The module action of $\mathcal{A}$ is given by pointwise scalar multiplication. Then the module homomorphisms $T \in \mathrm{Hom}(\mathcal{K}_-, \mathcal{K}_+)$ are operators of multiplication by an $L^\infty_{m\times n}$ function ($m \times n$ matrix valued L^∞ function) and those which map $\mathcal{K}_- = H^2_n$ into $\mathcal{J} = H^2_m$ correspond to $H^\infty_{m\times n}$ functions. Moreover, for $F \in L^\infty_{m\times n}, \|F\|_\infty = \|M_F|H^2_n\|$. The generalized Beurling-Lax theorem from Ball-Helton [1983] gives the BL Axiom for this case. Then the ABINT problem for this case is: given $K \in L^\infty_{m\times n}$, find $F \in K + H^\infty_{m\times n}$ with $\|F\|_\infty \leq 1$. This is the matrix Nehari problem (see Nehari [1957] for the scalar case and Adamian-Arov-Krein [1978] for the matrix case). The operator Γ in this case is the matrix Hankel operator $\mathcal{H}_K : f \in H^2_n \to P_{H^{2\perp}_m}(Kf)$. This problem comes up in engineering (see e.g. Francis-Helton-Zames [1984], Glover [1984], Francis-Doyle [1987], Helton [1987]) and engineers have been busy working on computational algorithms; we will have more to say on this in Section 4.

Example 2. This is the same as Example 1, except that we take $\mathcal{K}_- = \varphi^{-1}H^2_n$ and $\mathcal{J} = BH^2_m$ where $\varphi^{\pm 1} \in L^\infty_{n\times n}, B^{\pm 1} \in L^\infty_{m\times m}$. When $K \in H^\infty_{m\times n}$ and B, φ are matrix Blashcke products, we pick up the tangential Nevanlinna-Pick interpolation problem (see Fedchina [1975a, 1975b]). This is studied at length in Ball-Helton [1983, 1986a]. There exist interesting applications of this problem and variations of it to circuit theory (Youla-Saitoh [1967], Helton [1978, 1980], Ball-Helton [1982a]) and to systems theory (Zames [1981], Zames-Francis [1983], Khargonekar-Tannenbaum [1985], Tannenbaum [1982]). Note that if we assume B and φ are unitary-valued and consider $B^{-1}K\varphi$ in place of K, Example 2 can be reduced to Example 1.

Example 3. This example seemingly has nothing to do with analytic functions. We take for $\mathcal{A}$ the algebra Ω_u of $n \times n$ upper triangular matrices. We set $\mathcal{K}_-$ also equal to Ω_u, $\mathcal{K}_+$ equal to the set M_n of all $n \times n$ matrices and $\mathcal{J} = \Omega_u \subset M_n = \mathcal{K}_+$. All these spaces of matrices are Hilbert spaces in the Hilbert-Schmidt norm (also known as the Frobenius norm). The set M_n is an Ω_u-module where the module multiplication is matrix multiplication on the right:

$$(a, h) \in \Omega_u \times M_n \to ha \in M_n.$$

[Strictly speaking, this makes M_n an anti-module since $M_{a_2}M_{a_1} = M_{a_1 a_2}$ but this causes no difficulties.] The module homomorphisms from M_n into itself are the operators of left multiplication by an $n \times n$ matrix. The module homomorphisms $\mathcal{E}$ which map $\Omega_u = \mathcal{K}_-$ to itself ($= \mathcal{J}$) are such operators with multiplier in Ω_u. Also the matrix spectral norm $\|F\|$ (its norm as an operator on $\mathbb{C}^n$) is the same as its operator norm as a left multiplier on Ω_u. The BL Axiom for this setting was verified in Ball-Gohberg [1986a] (see also Ball-Gohberg [1985]). The proof is really again a variation of the Halmos wandering subspace argument; an earlier version of it for the Hilbert space case can be found in Arveson [1975]. The ABINT problem for this case is the following matrix completion problem: We specify the lower triangular entries $k_{i,j} (1 \leq j < i \leq n)$ and ask if we can find upper triangular entries $k_{ij}(1 \le i \le j \le n)$ so that the resulting matrix $K = [k_{ij}]_{1 \le i,j \le n}$ has spectral norm $\|K\|$ at most 1. Theorem 3.1 for this setting was obtained in Ball-Gohberg [1986a]. The analysis can be generalized to handle complicated block upper triangular structure as well. The condition $\|\Gamma\| \le 1$ for this case can be viewed as the Arveson distance formula (see Arveson [1975]) for the finite-dimensional case. The generalized Beurling-Lax theorem for this setting also has connections with generalized LU and QR factorization of matrices (Ball-Gohberg [1986a-c]).

As a generalization of this example, one might consider more general sparsity patterns than block upper triangular. By a *sparsity pattern* we mean a subset $\mathcal{S}$ of indices $\{(i,j) : 1 \le i,j \le n\}$, where the entries F_{ij} of an $n \times n$ matrix F are left unspecified. Given a specification of numbers $\{k_{ij} : (i,j) \notin \mathcal{S}\}$ one can then ask if there exist a matrix $F = [F_{ij}]_{1 \le i,j \le n}$ with $\|F\| \le 1$ such that $F_{ij} = k_{ij}$ for $(i,j) \notin \mathcal{S}$. By the result of Johnson-Rodman [1986] (see also Kraus-Larson [1986] for an infinite dimensional version), the obvious necessary condition (that all rectangular submatrices formed for locations where the entries k_{ij} are specified be contractions) is also sufficient for all possible choices of initial data $\{k_{ij} : (i,j) \notin \mathcal{S}\}$ for a given $\mathcal{S}$, if and only if, after a reordering of the rows and of the columns the pattern $\mathcal{S}$ can be brought to a block upper triangular form. This suggests that the BL-axiom for matrices breaks down as soon as one moves away from algebras $\mathcal{A}$ other than the upper triangular matrices.

Example 4. This example combines Examples 1 and 3. We take as the algebra $\mathcal{A}$ the algebra $H^{\infty}_{0\Omega_u}$ of $n \times n$ matrix H^{∞}-functions A such that the value $A(0)$ at the origin is upper triangular ($A(0) \in \Omega_u$). For $\mathcal{K}_-$ we take $H^2_{0\Omega_u}$, the space of $n \times n$ matrix-valued H^2-functions with value at the origin in Ω_u. Take $\mathcal{K}_+$ to be $L^2_{n \times n}$, the space of all $n \times n$ matrix-valued L^2-functions. The Hilbert space norm on $L^2_{n\times n}$ (and on $H^2_{0\Omega_u}$) is given by integration over the circle of the square of the pointwise Hilbert-Schmidt norm. We let $\mathcal{J} = H^2_{0\Omega_u}$ as well. All these spaces ($\mathcal{K}_-, \mathcal{K}_+$, and $\mathcal{J}$) are modules over $\mathcal{A} = H^{\infty}_{0\Omega_u}$ where the module multiplication is taken to be right pointwise matrix multiplication. A module homomorphism from $\mathcal{K}_-$ into $\mathcal{K}_+$ is multiplication on the left by a function in $L^{\infty}_{n\times n}$. Those which map $\mathcal{K}_-$ into $\mathcal{J}$ are those with the multiplier in $H^{\infty}_{0\Omega_u}$. The operator norm of a left multiplication operator L_F on $H^{\infty}_{0\Omega_u}$ agrees with the $L^{\infty}_{n\times n}$ norm of the multiplier F. The BL Axiom for this case is due to William Helton and myself and is announced in Section 8 of Helton [1987]. The ABINT problem for this case is that of finding the $L^{\infty}_{n\times n}$ distance of a function $K \in L^{\infty}_{n\times n}$ to the subspace $H^{\infty}_{0\Omega_u}$; here also it is possible to replace *upper triangular* with *block* *upper triangular*. This distance problem (without the parametrization of all solutions) has been solved in very general asymmetric algebra settings by McAsey [1980, 1981] and Solel [1985]. The problem comes up when one tries to minimize the sensitivity of a linear discrete-time plant with a stabilizing periodic compensator and can also be solved (at least apart from the linear fractional parametrization) by an iterative procedure (see Feintuch-Khargonekar-Tannenbaum [1986], Feintuch and Francis [1984], Khargonekar-Poola-Tannenbaum [1985]).

Example 5. We choose $\mathcal{A}$ to be again simply H^∞ as in Example 1, but

$$\mathcal{K}_- = \begin{bmatrix} H^2 \\ L^2 \end{bmatrix}, \mathcal{K}_+ = \begin{bmatrix} L^2 \\ L^2 \end{bmatrix}, \mathcal{J} = \begin{bmatrix} H^2 \\ 0 \end{bmatrix}.$$

The module action is pointwise scalar multiplication. A module homomorphism from $\mathcal{K}_-$ to $\mathcal{K}_+$ is given by multiplication (on the left) by a matrix function F in $L^\infty_{2\times 2}$. Those which map $\mathcal{K}_-$ into $\mathcal{J}$ are those with multiplier in

$$\begin{bmatrix} H^\infty & 0 \\ 0 & 0 \end{bmatrix}.$$

Also the operator norm of a left multiplication operator L_F agrees with the $L^\infty_{2\times 2}$ norm $\|F\|$ of its symbol. Verification of the BL Axiom for this case is not too difficult. The ABINT problem is: given $K \in L^\infty_{2\times 2}$, find its $L^\infty_{2\times 2}$ distance to the subspace

$$\begin{bmatrix} H^\infty & 0 \\ 0 & 0 \end{bmatrix}.$$

Matrix versions of this problem can be handled by Theorem 3.1 as well; we only took the simplest case to keep the notation to a minimum. This problem is also related to sensitivity minimization for linear systems and has been solved before by other methods (see Feintuch-Francis [1986]). Computational algorithms are presented in Vidyasager [1985] and Francis-Doyle [1987]). Nir Cohen and I have worked out the computational consequences of the point of view presented here (Ball-Cohen [1987]).

Example 6. This example contains all the previous ones as special cases. We take $\mathcal{A}$ again to be $H^\infty_{0\Omega_u}$, but

$$\mathcal{K}_- = \begin{bmatrix} H^\infty_{0\Omega_u} \\ L^2_{m\times n} \end{bmatrix}, \mathcal{K}_+ = \begin{bmatrix} L^2_{n\times n} \\ L^2_{m\times n} \end{bmatrix}, \text{ and } \mathcal{J} = \begin{bmatrix} H^2_{0\Omega_u} \\ 0 \end{bmatrix}.$$

Each of $\mathcal{K}_-, \mathcal{K}_+, \mathcal{J}$ is an $\mathcal{A}$-module with the multiplication given by pointwise right matrix multiplication. The module homomorphisms are left multiplications by $L^\infty_{(n+m)\times n}$ functions, and the error class $\mathcal{E}$ corresponds to multipliers in

$$\begin{bmatrix} H^\infty_{0\Omega_u} & 0 \\ 0 & 0 \end{bmatrix}.$$

The ABINT problem for this case is to compute the $L^\infty_{(m+n)\times n}$-distance of a given function K in $L^\infty_{(n+m)\times n}$ to the subspace

$$\begin{bmatrix} H^\infty_{0\Omega_u} & 0 \\ 0 & 0 \end{bmatrix}.$$

Of course one must also work out the BL Axiom for this case. Israel Gohberg, William Helton and I plan to publish details soon.

Example 7. This time take

$$\mathcal{A} = \begin{bmatrix} H^\infty & H^\infty \\ 0 & H^\infty \end{bmatrix}, \mathcal{K}_- = \begin{bmatrix} H^2 & H^2 \\ 0 & H^2 \end{bmatrix},$$

$$\mathcal{K}_+ = \begin{bmatrix} L^2 & L^2 \\ L^2 & L^2 \end{bmatrix} \text{ and } \mathcal{J} = \begin{bmatrix} H^2 & H^2 \\ 0 & H^2 \end{bmatrix}.$$

Note that this is like Example 4 except we require upper triangularity to hold pointwise everywhere rather than just at the origin. $\mathcal{K}_-, \mathcal{K}_+$, and $\mathcal{J}$ are all $\mathcal{A}$-modules with pointwise right matrix multiplication, module homomorphisms are left multiplications by $L^\infty_{2\times 2}$-functions and the error class $\mathcal{E}$ corresponds to multipliers in

$$\begin{bmatrix} H^\infty & H^\infty \\ 0 & H^\infty \end{bmatrix}.$$

Moreover, the operator norm of a homomorphism is the same as the $L^\infty_{2\times 2}$-norm of its multiplier. The ABINT problem is to compute the $L^\infty_{2\times 2}$ norm of a given function K in $L^\infty_{2\times 2}$ to the subspace

$$\begin{bmatrix} H^\infty & H^\infty \\ 0 & H^\infty \end{bmatrix}$$

of upper triangular H^∞-functions. However, for this case, as William Helton and I once found (unpublished), it turns out that the BL Axiom fails, so Theorem 3.1 does not apply. In the proof of Theorem 3.1 there may well exist no $\mathcal{M}$-maximal negative subspaces which are also submodules. It is probably not the case that the existence problem is settled simply by the norm of the operator Γ. It would be interesting to understand this example better.

Example 8. Choose $\mathcal{A}$ to be the algebra $H^\infty(R)$ of bounded analytic functions on a finitely connected domain R. The Hardy space $H^2(R)$ can be defined (see Abrahamse and Douglas [1976]), and is a module over $H^\infty(R)$. $H^2(R)$ can be identified as a subspace of $L^2(\partial R)$ (square integrable functions over the boundary of R with respect to harmonic measure for some point $t \in R$). In short, we consider the analogue of Example 1 with the region R replacing the underlying domain $\mathcal{D}$. The Beurling-Lax theorem (Hilbert space version) exists but has a more complicated form due to the existence of multi-valued analytic functions on R (see Abrahamse-Douglas [1976]). There is evidence that this complication causes the failure of the Krein space version of the Beurling-Lax theorem (i.e. the BL Axiom for this case). The existence criterion for the analogues of the Nehari and Nevanlinna-Pick problems has been obtained (see Abrahamse [1979] and Ball [1979]) by a dual extremal method close to the commutant lifting approach of Sarason [1967]. However, Theorem 3.1 does not apply due to the failure of the BL Axiom and one does not get the linear fractional parametrization of all the solutions.

4. Computational Aspects

For practical applications one would like to be able to compute the symbol Θ of the linear fractional map in Theorem 3.1 explicitly in terms of the original data of the problem. Nevanlinna's adaptation of the Schur algorithm (and its matrix generalization by Dewilde and Dym [1981, 1984] and more recently by Limebeer and Anderson [preprint]) achieved this for the Nevanlinna-Pick interpolation problem but only in an implicit recursive way; in some statistical applications (as in Dewilde-Dym [1981, 1984]) where one is updating information at successive moments in time, this is exactly what is needed. However, for other applications one may want a closed form explicit formula in terms of the original interpolation data. For matrix interpolation problems such formulas have been obtained by Soviet authors (see Kovalishina-Potapov [1982], Krein-Nudelman [1977]) and also by Ball-Ran [1987c]; the latter used the description of the linear fractional map in Theorem 3.1 as the starting point for the computation of its symbol Θ. The paper Delsarte-Genin-Kamp [1979] relating Pick matrices to Toeplitz matrices and matrix orthogonal polynomials should also be insightful for computational purposes.

In this section we discuss in detail the computation of Θ for the case of the matrix Nehari problem (Example 1 from Section 3). We are given a matrix function $K \in L^\infty_{m\times n}$ for which we assume that the Hankel operator H_K: $H^2_n \to H^{2\perp}_m (f \to P_{H^{2\perp}_m}(Kf))$ has $\|H_K\| < 1$ and we wish to describe all $F \in K + H^\infty_{m\times n}$ satisfying $\|F\|_\infty \leq 1$. Adamian-Arov-Krein [1978] obtained an explicit formula for the linear fractional map symbol Θ in terms of the Hankel operator H_K; this was later refined by Dym-Gohberg [1983a,

1983b] for a Banach algebra setting. In this lecture we state and verify this very pretty formula of Adamian-Arov-Krein by using the invariant subspace characterization of the symbol Θ for the linear fractional map given by Theorem 3.1.

To describe the result we need a bit more notation. We define operators $\widehat{e}_k\colon L^2_m \to \mathbb{C}^m$ and $e_k\colon L^2_n \to \mathbb{C}^n$ for k an integer, and $Z\colon H^2_n \to H^2_n, Z_\star : H^{2\perp}_m \to H^{2\perp}_m, \gamma : \mathbb{C}^n \to \mathbb{C}^n, \gamma_\star : \mathbb{C}^m \to \mathbb{C}^m, M_z : L^2_n \to L^2_n, \widehat{M}_z : L^2_m \to L^2_m$ by

$$\begin{aligned}
&\widehat{e}_k\colon \sum_{j=-\infty}^{\infty} z^j f_j \to f_k \\
&e_k\colon \sum_{j=-\infty}^{\infty} z^j g_j \to g_k \\
&Z = (I - H_K^\star H_K)^{-1} \\
&Z_\star = (I - H_K H_K^\star)^{-1} \\
&\Gamma = (e_0 Z e_0^\star)^{-1/2} \\
&\Gamma_\star = (\widehat{e}_{-1} Z_\star \widehat{e}^\star_{-1})^{-1/2} \\
&M_z\colon g(z) \to z g(z) \\
&\widehat{M}_z\colon f(z) \to z f(z).
\end{aligned}$$

Note that the adjoint maps $\widehat{e}_K^\star$ and $e_K^\star$ are given by

$$\begin{aligned}
&\widehat{e}_k^\star : d \in \mathbb{C}^m \to z^k d \in L^2_m \\
&e_k^\star : c \in \mathbb{C}^n \to z^k c \in L^2_n
\end{aligned}.$$

Theorem 4.1. *(see Adamian et al. [1978] and Dym-Gohberg [1983]) Suppose $K \in L^\infty_{m\times n}$ is such that $\|H_K\| < 1$. Let $\Theta\colon \mathbb{C}^{m+n} \to L^2_{m+n}$ be the map*

$$\Theta = \begin{bmatrix} \widehat{M}_z Z_\star \widehat{e}^\star_{-1}\gamma_\star & H_K Z e_0^\star \gamma \\ M_z H_k^\star Z_\star \widehat{e}^\star_{-1}\gamma_\star & Z e_0^\star \gamma \end{bmatrix} \tag{4.1}$$

and let

$$\Theta(z) = \begin{bmatrix} \Theta_{11}(z) & \Theta_{12}(z) \\ \Theta_{21}(z) & \Theta_{22}(z) \end{bmatrix}$$

be the associated $L^2_{(m+n)\times(m+n)}$ matrix function defined by

$$\Theta(z)c = (\Theta c)(z)c \in \mathbb{C}^{m+n}.$$

Then the set of all H^∞-approximants $F \in K + H^\infty_{m\times n}$ to K having $\|F\|_\infty \le 1$ coincides with the collection of all matrix functions F of the form

$$F = (\Theta_{11} G + \Theta_{12})(\Theta_{21} G + \Theta_{22})^{-1}$$

for some $G \in H^{\infty}_{m\times n}$ *with* $\|G\|_{\infty} \leq 1$.

Proof: By Theorem 3.1 Θ is characterized by any matrix function satisfying the two conditions

(i) $\Theta H^2_{m+n} = \begin{bmatrix} K \\ I \end{bmatrix} H^2_n + \begin{bmatrix} H^2_m \\ 0 \end{bmatrix}$ and

(ii) $\Theta(z)^\star J \Theta(z) = J$ for $|z| = 1$.

[Actually we are being a little sloppy here; without an extra assumption on K (such as continuous differentiability, for example), Θ may be unbounded and the left side of (i) should be amended to "the L^2_{m+n}-closure of ΘH^{∞}_{m+n}." This, however, does not affect any of the argument to follow.] By the proof of the Krein space Beurling-Lax theorem from Ball-Helton [1983], Θ is characterized, up to a right constant invertible factor, by the condition that it map the constants $\mathbb{C}^{m+n}$ injectively onto the wandering subspace

$$\mathcal{L} = \mathcal{M} \cap \begin{bmatrix} \widehat{M}_z & 0 \\ 0 & M_z \end{bmatrix} \mathcal{M}^{\perp_J}$$

where

$$\mathcal{M} = \begin{bmatrix} K \\ I \end{bmatrix} H^2_n + \begin{bmatrix} H^2_m \\ 0 \end{bmatrix},$$

and $\mathcal{M}^{\perp_J}$ is the J-orthogonal complement of $\mathcal{M}$ in L^2_{m+n}. It is straightforward to see that

$$\mathcal{M} = \begin{bmatrix} H_K \\ I \end{bmatrix} H^2_n + \begin{bmatrix} H^2_m \\ 0 \end{bmatrix} \tag{4.2}$$

$$\mathcal{M}^{\perp_J} = \begin{bmatrix} I \\ H^\star_K \end{bmatrix} H^{2^\perp}_m + \begin{bmatrix} 0 \\ H^{2\perp}_n \end{bmatrix} \tag{4.3.}$$

From the form of Θ in (4.1), it is immediate that the first column maps $\mathbb{C}^m$ into

$$\begin{bmatrix} \widehat{M}_z & 0 \\ 0 & M_z \end{bmatrix} \mathcal{M}^{\perp_J}$$

and that the second column maps $\mathbb{C}^n$ into $\mathcal{M}$. To see that the first column maps $\mathbb{C}^m$ into $\mathcal{M}$, we must show:

$$H_K M_z H^*_K Z_* \widehat{e}^*_{-1} = P_{H^{2\perp}} \widehat{M}_z Z_* \widehat{e}^*_{-1}. \tag{4.4}$$

Using the identity $H_K M_z | H^2_n = P_{H^{2\perp}_m} \widehat{M}_z H_K$, we see the left side of (4.4) equals

$$P_{H^{2\perp}_m} \widehat{M}_z H_K H^\star_K Z_\star \widehat{e}^\star_{-1}.$$

From the definition of $Z_\star$ as $(I - H_K H_K^\star)^{-1}$ we get $H_K H_K^\star Z_\star = -I + Z_\star$ on $H_m^{2\perp}$ and the above becomes

$$P_{H_m^{2\perp}} \widehat{M}_z(-I + Z_\star)\widehat{e}_{-1}^\star = P_{H_m^{2\perp}} \widehat{M}_z Z_\star \widehat{e}_{-1}^\star$$

and (4.4) is verified. To see that the second column of Θ maps $\mathbb{C}^n$ into

$$\begin{bmatrix} \widehat{M}_z & 0 \\ 0 & M_z \end{bmatrix} \mathcal{M}^{\perp_J}$$

we must show

4.5 $$H_K^\star \widehat{M}_{z^{-1}} H_K Z e_0^\star = P_{H_n^2} M_{z^{-1}} Z e_0^\star.$$

This is just a dual version of (4.4); it follows in the same way as (4.4) using the identities $H_K^\star \widehat{M}_{z^{-1}} | H_m^{2\perp} = P_{H_n^2} M_{z^{-1}} H_K^\star$,

$$H_K^\star H_K Z = -I + Z \text{ and } P_{H_n^2} M_{z^{-1}} e_0^\star = 0.$$

This completes the verification that Θ as in (4.1) maps $\mathbb{C}^{m+n}$ into $\mathcal{L}$. By a computation to follow Θ is also injective; since we know $\dim \mathcal{L} = m+n$, Θ is also onto $\mathcal{L}$. By the usual wandering subspace argument, we get that $\Theta(z)H_{m+n}^2 = \mathcal{M}$ and that $\Theta(z)(H_{m+n}^2)^{\perp_J} = \mathcal{M}^{\perp_J}$, and in particular condition (i).

To guarantee (ii), we need only check that $\Theta : \mathbb{C}^{m+n} \to L_{m+n}^2$ as given by (4.1) is a J-isometry, namely

$$\Theta^\star \begin{bmatrix} I_{L_m^2} & 0 \\ 0 & -I_{L_n^2} \end{bmatrix} \Theta = \begin{bmatrix} I_m & 0 \\ 0 & -I_n \end{bmatrix}.$$

This amounts to the three identities:

4.6 $$\widehat{e}_{-1} Z_\star \widehat{M}_z^{-1} \widehat{M}_z Z_\star \widehat{e}_{-1}^\star - \widehat{e}_{-1} Z_\star H_K M_{z^{-1}} M_z H_K^* Z_* \widehat{e}_{-1}^* = \widehat{e}_{-1} Z_\star \widehat{e}_{-1}^\star$$

4.7 $$e_0 Z H_K^\star H_K Z e_0^\star - e_0 Z^2 e_0^\star = -e_0 Z e_0^\star$$

and

4.8 $$\widehat{e}_{-1} Z_\star \widehat{M}_{z^{-1}} H_K Z e_0^\star - \widehat{e}_{-1} Z_\star H_K P_{H_n^2} M_{z^{-1}} Z e_0^\star = 0.$$

The left side of (4.6) immediately collapses to

$$\widehat{e}_{-1} Z_\star^2 \widehat{e}_{-1}^\star - \widehat{e}_{-1} Z_\star H_K H_K^\star Z_\star \widehat{e}_{-1}^\star.$$

From the definition of $Z_\star$ we get

$$Z_\star H_K H_K^\star = H_K H_K^\star Z_\star,$$

so the left side of (4.6) becomes

$$\widehat{e}_{-1}[I - H_K H_K^\star] Z_\star^2 \widehat{e}_{-1}^\star = \widehat{e}_{-1} Z_\star \widehat{e}_{-1}^\star$$

as desired. Equation (4.7) follows in a similar way from

$$Z H_K^\star H_K Z - Z^2 = (H_K^\star H_K - I) Z^2 = -Z.$$

To verify (4.8), we need to show that $Z_\star \widehat{M}_{z^{-1}} H_K Z e_0^\star - Z_\star H_K P_{H_n^2} M_{z^{-1}} Z e_0^\star$ has range in $\widehat{M}_{z^{-1}} H_m^{2\perp}$. This follows if we show

$$\widehat{M}_{z^{-1}} H_K Z e_0^\star - H_K P_{H_n^2} M_{z^{-1}} Z e_0^\star = (I - H_K H_K^\star) \widehat{M}_{z^{-1}} H_K Z e_0^\star. \tag{4.9}$$

Clearly (4.9) is equivalent to

$$H_K P_{H_n^2} M_{z^{-1}} Z e_0^\star = H_K H_K^\star \widehat{M}_{z^{-1}} H_K Z e_0^\star.$$

From the intertwining $H_K^\star \widehat{M}_{z^{-1}} = P_{H_n^2} M_{z^{-1}} H_K^\star$, this in turn becomes

$$H_K P_{H_n^2} M_{z^{-1}} Z e_0^\star - H_K P_{H_n^2} M_{z^{-1}} H_K^\star H_K Z e_0^\star = 0.$$

To verify this, compute

$$H_K P_{H_n^2} M_{z^{-1}} (I - H_K^\star H_K) Z e_0^\star = H_K P_{H_n^2} M_{z^{-1}} e_0^\star = 0.$$

This completes the verification of all properties required of Θ. ■

Adamian-Arov-Krein did not derive their formula from these considerations, but rather by the now famous method of one-step extensions. Given their answer, we were able to verify above its validity directly by using the Beurling-Lax-Halmos point of view from Theorem 3.3. Nevertheless, in principle it is possible to derive a formula like (4.1) directly using Theorem 3.1 as a starting point. The steps are as follows:

Step 1. Find a preliminary map Θ_0 which maps $\mathbb{C}^{m+n}$ onto $\mathcal{L}$; this amounts to computing a parametrization of $\mathcal{L}$.

Step 2. Compute the operator $W = \Theta_0^\star J \Theta_0 : \mathbb{C}^{m+n} \to \mathbb{C}^{m+n}$.

Step 3. Compute the J-Cholesky factorization of W: $W = \widehat{\gamma}^\star J \widehat{\gamma}$ for some $\widehat{\gamma}$ on $\mathbb{C}^{m+n}$.

Necessarily, $\widehat{\gamma}$ is inverible.

Step 4. Set Θ equal to $\Theta_0 \widehat{\gamma}^{-1}$.

If one were fortuitous enough to come up with

$$\Theta_0 = \begin{bmatrix} \widehat{M}_z Z_\star \widehat{e}_{-1}^\star & H_K Z e_0^\star \\ M_z H_K^\star Z_\star \widehat{e}_{-1}^\star & Z e_0^\star \end{bmatrix}$$

for Step 1, then Step 2 would produce

$$W = \begin{bmatrix} \widehat{e}_{-1} Z_\star \widehat{e}_{-1}^\star & 0 \\ 0 & -e_0 Z e_0^* \end{bmatrix}$$

(if one were also fortuitous enough to note all the identities in the above derivation). Since W is diagonal, Step 3 would then be easy; we would get

$$\widehat{\gamma} = \begin{bmatrix} \gamma_*^{-1} & 0 \\ 0 & \gamma^{-1} \end{bmatrix}.$$

Then, finally, Step 4 would produce the Θ as given in (4.1). However, without peeking, as the result of Step 1, I was able to come up with

$$\Theta_0 = \begin{bmatrix} \widehat{M}_z Z_\star \widehat{e}_{-1}^\star & \widehat{M}_z Z_\star \widehat{M}_{z^{-1}} H_K e_0^\star \\ M_z H_K^\star Z_\star \widehat{e}_{-1}^\star & M_z H_K^\star Z_\star \widehat{M}_{z^{-1}} H_K e_0^\star + e_0^\star \end{bmatrix}. \tag{4.10}$$

After a laborious computation, Step 2 produces

$$W = \begin{bmatrix} \widehat{e}_{-1} Z_\star \widehat{e}_{-1}^\star & \widehat{e}_{-1} Z_\star \widehat{M}_{z^{-1}} H_K e_0^\star \\ e_0 H_K^\star P_{H_m^{2\perp}} \widehat{M}_z Z_\star \widehat{e}_{-1}^\star & -I_n \end{bmatrix}.$$

From the general identity

$$\begin{bmatrix} A & B \\ B^\star & -I \end{bmatrix} = \begin{bmatrix} I & B \\ 0 & I \end{bmatrix} \begin{bmatrix} A + B^\star B & 0 \\ 0 & -I \end{bmatrix} \begin{bmatrix} I & 0 \\ B^\star & I \end{bmatrix},$$

we see that Step 3 yields

$$\widehat{\gamma} = \begin{bmatrix} (A + BB^\star)^{1/2} & 0 \\ B^\star & I \end{bmatrix}, \tag{4.11}$$

where

$$A = \widehat{e}_{-1} Z_* \widehat{e}_{-1}^*, B = \widehat{e}_{-1} Z_* \widehat{M}_{z^{-1}} H_K e_0^*,$$

Finally, Step 4 gives

4.12 $$\Theta = \Theta_0 \widehat{\gamma}^{-1} \text{ with } \Theta_0 \text{ as in (4.10) and } \widehat{\gamma} \text{ as in (4.11).}$$

We have shown that both choices (4.1) and (4.12) for Θ are valid, although the Adamian-Arov-Krein choice is certainly more elegant; it follows that these two Θ's differ only by a right factor which is a J-unitary operator on $\mathbb{C}^{m+n}$.

Notes:

1. It is probably the case that an elegant formula like (4.1) exists for the other examples of the abstract interpolation problem presented at the end of the last section. These remain to be worked out.

2. In connection with sensitivity minimization and model reduction problems (see Doyle-Francis [1987] for a survey) from linear systems theory, one assumes that one knows a *realization* of K as the transfer function of a linear system

4.13 $$K(z) = C(zI - A)^{-1}B$$

where the matrices A, B, C have sizes $p \times p$, $p \times n$, and $m \times p$ respectively. One then seeks an explicit formula for the symbol Θ of the linear fractional map, but in terms of the matrices A, B, C appearing in the realization (4.13). Also, one would like to have the matrix function $\Theta(z)$ also realized as the transfer function of a linear system

$$\Theta(z) = \underline{D} + \underline{C}(zI - \underline{A})^{-1}\underline{B}$$

as this has a nice physical interpretation. This problem was solved completely by Glover [1984] for the continuous time setting (where the left half plane replaces the unit disk). One can then get an analogous result for the discrete-time case (the case discussed here) by a linear fractional change of variable. Andre Ran and I recently obtained two more derivations of these results, both of which used the Beurling-Lax characterization of Θ in Theorem 3.1 as the starting point; the first [1986, 1987a] reduced the problem to a now well-understood matrix Wiener-Hopf factorization problem while the second [1987b] reformulated the problem as one of recovering a function from its spectral data (the *inverse spectral problem*).

5. Variations on the Interpolation Problem

Here we discuss two other types of interpolation problems important for applications and not covered in the previous sections

Boundary Interpolation problems

We consider only the simplest scalar case of a boundary version of the Nevanlinna-Pick interpolation problem. We now are given an n-tuple $\{z_i\}$ of points on the boundary of the unit disk together with an n-tuple of complex numbers $\{w_i\}(1 \leq i \leq n)$. We seek a function f in the class $\mathcal{B}$ of functions analytic on the unit disk with modulus bounded by 1 such that $f(z_i) = w_i$ for $1 \leq i \leq n$. In general, for z_i a point with $|z_i| = 1$ and $f \in \mathcal{B}, f(z_i)$ is defined as a radial limit $\lim_{r \nearrow 1} f(rz_i)$. An obvious necessary condition for such an f to exist is that $|w_i| \leq 1$ for $1 \leq i \leq n$. This turns out to be also sufficient. This was observed by Khargonekar and Tannenbaum [1985] by using the Pick matrix condition on a disk slightly larger than the unit disk and taking a limit. The problem in this form comes up in some control theoretic contexts (see Khargonekar-Tannenbaum [1955]). We get a problem of more mathematical interest if we impose an additional constraint on the radial derivative at each point z_i. The full problem is: Find $f \in \mathcal{B}$ such that $f(z_i) = w_i$ and $z_i \overline{w}_i f'(z_i) \leq p_i$ for $1 \leq i \leq n$ where $\{z_i\}$ and $\{w_i\}$ are given n-tuples of complex numbers of modulus 1 and $\{p_i\}$ is an n-tuple of nonnegative real numbers. The inequality in the derivative constraints guarantees that the set of interpolating functions is weak-* closed in H^∞; this makes the problem more natural mathematically. This type of boundary interpolation problem arises in circuit theory, in particular in the study of classical LC ladder filters (see e.g. Helton [1977]).

This problem was already studied by Nevanlinna [1919]. He showed that the Schur algorithm procedure given in Section 1 can be adapted for this case as well. The symbol Θ for the linear fractional map which parametrizes all the solutions (when they exist) in this case is a rational 2×2 matrix function with a pole and a zero at each interpolating point z_i on the boundary; at all other boundary points $\Theta(z)$ is a J-unitary matrix

$$\left(J = \begin{bmatrix} 1 & 0 \\ 0 & -1 \end{bmatrix} \right).$$

[By definition, a point z_0 is a pole of the matrix function Θ if it shows up as a pole in some matrix entry of Θ, and z_0 is a zero if it is a pole of the matrix function Θ^{-1}; here we assume that $\det \Theta$ does not vanish identically.] The associated Pick matrix for the problem is

$$\Lambda = [\Lambda_{ij}]_{1 \leq i,j \leq n}$$

where

$$\Lambda_{ij} = \begin{cases} \frac{1-w_j\overline{w}_i}{1-z_j\overline{z}_i} & \text{if } i \neq j \\ p_i & \text{if } i = j. \end{cases}$$

In Ball-Helton [1986c] it was shown how this problem can be handled using the same Grassmannian approach as was discussed in Section 2. Indeed, let k be any H^∞-function which satisfies the interpolation constraints

$$k(z_i) = w_i, z_i\overline{w}_i k'(z_i) = p_i$$

for $1 \leq i \leq n$ and let $b(z)$ be the polynomial

$$b(z) = \prod_{i=1}^{n}(z - z_i).$$

Then a function f in H^∞ satisfies all the interpolation constraints with equality if and only if $f \in k + b^2 H^\infty$, and with only inequality on the derivative constraints if and only if f is in the H^∞-closure of $k + b^2 H^\infty$. Instead of looking for interpolants f in $\mathcal{B}$, we seek their graph subspaces. To set this up, it turns out that the right space to work with is

$$\mathcal{M} = \begin{bmatrix} b & kb^{-1} \\ 0 & b^{-1} \end{bmatrix} H_2^2.$$

Note that $\mathcal{M}$ is not contained in L_2^2 due to the pole in b^{-1} at each z_j (which are not points on the unit circle). Instead, we have $\mathcal{M} \subset \mathcal{K}$ where

$$\mathcal{K} = \begin{bmatrix} b & kb^{-1} \\ 0 & b^{-1} \end{bmatrix} L_2^2.$$

A key point is that the matrix function

$$M_{k,b} = \begin{bmatrix} |b|^2 & \overline{b}kb^{-1} \\ \overline{b}^{-1}\overline{k}b & \overline{b}^{-1}(|k|^2 - 1)b \end{bmatrix} = \begin{bmatrix} \overline{b} & 0 \\ \overline{b}^{-1}\overline{k} & \overline{b}^{-1} \end{bmatrix} \begin{bmatrix} 1 & 0 \\ 0 & -1 \end{bmatrix} \begin{bmatrix} b & kb^{-1} \\ 0 & b^{-1} \end{bmatrix}$$

and its inverse are bounded on the unit circle. This implies that $\mathcal{K}$ is a Krein space with inner product

$$[F, F]_{\mathcal{K}} =< M_{k,b} \begin{bmatrix} b & kb^{-1} \\ 0 & b^{-1}cr \end{bmatrix}^{-1} F, \begin{bmatrix} b & kb^{-1} \\ 0 & b^{-1} \end{bmatrix}^{-1} F >_{L_2^2}.$$

Formally the inner product is simply

$$[F, F]_{\mathcal{K}} =< \begin{bmatrix} 1 & 0 \\ 0 & -1 \end{bmatrix} F, F >_{L_2^2}.$$

This is only formal since in general $F \notin L^2_2$. Even so, $\infty - \infty$ cancellation causes this inner product to make sense for all $f \in \mathcal{K}$. We now follow the program given in Section 2. The graph space associated with a function $f \in H^\infty$ we take to be

$$\mathcal{G}_f = \begin{bmatrix} f \\ 1 \end{bmatrix} b^{-1} H^2.$$

In general, a subspace $\mathcal{G} \subset \mathcal{K}$ is a graph space (of this form with its angle operator having domain equal to $b^{-1}H^2$) if and only if

(i)

$$\mathcal{G} \dot{+} \begin{bmatrix} b^{-1}L^2 \\ 0 \end{bmatrix} = \begin{bmatrix} b^{-1}L^2 \\ b^{-1}H^2 \end{bmatrix}.$$

Note that the spaces

$$\begin{bmatrix} 0 \\ b^{-1}H^2 \end{bmatrix} \text{ and } \begin{bmatrix} b^{-1}L^2 \\ 0 \end{bmatrix}$$

are not in the Krein space $\mathcal{K}$. Nevertheless, we may consider $\mathcal{K}$ as a linear manifold in the linear space

$$\begin{bmatrix} b^{-1}L^2 \\ b^{-1}H^2 \end{bmatrix}$$

and (i) is a well-defined, purely algebraic, condition. Graph spaces whose angle operator is a multiplication operator again are characterized by

(ii)

$$M_z \mathcal{G} \subset \mathcal{G}$$

where M_z is as usual multiplicaion by the function z. The angle operator M_f has a multiplier f in the unit ball $\mathcal{B}$ of H^∞ if and only if in addition

(iii) $\mathcal{G}$ is a negative subspace with respect to $[\cdot,\cdot]_{\mathcal{K}}$.

The condition that the multiplier f of the angle operator M_f satisfy the interpolation constraints with equality $f \in k + b^2 H^\infty$ translates to

(iv) $\mathcal{G} \subset \mathcal{M}$ where

$$\mathcal{M} = \begin{bmatrix} b^{-1} & kb^{-1} \\ 0 & b^{-1} \end{bmatrix} H^2_2.$$

If we intersect both sides of (i) with $\mathcal{M}$, we see that (i) and (iv) can be combined to the direct sum condition

$$\mathcal{G} \dot{+} \begin{bmatrix} bH^1 \\ 0 \end{bmatrix} = \mathcal{M}. \tag{5.1}$$

We also want to impose condition (iii) that $\mathcal{G}$ be $[\cdot,\cdot]_{\mathcal{K}}$-negative. We are in a position to apply Lemma 1.1; the catch, however, is that

$$\begin{bmatrix} bH^2 \\ 0 \end{bmatrix},$$

while being a strictly positive subspace of $\mathcal{M}$, fails to be a regular subspace. Thus Lemma 1.1 must be generalized to this more delicate situation. With the assumption that

$$\mathcal{P} = \begin{bmatrix} bH^2 \\ 0 \end{bmatrix}$$

is $\mathcal{M}$-maximal positive, we do get that regular $\mathcal{M}$-maximal negative subspaces $\mathcal{G}$ satisfy (5.1) but in general, the manifold

$$\mathcal{G} + \begin{bmatrix} bH^2 \\ 0 \end{bmatrix}$$

can be a dense proper linear submanifold of $\mathcal{M}$. In this case it can be shown that the corresponding angle operator f for $\mathcal{G}$ satisfies the interpolation conditions with inequalities on the derivatives.

Assume that $\mathcal{P}$ is maximal positive and that we are in the generic situation where $\mathcal{M}$ is regular. Then $\mathcal{M}$ again has a Beurling-Lax representation

$$\mathcal{M} = \Theta H_2^2$$

for a rational matrix function $\Theta(z)$ satisfying

$$\Theta(z)^* J \Theta(z) = J.$$

In this case of course Θ must have poles and zeros on the unit circle. As before, Θ defines a linear fractional map

$$g \to (\Theta_{11} g + \Theta_{12})(\Theta_{21} g + \Theta_{22})^{-1}$$

which parametrizes all solutions of the interpolation problem (with only inequalities in the derivative constraints). The existence criterion, that the subspace $\mathcal{P}$ be maximal positive, again can be shown to be equivalent to the positive semidefiniteness of the Pick matrix. For details, see Ball-Helton [1986c].

Other types of boundary interpolation problems appear in the literature. Abrahamse and Fisher [1980] considered the problem of interpolating finitely many arcs on the unit circle with a Blaschke product. Löwner [1943] studied the problem of finding a function mapping the upper half plane into itself whose restriction to a finite interval on the line is a prescribed smooth function; Rosenblum and Rovnyak [1975, 1980] later gave operator-theoretic treatments of this problem.

5.2 More General Ranges

Another variation on the Nevanlinna-Pick interpolation problem is the following. We are given an n-tuple $\{z_i\}$ of distinct complex numbers in the open unit disk D, an n-tuple $\{w_i\}$ of complex numbers and a region Ω in the complex plane. The problem is to find a function f analytic on the disk such that $f(z_i) = w_i$ for $l \leq i \leq n$ and such that $f(D) \subset \lambda$. When $\lambda = D$ this is just Nevanlinna-Pick interpolation. When $\lambda = \mathbb{C}$, this is often called Lagrangian interpolation; then the problem is always solvable and one can take f to be analytic on all of $\mathbb{C}$ (in fact, a polynomial). When λ is a simply connected region, not the whole complex plane, by the Riemann mapping theorem there exists a conformal map $\psi : D \to \lambda$. Then f interpolates $\{z_i\}$ and $\{\psi^{-1}(w_i)\}$ and maps D into D. By th Pick matrix condition there exists such a function g if and only if the matrix

5.2
$$\left[\frac{1 - \overline{\psi^{-1}(w_j)}\psi^{-1}(w_i)}{1 - \overline{z}_j z_i}\right]_{1 \leq i,j \leq n}$$

is positive semi-definite. Given any such function g, $f = \psi \circ g$ then solves the original interpolation problem. Thus (5.2) gives a necessary and sufficient condition for the problem to be solvable for the case that λ is a proper simply connected region in the complex plane.

This problem (with λ simply connected) has come up in connection with the robust stabilization problem for linear systems (see Tannenbaum [1986] and Khargonekar-Tannenbaum [1985]). When λ is such that the Riemann map ψ can be computed explicitly, then the Pick matrix above is explicit. For multi-input multi-output systems, there arises a matrix analogue of this problem; here the results are less satisfactory.

We now consider the case where λ is finitely connected–say λ has N holes. Then we no longer have a conformal map but instead a covering map $\psi : D \to \lambda$. The inverse map $\psi^{-1} : \lambda \to D$ is now multivalued; for each point $w \in \lambda$ there exist countably many choices for $\psi^{-1}(w)$. If $f : D \to \lambda$ is analytic, then there is an analytic function $g : D \to D$ such that $f = \psi \circ g$; indeed, this is the basic property of the covering map ψ. If f also interpolates $\{z_i\}$ and $\{w_i\}$, then g interpolates $\{z_i\}$ and $\{\psi^{-1}(w + i)\}$ for some choice of inverse image points $\psi^{-1}(w_i), \ldots, \psi^{-1}(w_n)$. Thus, for the appropriate choice of n-tuple $\{\psi^{-1}(w_i)\}(1 \leq i \leq n)$, the Pick matrix

$$\left[\frac{1 - \overline{\psi^{-1}(w_j)}\psi^{-1}(w_i)}{1 - \overline{z}_j z_i}\right]_{1 \leq i,j \leq n}$$

is positive semidefinite. Conversely, if g interpolats $\{z_i\}$ and some n-tuple of the form $\{\psi^{-1}(w_i)\}$, then $f = \psi \circ g$ solves the original interpolation problem. Thus we see that our problem has a solution if and only if at least one of countably many different Pick matrices is positive semidefinite–a rather unsatisfactory solution.

To be specific, let us take $\lambda = D/\{0\}$. The interpolation problem then is the standard Nevanlinna-Pick interpolation with the extra constraint imposed that the interpolant f should not vanish on the disk. This problem turned up in a couple of stabilization problems in systems theory (see Ball-Helton [1979] and Tannenbaum [1982]). In this case it is easier to work with the covering map $z \to e^{-z} = \psi(z)$ of the right half plane (RHP) onto λ rather than of D onto λ. Then inverse map ψ^{-1} is given by $\psi^{-1}(w) = -\ell n w - 2\pi\sqrt{-1}k (k = 0, \pm 1, \pm 2, \ldots)$. The condition in general that there exist an analytic function $g : D \to RHP$ which interpolates $\{z_i\}$ and $\{\eta_i\}$ is that the matrix

$$\left[\frac{\overline{\eta}_j + \eta_i}{1 - \overline{z}_j z_i}\right]_{1 \le i,j \le n}$$

be positive semidefinite. Thus the condition that there exist an analytic $f : D \to D\backslash\{0\}$ which interpolates $\{z_i\}$ and $\{w_i\}$ is that for some choice of n integers $\{k_1, \ldots, k_n\}$ the matrix

$$\left[\frac{-\overline{\ell n w_j} - \ell n w - i + a\pi\sqrt{-1}(k_j - k_i)}{1 - \overline{z}_j z_i}\right]_{1 \le i,j \le n}$$

be positive semidefinite. In practice, it should be possible to test only finitely many of these Pick matrices. A detailed study of this type of interpolation problem using constructive variational tchniques has been done by Scales [1982].

NOTES: These are just two of the variations that could have been mentioned. Others which also have applications in engineering are: control of the matrix singular values rather than the norm on the unit circle (see Helton [1980] and Ball-Helton [1982a]), demanding J-contractive matrix values on the unit disk rather than contractive values (see Ball-Helton [1982a]) and the imposition of additional symmetry constraints on the interpolants (see Ball-Helton [1986a]). There has also been work by Dym-Gohberg [1982a-b, 1983a-b] and Ball [1984] on Wiener-Hopf factorization in connection with interpolation problems. Dym and Gohberg also study maximization of entropy. Analogous results (Dym-Gohberg [1982a], Ball-Gohberg [1986c]) also hold for finite matrices.

References

1. Abrahamse, M. B., The Pick interpolation theorem for finitely connected domains, Mich. Math. J. **26** (1979) , 195–203.
2. _______ and R. G. Douglas, A class of subnormal operators related to multiply connected domains, Advances in Math. **19** (1976), 106–148.
3. _______ and S. D. Fischer, Mapping intervals to intervals, Pacific J. Math. **91** (1971), 13–27.
4. _______, Infinite block Hankel matrices and related extension problems, AMS Transl. **111** (1978), 133–156.
5. Adamjan, V. M., D. Z. Arov, and M. G. Krein, Analytic properties of Schmidt pairs for a Hankel operator and the generalized Schur-Takagi problem, Math. U.S.S.R. Sbornik **15** (1971), 31-72.
6. Arveson, W. B., Interpolation problems in nest algebras, J. Funct. Anal. **20** (1975), 208–233.
7. Ball, J. A., A lifting theorem for operator models of finite rank on multiply connected domains, J. Operator Theory **1** (1979), 3–25.
8. _______, A non-Euclidean Lax-Beurling theorem with applications to matrical Nevanlinna–Pick interpolation, in Toeplitz Centennial: Toeplitz Memorial Conference in Operator Theory (ed. I. Gohberg), vol. 0T3, (1982), Birkhauser, 67–84.
9. _______, Interpolation problems of Pick–Nevanlinna and Loewner types for meromorphic matrix functions, *Integral Equations and Operator Theory*, **6** (1983), 804–840.
10. _______, Invariant subspace representations, unitary interpolants and factorization indices, in *Topics in Operator Theory Systems and Networks* (ed. H. Dym and I. Gohberg), vol. 0T 12 (Birkhauser) (1984), 11-38.
11. Ball, J. A. and N. Cohen, Sensitivity minimization in an H^∞ norm: parametrization of all suboptimal solutions, Int. J. Control **46** (1987), 785-816.
12. Ball, J. A. and I. Gohberg, A commutant lifting theorem for triangular matrices with diverse applications, Integral Equations and Operator Theory **8** (1985), 205–267.
13. _______, Shift invariant subspaces, factorization and interpolation for matrices I: The canonical case: Linear Algebra and Applications, **74** (1986a), 87-150.
14. _______, Pairs of shift invariant subspaces of matrices and noncanonical factorization, Linear and Multilinear Algebra, **20** (1986b), 27-61.

15. ———, Classification of shift invariant subspaces of matrices with Hermitian form and completion of matrices, in *Operator Theory and Systems*, (ed. H. Bart, I. Gohberg, and M. A. Kaashoek), OT 19 Birkhäuser (Basel-Boston-Stuttgart) (1986c), pp. 23-85.

16. Ball, J. A. and J. W. Helton, Interpolation with outer functions and gain equalization in amplifiers, in Proc. Third Int. Symp. on Mathematical Theory of Networks and Systems, Delft, Western Periodicals (N. Hollywood, California), 1979, 41–49.

17. ———, Lie groups over the field of rational functions, signed spectral factorization, signed interpolation and amplifier design, J. Operator Theory **8** (1982a), 19–64.

18. ———, Factorization results related to shifts in an indefinite metric, Integral Equations and Operator Theory **5** (1982b), 632–658.

19. ———, A Beurling–Lax theorem for the Lie group $U(m,n)$ which contains most classical interpolation, J. Operator Theory **9** (1983), 107–142.

20. ———, Beurling–Lax representations using classical Lie groups with many applications II: $GL(n,C)$ and Wiener–Hopf factorization, Integral Eequations and Operator Theory **7** (1984), 291–309.

21. ———, Beurling–Lax representations using classical Lie groups with many applications III: groups preserving forms, Amer. J. Math., **108** (1986a), 95–174.

22. ———, Beurling–Lax representations using classical Lie groups with many applications IV: $GL(n,R), U^*(2n), SL(n,C)$ and a solvable group, J. Functional Analysis, **69** (1986b), 178-206.

23. ———, Interpolation problems of Pick–Nevanlinna and Loewner types for meromorphic matrix functions: parametrization of the set of all solutions, Integral Equations and Operator Theory, **9** (1986c), 155-203.

24. Ball, J. A. and A. C. M. Ran, Hankel norm approximation of a rational matrix function in terms of its realization, in *Modelling, Identification and Robust Control* (ed. C. Byrnes and A. Lindquist) North-Holland (Amsterdam) (1986), pp. 285-296.

25. ———, Optimal Hankel norm model reductions and Wiener-Hopf factorization I: The canonical case, SIAM J. Control and Opt., **25** (1987a), 362-382.

26. ———, Global inverse spectral problems for rational matrix functions, Linear Algebra and Applications, **86** (1987b), 237-282.

27. ———, Local inverse spectral problems for rational matrix functions, Integral Equations and Operator Theory, **10** (1987c), 349-415.

28. Bognar, J.. *Indefinite Inner Product Spaces*, Springer–Verlag, 1974.

29. Caratheodory, C., Über den Variabilitatsbereich der Koeffizienten von Potenzreihen, die gegebene Werte nicht annehmen, Math. Ann. **64** (1907), 95–115.

30. Delsarte, Ph., Y. Genin, and Y. Kamp, The Nevanlinna–Pick problem for matrix valued functions, SIAM J. Appl. Math. **36** (1979), 47–61.

31. Dewilde, P. and H. Dym, Lossless chain scattering matrices and optimum linear prediction: The vector case, Circuit Theory and Appl. **9** (1981), 135–175.

32. ———, Lossless inverse scattering for digital filters, IEEE Trans. Inf. Theory **30** (1984), 644–662.

33. Dym, H. and I. Gohberg, Extensions of triangular operators and matrix functions, Indiana Univ. Math. J. **31** (1982a), 579–606.

34. ———, Extensions of matrix valued functions and block matrices, Indiana Univ. Math. J. **31** (1982b), 733–765.

35. ———, Unitary interpolants, factorization indices and infinite Hankel block matrices, J. Funct. Anal. **54** (1983a), 229–289.

36. ———, Hankel integral operators and isometric interpolants on the line, J. Funct. Anal. **54** (1983b), 290–307.

37. Fedchina, I. I., Tangential Nevanlinna–Pick problem with multiple points (Russian), Akad. Nauk Armjan SSR Dokl. **61** (1975a), 214–218.

38. ———, Description of solutions of the tangential Nevanlinna–Pick problem (Russian), Akad. Nauk Armjan SSR Dokl. **60** (1975b), 37–42.

39. Feintuch, A. and B. A. Francis, Uniformly optimal control of linear time-varying systems, Systems and Control Letters **5** (1984), 67–71.

40. Feintuch, A., P. Khargonekar, and A. Tannenbaum, On the sensitivity minimization problem for linear time-varying systems, SIAM J. Control and Opt., **24** (1986), 1076-1085.

41. Francis, B. A. and J. C. Doyle, Linear control theory with an H^∞ optimality criterion, SIAM J. Control and Opt. **25** (1987), 815–844.

42. Francis, B. A., J. W. Helton, and G. Zames, H^∞–optimal feedback controllers for linear multivariable systems, IEEE Trans. Aut. Control, **AC-29** (1984), 888–900.

43. Garnett, J. B., *Bounded Analytic Functions*, Academic Press, 1981.

44. Glover, K., All optimal Hanekl–norm approximations of linear multivariable systems and their L^∞–error bounds, Int. J. Control **39** (1984), 1115–1193.

45. Gohberg, I., P. Lancaster, and L. Rodman, Matrices and Indefinite Scalar Products, vol. 0T8, 1983, Birkhauser.

46. Helton, J. W., A simple test to determine gain bandwidth limitations, Proc. IEEE Int. Conf. Circuit Theory, 1977, Phoenix.

47. ———, Orbit structure of the Möbius transformations semi–group acting on H^∞ (broadband matching), in *Topics in Functional Analysis, Advances in Math. Suppl. Studies* **3** (1978), Academic Press, 129–257.

48. ———, The distance of a function to H^∞ in the Poincare metric; electrical power transfer, J. Funct. Anal. **38** (1980), 273–314.

49. ———, Operator Theory, Analytic Functions, Matrices and Electrical Engineering, CBMS Regional Conference Series in Mathematics, No. 68, Amer. Math. Soc. (Providence), 1987.

50. Hoffman, K., *Banach Spaces of Analytic Functions*, Prentice–Hall, 1965.

51. Johnson, C. and L. Rodman, Completion of partial matrices to contractions, J. Functional Analysis, **69** (1986), 260-267.

52. Khargonekar, P. and K. Poola, Uniformly optimal control of linear time-invariant plants: nonlinear time-varying controllers, Systems and Control Letters **6** (1986), 303-308.

53. Khargonekar, P. K. Poola, and A. Tannenbaum, Robust control of linear time–invariant plants using periodic compensation, IEEE Trans. Aut. Control, AC-**30** (1985), 1088-1096.

54. Khargonekar, P. and A. Tannenbaum, Noneuclidean metrics and the robust stabilization of systems with parameter uncertainty, IEEE Trans. Aut. Control **AC-30** (1985), 1005–1013.

55. Kovalishina, I. V. and V. P. Potapov, Integral representations of Hermitian positive functions, Private translation by T. Ando, 1981, Sappora, Japan.

56. Kraus, J. and D. R. Larson, Reflexivity and distance formulae, Proc. London Math. Soc., to appear.

57. Krein, M. G. and A. A. Nudelman, *The Markov Moment Problem and Extremal Problems*, Transl. Math. Monographs, vol. **50**, 1977, Amer. Math. Soc.

58. Limebeer, D. J. N. and B. D. O. Anderson, An interpolation theory approach to H^∞ controller degree bounds, Linear Alg. and Appl., to appear.

59. Löwer, K., Über monotone Matrixfunktionen, Math. Z. **38** (1934), 177–216.

60. Marshall, D. E., An elementary proof of the Pick–Nevanlinna interpolation theorem, Mich. Math. J. **21** (1974), 219–223.
61. McAsey, M., Canonical models for invariant subspaces, Pac. J. Math **91** (1980), 377-395.
62. ———, Invariant subspaces of nonselfadjoint crossed products, Pac. J. Math. **96** (1981), 457-473.
63. Sz.–Nagy, B. and A. Koranyi, Relations d'un problem de Nevanlinna et Pick avec de theorie des operateurs de l'espace Hilbertien, Acta Math. Acad. Sci. Hungar. **7** (1956), 295–302.
64. ———, Operator theoretische Behandlung und Veralgemeinerung eines Problemkreises in der komplexen Funktionentheorie, Act Math. **100** (1958), 608–633.
65. Nehari, Z., On bounded blinear forms, Ann. of Math **65** (1957), 153–162.
66. Nevanlinna, R., Über beschrankte Funktionen, die in gegebene Punkten vorgeschriebene Werte annehmen, Ann. Acad. Sci. Fenn. (1919), 1–71.
67. ———, Über beschrankte analytische Funktionen, Ann. Acad. Sci. Fenn. #7 **32** (1929).
68. Pick, G., Über die Beschrankungen analytischer Funktionne welche durch vorgegebene Funktionswerte bewirkt wird, Math. Ann. **77** (1916), 7–23.
69. Rosenblum, M. and J. Rovnyak, Restrictions of analytic functions I, Proc. Amer. Math. Soc. **48** (1975), 113–119.
70. ———, An operator–theoretic approach to theorems of the Pick–Nevanlinna and Loewner types I, Integral Equation and Operator Theory **3** (1980), 408–436.
71. ———, *Hardy Classes and Operator Theory*, Oxford University Press, 1985.
72. Sarason, D., Generalized interpolation in H^∞, Trans. Amer. Math. Soc. **127** (1967), 179–203.
73. ———, Operator–theoretic aspects of the Nevanlinna–Pick interpolation problem, in Operators and Function Theory (ed. S. C. Power), D. Reidel (Dordrecht, Holland) (1985), 279–314.
74. Scales, W. A., Interpolation with meromorphic functions of minimal norm, Dissertation, University of California at San Diego (1982).
75. Schur, I., Über Potenzreihen die im Innern des Einheits Kreisses beschränkt sind II., J. reine. angew. Math. **148** (1918), 122–145.
76. Solel, B., Analytic operator algebras (factorization and an expectation), Trans. Amer. Math. Soc. **287** (1985), 799-817.

77. Takagi, T.; On an algebraic problem related to an analytic theorem of Caratheodory and Fejer, Japan J. Math. **1** (1924), 83–93.
78. Tannenbaum, A., Modified Nevanlinna–Pick interpolation and feedback stabilization of linear plants with uncertainty in the gain factor, Int. J. Control **36** (1982), 331-336.
79. ———, On the multivariable gain margin problem, Automatica **22** (1986), 381-383.
80. Vidyasagar, M., *Control System Snythesis: A Factorization Approach*, MIT Press, Cambridge, Massachusetts, 1985.
81. Youla, D. C. and M. Saitoh, Interpolation with positive real functions, J. Franklin Institute **284** (1967), 77–108.
82. Zames, G., Feedback and optimal sensitivity: model reference transformations, multiplicative seminorms, and approximate inverses, IEEE Trans. Aut. Control **AC-23** (1981), 301–320.
83. Zames, G. and B. A. Francis, Feedback, minimax sensitivity, and optimal robustness, IEEE Trans. Aut. Control **AC-28** (1983), 585–601.

Joseph A. Ball
Department of Mathematics
Virginia Polytechnic Institute and State University
Blacksburg, Virginia 24061
U.S.A.

A Fourier Analysis Theory of Abstract Spectral Decompositions

by

Earl Berkson

These lecture notes correspond to a series of four talks given March 24-28, 1986, as part of Indiana University's Special year in Operator Theory. I am grateful to the Director of the Special Year, Professor John B. Conway, and to all my colleagues at Indiana University for inviting me to present this account of spectral decomposability.

1. Spectral Families of Projections and Well-bounded Operators: Motivation and Fundamental Theory

In this paper I shall describe an ongoing joint research program with T. A. Gillespie and, more recently, P. S. Muhly as well. The aim of this program is the development of an abstract operator theory which accounts for and treats the weakened forms of orthogonality permeating modern analysis outside the Hilbert space setting (e.g., in Fourier analysis and multiplier theory). Typically, these weakened forms of orthogonality cannot be implemented by projection-valued measures. Let us illustrate this state of affairs with a few examples. A famous theorem of Marcel Riesz asserts that for $f \in L^p(\mathbf{T})$ $(1 < p < \infty)$ the Fourier series of f converges to f in the mean of order p. For $p = 2$, this Fourier inversion theorem merely states the completeness of the trigonometric system, and is an easy consequence of the rich theory of Hilbert space orthogonality combined with any one of a variety of standard tools (e.g., $(C,1)$-summability or the Weierstrass Approximation Theorem). For $p \neq 2$, Marcel Riesz's theorem bears a tantalizing formal resemblance to its $L^2(\mathbf{T})$-version, but the theorem is no longer so elementary, and the usual methods for $p = 2$ are of no avail, because the space $L^p(\mathbf{T})$ lacks the intrinsic orthogonality structure enjoyed by Hilbert space. Indeed, slavish imitation of the Hilbert space methods is doomed to failure, since, as is well-known, for $p \neq 2$, $L^p(\mathbf{T})$ contains functions whose Fourier series fail to converge unconditionally in $L^p(\mathbf{T})$ [17, p. 12]. In particular, the projections on the individual elements of the trigonometric system $\{e^{in\vartheta}\}_{n=-\infty}^{\infty}$ will no longer give rise to a spectral measure. In order to account for the features of Hilbert space orthogonality that survive the transition to $L^p(\mathbf{T})$, we can seek a *Hilbert space minded* abstract operator-theoretic rationale that simultaneously provides for Fourier series convergence in all the spaces $L^p(\mathbf{T})$, $1 < p < \infty$. The L^2-convergence of Fourier series can be established by applying Stone's Theorem for unitary groups to the translation group acting on $L^2(\mathbf{T})$. The spectral measure provided by Stone's Theorem will, in this instance, be supported on the integers, where its values are the projections on the individual elements of the trigonometric system. The trigonometric system automatically becomes an orthonormal basis thereby. In (2.8) below we shall discuss a Banach space analogue of Stone's Theorem that handles the $L^p(\mathbf{T})$-case (see the remarks in (2.24)). Some delicacy is required, since an appropriate generalization of Stone's Theorem must necessarily be formulated without recourse to spectral measures in view of the conditional convergence of Fourier series outside of the L^2-setting.

Harmonic analysis on a group can be viewed as the study of translation operators, since the latter embody the group structure in alternate form. In particular, the weakened forms of orthogonality in $L^p(\mathbf{R})$ connected with the *partial sum* projection operators, such as Littlewood-Paley theory, trace back to convolution with the Hilbert kernel, and hence ultimately to translations. It is thus desirable to have an abstract theory that guarantees spectral decompositions for translation operators. Suppose that G is a locally compact abelian group and $x \in G$. For $1 < p < \infty$, let $R_x^{(p)}$ be the translation operator on $L^p(G)$ corresponding to x. If $p = 2$, then $R_x^{(p)}$ is unitary. The case $p \neq 2$ is quite different, for in this setting it is known (see [20]) that if x has infinite order, then $R_x^{(p)}$ is not a spectral operator in the sense of Dunford-and so $R_x^{(p)}$ cannot have an associated spectral measure. We shall see later on that if V is an invertible power-bounded operator on $L^p(\mu)$ ($1 < p < \infty$, μ an arbitrary measure), then V can be expressed as a Riemann-Stieltjes integral with respect to a projection-valued function of one variable. In particular, this result will apply to the translation operators $R_x^{(p)}$.

The preceding discussion will serve as an abbreviated motivation for developing an operator-theoretic approach to weakened forms of orthogonality. Further examples could be drawn from diverse topics to which our theory applies, including ergodic flows in commutative and non-commutative analysis, the Paley-Wiener theorem, generalized analyticity and invariant subspaces à la Helson, and harmonic conjugation. Our theory begins with the relevant notion for spectral decompositions.

Definition. Let $\mathcal{B}(Y)$ be the algebra of bounded operators on a Banach space Y, and let I denote the identity operator on Y. A *spectral family of projections in Y* is a uniformly bounded, projection-valued function $E(\bullet) : \mathbf{R} \to \mathcal{B}(\mathbf{Y})$ such that:

(i) $E(s)E(t) = E(t)E(s) = E(s)$ for $s \le t$;

(ii) for each $s \in \mathbf{R}$, $E(\bullet)$ is right continuous at s in the strong operator topology, and $E(\bullet)$ has a strong left-hand limit at s (denoted $E(s^-)$);

(iii) $E(s) \to I$ (resp., $E(s) \to 0$) in the strong operator topology as $s \to +\infty$ (resp., $s \to -\infty$).

If there is a compact interval $[a,b]$ such that $E(s) = 0$ for $s < a$ and $E(s) = I$ for $s \ge b$, we shall say that $E(\bullet)$ is *concentrated* on $[a,b]$.

Suppose that $\mathcal{P}$ is a spectral measure on the Borel subsets of $\mathbf{R}$ taking values in $\mathcal{B}(Y)$. Define $\varepsilon(\bullet) : \mathbf{R} \to \mathcal{B}(\mathbf{Y})$ by putting $\varepsilon(t) = \mathcal{P}((-\infty,t])$. It is easily seen that $\varepsilon(\bullet)$ is a spectral family of projections. On the other hand, if Y has a basis $\{x_n\}_{n=1}^{\infty}$, and Q_n denotes the projection on the nth basis vector, then we can form a spectral family $\mathcal{G}(\bullet)$ that has a *jump* of Q_n at n ($n = 1,2,\ldots$), and is constant on each of the intervals complementary to the set $\mathbf{N}$ of positive integers. If the basis $\{x_n\}$ fails to be unconditional, then the spectral family $\mathcal{G}(\bullet)$ cannot arise from a spectral measure in contrast to the spectral family $\varepsilon(\bullet)$ described above. This shows that the properties used to define a spectral family constitute a weakening of the notion of spectral measure. The definition of spectal family will turn out to be just strong enough, but not too strong, for our purposes.

For spectral representations of operators, we shall require a notion of integration with respect to a spectral family. Suppose first that $\mathcal{H}$ is a Hilbert space, A is a bounded, self-adjoint operator on $\mathcal{H}$ with spectrum $\sigma(A)$ contained in $[a,b]$, and $\mathcal{P}$ is the spectral measure of A. Let $\varepsilon(\bullet)$ be the spectral family corresponding to $\mathcal{P}$ as above, $\varepsilon(t) = \mathcal{P}((-\infty,t])$. In the earlier Hilbert space literature, the spectral theorem representation for A was expressed in terms of a Riemann-Stieltjes integral with respect to $\varepsilon(\bullet)$-specifically,

$$A = a\varepsilon(a) + \int_a^b t \, d\varepsilon(t).$$

This formulation of the spectral theorem, which suppresses explicit mention of spectral measures, points the way to a successful theory of integration with respect to spectral families (and, as will be seen from Theorem (1.13), to a Banach space generalization of self-adjoint operators).

We shall now outline the integration theory of spectral families that evolved in the works of G. L. Krabbe and of P. G. Spain, and is described in detail in [13, Chapter

17]. Let $J = [a,b]$ be a compact interval, and let $BV(J)$ denote the Banach algebra consisting of all complex-valued functions on J having bounded variation (the operations in $BV(J)$ are taken pointwise, and the Banach algebra norm $\|\cdot\|_J$ in $BV(J)$ is defined by $\|f\| = |f(b)| + \operatorname{var}(f,J)$). Let Π denote the set of all partitions of $[a,b]$, partially ordered and directed by refinement. Suppose that $E(\cdot)$ is a spectral family in the Banach space Y, and $E(\cdot)$ is concentrated on J. The key tool for the integration theory of $E(\cdot)$ is the following technical lemma (see [13, Lemma 17.2] for its elementary proof).

Lemma 1.1. *For $x \in Y$ and $u = (\lambda_0,\lambda_1,\ldots \lambda_n) \in \Pi$, put*

$$\omega(u,x) = \max_{1\le j\le n} \sup\{\|E(\lambda)x - E(\lambda_{j-1})x\| : \lambda \in [\lambda_{j-1},\lambda_j)\}.$$

Then for each $x \in Y$, $\lim_{u\in\Pi} \omega(u,x) = 0$.

For $g \in BV([a,b])$ and $u = (\lambda_0,\lambda_1,\ldots,\lambda_n) \in \Pi$, put

$$\mathcal{S}(g,u) = g(a)E(a) + \sum_{j=1}^{n} g(\lambda_j)\{E(\lambda_j) - E(\lambda_{j-1})\}. \tag{1.2}$$

Rearrangement of the terms on the right in the manner of integration by parts gives

$$\mathcal{S}(g,u) = g(b)E(b) - \sum_{j=1}^{n}\{g(\lambda_j) - g(\lambda_{j-1})\}E(\lambda_{j-1}). \tag{1.3}$$

In particular, $\|\mathcal{S}(g,u)\| \le \|g\|_J \sup\{\|E(\lambda)\| : \lambda \in \mathbf{R}\}$.

Lemma 1.4. *Let $u, v \in \Pi$ with $v \ge u$, and let $g \in BV(J)$. Then for $x \in Y$*

$$\|\mathcal{S}(g,v)x - \mathcal{S}(g,u)x\| \le \operatorname{var}(g,J)\omega(u,x).$$

Proof: Standard and elementary from (1.3). ■

It is evident from Lemmas (1.1) and (1.4) that for $g \in BV(J)$

$$\int_J^{\oplus} g\,dE \equiv \lim_{u\in\Pi} \mathcal{S}(g,u)$$

exists in the strong operator topology. The foregoing considerations readily give us the following fundamental proposition.

Proposition 1.5. *The mapping $g \to \int_J^{\oplus} g\,dE$ is an identity preserving algebra homomorphism of $BV(J)$ into $\mathcal{B}(Y)$ satisfying*

$$\left\| \int_J^{\oplus} g\,dE \right\| \le \|g\|_J \sup\{\|E(\lambda)\| : \lambda \in \mathbf{R}\} \qquad \text{for } \mathbf{g} \in \mathbf{BV(J)}. \tag{1.6}$$

Furthermore,

$$\left\| \left(\int_J^{\oplus} g\,dE \right) x - \mathcal{S}(g,u)x \right\| \le \operatorname{var}(g,J)\omega(u,x), \tag{1.7}$$

$$\text{for } g \in BV(J), \quad u \in \Pi, \quad x \in Y.$$

Remark. It is shown in [13, Theorems 17.4 and 17.8] that if $g \in BV(J)$ is continuous on J, then $\int_J^{\oplus} g\,dE$ is the strong limit of Riemann-Stieltjes sums obtained by replacing $g(\lambda_j)$, $j = 1,2,\ldots,n$ in (1.2) by values of g at arbitrary intermediate points of the corresponding partition subintervals.

One further inequality will be useful. Given $f \in BV(J)$, $g \in BV(J)$, $x \in Y$, $u \in \Pi$, we have

$$\left\| \left\{ \int_J^{\oplus} f\,dE - \int_J^{\oplus} g\,dE \right\} x \right\| \le \{\operatorname{var}(f,J) + \operatorname{var}(g,J)\}\omega(u,x) \tag{1.8}$$

$$+\|\mathcal{S}(f,u)x - \mathcal{S}(g,u)x\|.$$

This comes from writing

$$\int_J^{\oplus} f\,dE - \int_J^{\oplus} g\,dE = \left\{ \int_J^{\oplus} f dE - \mathcal{S}(f,u) \right\} - \left\{ \int_J^{\oplus} g\,dE - \mathcal{S}(g,u) \right\}$$
$$+ \mathcal{S}(f,u) - \mathcal{S}(g,u)\},$$

and utilizing (1.7). An an immediate consequence of (1.8) and Lemma 1.1, we obtain the following valuable convergence theorem for the integration theory.

Proposition 1.9. *Let $\{g_\alpha\}$ be a net in $BV(J)$, and let g be a complex-valued function on J such that*

(i) $\sup_\alpha \operatorname{var}(g_\alpha, J) < \infty$;
(ii) $g_\alpha \to g$ pointwise on J. Then $g \in BV(J)$ and $\left\{ \int_J^{\oplus} g_\alpha\,dE \right\}$ converges to $\int_J^{\oplus} g\,dE$ in the strong operator topology.

Having settled on spectral families of projections as the relevant vehicle for our study of weakened forms of orthogonality, we now turn our attention to finding conditions that guarantee, rather than assume, the existence of spectral families. A first step in this direction is to characterize when a bounded Banach space operator A will have a spectral representation formally analogous to that in the spectral theorem for self-adjoint operators-that is,

$$A = \int_J^{\oplus} \lambda \, dE(\lambda), \tag{1.10}$$

for a suitable spectral family $E(\cdot)$ and compact interval J. Some years ago this matter was settled for operators on reflexive spaces by J. R. Ringrose and D. R. Smart [26], [28]. In order to motivate their characterization, we shall first discuss some background suggestive of it.

Suppose that $\mathcal{U}$ is a unital Banach algebra and Y is a Banach space. A norm-continuous algebra homomorphism ϑ of $\mathcal{U}$ into $\mathcal{B}(Y)$ such that $\vartheta(1) = I$ will be called a *representation of* $\mathcal{U}$ *in* Y. Suppose that $\mathcal{U}_0$ is a unital Banach algebra of functions defined on a compact subset K of the complex plane $\mathbf{C}$, the operations in $\mathcal{U}_0$ being pointwise, and $\mathcal{U}_0$ containing the constants as well as the function $f_1(z) \equiv z$. If $T \in \mathcal{B}(Y)$, we shall use the terminology "$\mathcal{U}_0$ functional calculus for T" to mean a representation ϑ of $\mathcal{U}_0$ in Y such that $\vartheta(f_1) = T$. Throughout this section the algebra $\mathcal{U}_0$ will be one of the following: $BV(J)$ for a compact interval J; $AC(J)$, the closed subalgebra of $BV(J)$ consisting of the absolutely continuous functions on J, $AC(J)$ being endowed with the norm $\|\cdot\|_J$; $C(K)$ (resp., $B(K)$), the algebra of all continuous complex-valued functions (resp., all bounded Borel functions) defined on a compact subset $K \subseteq \mathbf{C}$, the norm in $C(K)$ (resp., $B(K)$) being the usual *sup* norm, $\|f\| \equiv \sup\{|f(z)| : z \in K\}$. It is a long established fact in Dunford's theory of spectral operators that a bounded operator S on a reflexive Banach space X is scalar-type spectral if and only if S has a $C(\sigma(S))$-functional calculus (see, for example, [15, Proof of Theorem 18] or [13, Theorem 5.21]). The nontrivial half of this equivalence is the *if* part, which is shown by using the Riesz representation theorem for the linear functionals on $C(\sigma(S))$ to obtain a projection-valued measure $\mathcal{G}$ such that

$$\Psi(f) = \int_{\sigma(S)} f \, d\mathcal{G} \qquad \text{for all } f \in C(\sigma(S)), \tag{1.11}$$

where Ψ is the given $C(\sigma(S))$-functional calculus. Since bounded Borel functions can be integrated against $\mathcal{G}$, (1.11) shows that Ψ can be extended to a $B(\sigma(S))$-functional calculus for S, which we denote by Ψ. Thus the spectral projections $\mathcal{G}(\alpha)$, where α ranges over the Borel sets, are simply the Ψ-images of the corresponding characteristic functions. In effect then, $\mathcal{G}$, the spectral measure of S, arises because the Riesz representation theorem forces the given $C(\sigma(S))$-functional calculus to extend to a $B(\sigma(S))$-functional calculus. Let us now consider the possibility of formulating, for a Banach space operator A, a functional calculus just sufficient to guarantee that (1.10) holds for a suitable spectral family $E(\cdot)$ and compact interval J. On aesthetic grounds the relevant algebra of functions $\mathcal{U}_0$ should contain as few nontrivial idempotents as possible. By analogy with the case of scalar-type operators just discussed, we would like our prospective functional calculus for A to extend to a larger class of functions $\mathcal{U}_1$; however, we do *not* want too many projections to result from this extension (in fact, it is desirable that the characteristic functions in $\mathcal{U}_1$ should essentially all come from intervals). Natural candidates for $\mathcal{U}_0$ and $\mathcal{U}_1$, respectively, are $AC(J)$ and $BV(J)$, where J is a compact interval. We are thus led to consider the following notion originated by Ringrose and Smart [26], [27], [28].

Definition. Let A be a bounded operator on a Banach space Y. We say that A is *well-bounded* provided there is a compact interval J such that A has an $AC(J)$-functional calculus, or, equivalently, such that, for some constant K,

1.12 $$\|p(A)\| \leq K\|p\|_J,$$

for all polynomials p with complex coefficients.

In the setting of the arbitrary Banach space Y, this definition of well-bounded operator is slightly too weak for our purposes, since, in particular, it does not automatically produce a spectral family of projections in Y (see [27, S6]). In order to obtain a spectral family for a well-bounded operator so that (1.10) holds, a suitable weak compactness assumption on the $AC(J)$-functional calculus must be imposed (see the notion of well-bounded operator of type (B), introduced in [3]). However, in our discussions we shall only be concerned with the theory of well-bounded operators on reflexive Banach spaces, where such difficulties do not arise. The various aspects of the general Banach space theory of well-bounded operators can be found in the respective references cited below. We now take up the spectral theorem of Ringrose and Smart for well-bounded operators on reflexive spaces.

Theorem 1.13. *Let A be a bounded operator on a reflexive Banach space X. Then A is well-bounded if and only if there is a spectral family $E(\cdot)$ in X such that for some compact interval J, $E(\cdot)$ is concentrated on J and $A = \int_J^{\oplus} \lambda \, dE(\lambda)$.*

Proof: The *if* assertion is easy. If the spectral family $E(\cdot)$ in the statement of the theorem exists, then application of Proposition 1.5 immediately gives us (1.12). The converse, which is the heart of the theorem, is not so easy. I shall utilize what is currently the simplest method of proof for it; this method was developed in [4, S2] as a by-product of other considerations. The proof will be carried out in successive stages, beginning with two preliminary lemmas.

Lemma 1.14. *Let Q be a well-bounded operator on a Banach space Y. If Q is quasinilpotent, then $Q = 0$.*

Proof: Pick $M > 0$ so that Q has an $AC([-M,M])$-functional calculus (which we shall denote by $f \to f(Q)$). If $f \in AC([-M,M])$ and f vanishes on an interval $(-\rho,\rho)$, then for each $z \in \mathbf{C}$ with $|z| < \rho/2$, let $h_z \in AC([-M,M])$ be defined by setting $h_z(t) = (z-t)^{-1}$ for $\rho \leq |t| \leq M$, and then taking h_z to be linear on $[-\rho,\rho]$. Obviously $(z-t)h_z(t)f(t) = f(t)$ for $|t| \leq M$, and so $h_z(Q)f(Q) = (z-Q)^{-1}f(Q)$ for $0 < |z| < \rho/2$. However, $\{h_z : |z| < \rho/2\}$ is easily seen to be a bounded subset of $AC([-M,M])$. So $(z-Q)^{-1}f(Q)$ has a removable singularity at $z = 0$. By Liouville's Theorem, $f(Q) = 0$. For each positive integer n, let $f_n \in AC([-M,M])$ be defined by taking $f_n(t) = t$ for $|t| \leq n^{-1}$, and making f_n constant on each of the two remaining subintervals of $[-M,M]$. Since $f_n(t) - t$ vanishes for $|t| < n^{-1}$, the preceding argument shows that $f_n(Q) = Q$. But $\|f_n\|_{[-M,M]} = 3/n$ for all n. Hence $Q = 0$.

The next lemma is a result of Fong and Lam [19, proof of Proposition 2.2].

Lemma 1.15. *Suppose that $\mathcal{A}$ is an algebra over $\mathbf{R}$ with identity $\mathcal{I}$, and $\mathcal{K}$ is a subset of $\mathcal{A}$ such that each of $\mathcal{K}$ and $\mathcal{I} - \mathcal{K}$ is closed under multiplication. Then every extreme point of $\mathcal{K}$ is an idempotent.*

Proof: Let x be an extreme point of $\mathcal{K}$. Since $\mathcal{K}$ and $\mathcal{I} - \mathcal{K}$ are closed under multiplication, $x^2 \in \mathcal{K}$ and $(2x - x^2) \in \mathcal{K}$. Since $x = 2^{-1}\{x^2 + (2x - x^2)\}$, and x is an extreme point, x must be idempotent.

We now resume the proof of the *only if* assertion of Theorem 1.13. Let $J = [a,b]$ be a compact interval such that A has an $AC(J)$-functional calculus Ψ. For $\lambda \in [a,b)$ and $0 < \delta < b - \lambda$, let $\mathcal{F}_{\lambda,\delta}$ be the set consisting of all real-valued $f \in AC(J)$ such that $f \equiv 1$ on $[a,\lambda]$, $f \equiv 0$ on $[\lambda+\delta,b]$, and f is decreasing on $[\lambda,\lambda+\delta]$. Let $\mathcal{K}_{\lambda,\delta}$ be the closure of $\{\Psi(f)\colon f \in \mathcal{F}_{\lambda,\delta}\}$ in the weak operator topology of $\mathcal{B}(X)$. Put $\mathcal{K}_\lambda = \cap\{\mathcal{K}_{\lambda,\delta}\colon 0 < \delta < b - \lambda\}$. Clearly each set $\mathcal{F}_{\lambda,\delta}$ is convex, and non-empty, and consists of unit vectors in $AC(J)$. With the aid of reflexivity of X, it follows that each $\mathcal{K}_{\lambda,\delta}$ is a convex, non-void, weakly compact subset of $\mathcal{B}(X)$. Since $\delta_1 < \delta_2$ implies $\mathcal{K}_{\lambda,\delta_1} \subseteq \mathcal{K}_{\lambda,\delta_2}$, it follows by compactness that $\mathcal{K}_\lambda$ is non-void. Obviously $\mathcal{K}_\lambda$ is weakly compact and convex. It is easy to see that for each δ, $\mathcal{K}_{\lambda,\delta}$ and $(I - \mathcal{K}_{\lambda,\delta})$ are commutative semigroups. Hence $\mathcal{K}_\lambda$ and $(I - \mathcal{K}_\lambda)$ are commutative semigroups. By Lemma 1.15 we have:

1.16 *Each extreme point of $\mathcal{K}_\lambda$ is an idempotent.*

We now proceed to show that each set $\mathcal{K}_\lambda$ for $\lambda \in [a,b)$ is a singleton set consisting of a projection operator. First we establish a lemma. Let us observe beforehand that the closure of the range of Ψ in the weak operator topology is commutative.

Lemma 1.17. *Let $\lambda \in [a,b)$ and $E \in \mathcal{K}_\lambda$ with $E^2 = E$. Then $A|EX$, the restriction of A to EX, satisfies $\sigma(A|EX) \subseteq [a,\lambda]$. Also, $\sigma(A \mid (I-E)X) \subseteq [\lambda,b]$.*

Proof: If $z \in \mathbf{C} \setminus J$, then $f_z(t) \equiv (z-t)^{-1}$ belongs to $AC(J)$, and so z is in the resolvent set of A and $\Psi(f_z) = (z-A)^{-1}$. Thus E commutes with $(z-A)^{-1}$, and $\sigma(A \mid EX), \sigma(A \mid (I-E)X)$ are subsets of J. Fix an arbitrary μ so that $\lambda < \mu \leq b$, and pick $\delta > 0$ so that $\lambda + \delta < \mu$. Let $g \in AC(J)$ be such that $g(t) = (\mu - t)^{-1}$ for $t \in [a,\lambda+\delta]$. Then for each $f \in \mathcal{F}_{\lambda,\delta}$, $(\mu - t)gf = f$. Hence for each $C \in \mathcal{K}_{\lambda,\delta}$, $(\mu - A)\Psi(g)C = C$. If in particular we take C to be E, then the first conclusion of the lemma is established. If $\lambda = a$, the second conclusion is trivial; so assume that $a < \lambda < b$. Fix an arbitrary β such that $a \leq \beta < \lambda$. Let $h \in AC(J)$ be such that $h(t) = (\beta - t)^{-1}$ for $t \in [\lambda,b]$. If $\delta > 0$ and $f \in \mathcal{F}_{\lambda,\delta}$, then $(\beta - t)(1-f)h = (1-f)$. It follows readily that $(\beta - A)(I-E)\Psi(h) = I - E$, and the remaining conclusion of the lemma is now apparent.

In order to show that $\mathcal{K}_\lambda$ $(a \leq \lambda < b)$ consists of a single projection, it suffices, by virtue of (1.16) and the Krein-Milman Theorem, to show that any two projections E, F in $\mathcal{K}_\lambda$ are equal. From Lemma 1.17 we see that, in particular, $\sigma(A \mid FX)$ and $\sigma(A \mid (I-E)X)$ do not separate the plane. By standard spectral theory (for example, by [13, Theorem 1.29]) this gives us

$$\sigma(A \mid F(I-E)X) \subseteq \sigma(A \mid FX) \cap \sigma(A \mid (I-E)X).$$

Application of Lemma 1.17 now shows that

$$\sigma(A \mid F(I-E)X) \subseteq \{\lambda\}. \tag{1.18}$$

However, well-boundedness of A implies well-boundedness of $(A - \lambda)$, and hence of the restriction $\{(A - \lambda) \mid F(I - E)X\}$. The latter operator is quasinilpotent by (1.18), and use of Lemma 1.14 gives us

$$1.19 \qquad Ax = \lambda x \qquad \text{for } x \in F(I - E)X\,.$$

Since the polynomials are dense in $AC(J)$, we infer from (1.19) that for $x \in F(I - E)X$ and $f \in AC(J)$, $\Psi(f)x = f(\lambda)x$. In view of the definition of $\mathcal{K}_\lambda$, this shows that $Ex = Fx = x$ for $x \in F(I - E)X$. Hence for all $y \in X$,

$$F(I - E)y = E\{F(I - E)y\} = 0\,.$$

So $F = FE$. Reversing the roles of F and E, we see that $E = EF = FE$. So $E = F$, as required, and each $\mathcal{K}_\lambda$ consists of a single projection operator.

Put $\mathcal{K}_\lambda = \{E(\lambda)\}$ for $a \le \lambda < b$, and define $E(\lambda) = I$ for $\lambda \ge b$, and $E(\lambda) = 0$ for $\lambda < a$. Two stages remain to be carried out in order to complete the proof of Theorem 1.13. First, we shall show that the function $E(\bullet)$: $\mathbf{R} \to \mathcal{B}(\mathbf{X})$ just defined is a spectral family. Then we shall establish that $A = \int_J^{\oplus} \lambda\, dE(\lambda)$. Denoting by $\mathcal{M}$ the closure in the weak operator topology of $\cup\{\mathcal{K}_{\lambda,\delta}\colon \lambda \in [a,b),\ 0 < \delta < b - \lambda\}$, let us observe beforehand that, since each $\mathcal{F}_{\lambda,\delta}$ consists of unit vectors in $AC(J)$, we have $\sup\{\|T\|\colon T \in \mathcal{M}\} \le \|\Psi\|$. In particular,

$$\sup\{\|E(\lambda)\|\colon \lambda \in \mathbf{R}\} \le \|\Psi\|\,,$$

and also, by virtue of the reflexivity of X, $\mathcal{M}$ is compact in the weak operator topology. If $\lambda \in [a,b)$, and $\{f_\delta\}$ belongs to the Cartesian product $\prod_\delta \mathcal{F}_{\lambda,\delta}$, then $\{\Psi(f_\delta)\}$ is contained in the weakly compact set $\mathcal{M}$. Moreover, since $\mathcal{K}_{\lambda,\delta_1} \subseteq \mathcal{K}_{\lambda,\delta_2}$ if $\delta_1 < \delta_2$, it is clear that any weakly convergent subnet of $\{\Psi(f_\delta)\}$ must have its limit contained in $\mathcal{K}_\lambda$ (which consists of $E(\lambda)$). It follows from the last two sentences that $\{\Psi(f_\delta)\}$ converges in the weak operator topology to $E(\lambda)$ as $\delta \to 0^+$. Suppose next that $a \le \lambda < \mu < b$. For $0 < \delta < b - \mu$, pick $f_\delta \in \mathcal{F}_{\lambda,\delta}$, $g_\delta \in \mathcal{F}_{\mu,\delta}$. Temporarily fix $\delta_0 \in (0, b - \mu)$. If $0 < \delta < b - \mu$, we have $f_\delta g_{\delta_0} \in \mathcal{F}_{\lambda,\delta}$. Hence $\lim_\delta[\Psi(f_\delta)\Psi(g_{\delta_0})] = E(\lambda)$ in the weak operator topology, and so $E(\lambda)\Psi(g_{\delta_0}) = E(\lambda)$. Letting $\delta_0 \to 0^+$, we obtain $E(\lambda)E(\mu) = E(\lambda)$. This establishes that $E(\bullet)$ satisfies the monotonicity requirement (i) in the definition of spectral family. To see that $E(\bullet)$ is a spectral family, it remains only to verify (ii) in the definition. Since X is reflexive, and $E(\bullet)$ is increasing and uniformly bounded, it follows immediately from [1, Corollary 2] that $E(\bullet)$ has a strong left-hand limit and a strong right-hand limit at each point of $\mathbf{R}$. We complete the demonstration that $E(\bullet)$ is a spectral family by showing that for each $s \in [a,b)$ the strong right-hand limit $E(s^+)$ coincides with $E(s)$. Since $E(\bullet)$ is increasing, $E(s) = E(s)E(s^+)$, and so it suffices to show that $E(s)E(s^+) = E(s^+)$. For $s < t < b$, $E(s^+)E(t) = E(s^+)$. From the latter equality, we infer with the aid of Lemma 1.17 that $\sigma(A \mid E(s^+)X) \subseteq [a,s]$. Hence

1.20 $$\sigma(A \mid E(s^+)\{I - E(s)\}X) \subseteq \{s\}.$$

Let $Z \equiv E(s^+)\{I - E(s)\}X$. It follows from (1.20) and Lemma 1.14 that $Az = sz$ for $z \in Z$. Hence $\Psi(f)z = f(s)z$ for $f \in AC(J)$, $z \in Z$. For $0 < \delta < b - s$, pick $f_\delta \in \mathcal{F}_{s,\delta}$. Thus, $\Psi(f_\delta)z = z$ and $\{\Psi(f_\delta)\}$ converges in the weak operator topology, as $\delta \to 0^+$, to $E(s)$. We infer that $E(s)z = z$ for $z \in Z$. Thus Z contains only 0, and it is now established that $E(\bullet)$ is a spectral family.

Fix $x \in X$ and $x^\star \in X^\star$, the dual space of X. The mapping $f \in AC(J) \to \langle \Psi(f)x, x^\star \rangle$ is a continuous linear functional on $AC(J)$. Since the mapping $f \in AC(J) \to (f', f(b))$ is a linear isometry of $AC(J)$ onto the direct sum $L^1[a,b] \oplus \mathbf{C}$, we see that there are $\alpha \in \mathbf{C}$ and a function $\Phi \in L^\infty[a,b]$ such that

1.21 $$\langle \Psi(f)x, x^\star \rangle = \int_a^b f'\Phi + \alpha f(b), \qquad \text{for } f \in AC(J).$$

Taking $f \equiv 1$, we get $\alpha = \langle x, x^\star \rangle$. Let $\lambda \in [a,b)$ and for $0 < \delta < b - \lambda$, let $F_\delta \in \mathcal{F}_{\lambda,\delta}$ be linear on $[\lambda, \lambda + \delta]$. Specializing (1.21) to F_δ and letting $\delta \to 0^+$, we find that

$$\langle E(\lambda)x, x^\star \rangle = - \lim_{\delta \to 0+} \delta^{-1} \int_\lambda^{\lambda+\delta} \Phi.$$

For almost all λ the right-hand side of this equation is $[-\Phi(\lambda)]$, and so $\langle E(\lambda)x, x^\star \rangle = -\Phi(\lambda)$ for almost all $\lambda \in [a,b)$. Using this in (1.21), we have

$$\langle \Psi(f)x, x^\star \rangle = \langle x, x^\star \rangle f(b) - \int_a^b f'(\lambda)\langle E(\lambda)x, x^\star \rangle \, d\lambda, \qquad \text{for } f \in AC(J).$$

Upon taking $f(\lambda) \equiv \lambda$ and integrating by parts, we obtain the desired conclusion that $A = \int_J^\oplus \lambda \, dE(\lambda)$. This completes the proof of Theorem 1.13. ∎

We next examine the features of this *spectral theorem* for well-bounded operators (see also [28], [13, Chapter 17], and [6, Propositions 2.11-2.13]).

Proposition 1.22. *Let A be a well-bounded operator on a reflexive Banach space X. The spectral family $E(\bullet)$ occurring in Theorem 1.13 is uniquely determined (and called the spectral family of A). The spectral family $E(\bullet)$ of A has the following properties.*

(i) *An operator $S \in \mathcal{B}(X)$ commutes with A if and only if S commutes with $E(\lambda)$ for all $\lambda \in \mathbf{R}$.*

(ii) *An open interval $\mathcal{I}$ is contained in $\rho(A)$, the resolvent set of A, if and only if $E(\bullet)$ is constant on $\mathcal{I}$.*

(iii) *$E(\bullet)$ is concentrated on a compact interval $\mathcal{K}$ if and only if A has an $AC(\mathcal{K})$-functional calculus; this is equivalent to the assertion that $\sigma(A) \subseteq \mathcal{K}$. In this case $A = \int_{\mathcal{K}}^{\oplus} \lambda\, dE(\lambda)$.*

(iv) *For each $\lambda \in \mathbf{R}$, $\{E(\lambda) - E(\lambda^-)\}$ is a projection operator whose range is $\{x \in X\colon Ax = \lambda x\}$.*

(v) *A closed subspace $\mathcal{N}$ of X is invariant under A if and only if $\mathcal{N}$ is invariant under $E(\lambda)$ for all $\lambda \in \mathbf{R}$.*

Proof: To prove uniqueness suppose that $E_k(\bullet)$ $(k = 1, 2)$ are spectral families as in the statement of Theorem 1.13 with corresponding compact intervals J_k. If λ_0 is any real number, we can pick a compact interval $J_0 = [a_0, b_0]$, which has λ_0 in its interior and contains J_1 and J_2 as subsets. Thus

$$A = \int_{J_0}^{\oplus} \lambda\, dE_k(\lambda)$$

for $k = 1, 2$. Choose a sequence $\{f_n\}$ of polynomials such that $\{f_n\}$ is uniformly bounded in $AC(J_0)$ and $\{f_n\}$ tends pointwise on J_0 to the characteristic function of $[a_0, \lambda_0]$. By Proposition 1.9, for $k = 1, 2$,

$$\int_{J_0}^{\oplus} f_n(\lambda)\, dE_k(\lambda) \to E_k(\lambda_0)$$

in the strong operator topology as $n \to \infty$. In other words, $f_n(A) \to E_k(\lambda_0)$ in the strong operator topology. This establishes the uniqueness assertion and also conclusions (i), (v). The integration with respect to $E(\bullet)$ of suitable functions of bounded variation provides a natural proof of (ii) (see [6, Proposition 2.13]). We next consider (iii). Suppose that $E(\bullet)$ is concentrated on the compact interval $\mathcal{K}$ and J is a compact interval as in the statement of Theorem 1.13. Let J_0 be a compact interval containing $\mathcal{K}$ and J in its interior. Thus

$$A = \int_{J_0}^{\oplus} \lambda\, dE(\lambda).$$

Since $E(\bullet)$ is concentrated on $\mathcal{K}$,

$$\int_{J_0}^{\oplus} \lambda \, dE(\lambda) = \int_{\mathcal{K}}^{\oplus} \lambda \, dE(\lambda) = A.$$

It follows by Proposition 1.5 that

$$\|p(A)\| \leq \sup\{\|E(\lambda)\| : \lambda \in \mathbf{R}\} \, \|\mathbf{p}\|_{\mathcal{K}},$$

for every polynomial p-whence A has an $AC(\mathcal{K})$-functional calculus. Conversely, if A has an $AC(\mathcal{K})$-functional calculus, then the proof of the *only if* assertion of Theorem 1.13 constructs the spectral family of A as a spectral family concentrated on $\mathcal{K}$. Moreover, if $z \in \mathbf{C} \setminus \mathcal{K}$, then $f_z(\lambda) \equiv (z - \lambda)^{-1}$ is in $AC(\mathcal{K})$, and its image under the $AC(\mathcal{K})$-functional calculus is $(z - A)^{-1}$. Hence $\sigma(A) \subseteq \mathcal{K}$. To complete the proof of (iii) it suffices to show that if $\sigma(A) \subseteq \mathcal{K}$, then $E(\bullet)$ is concentrated on $\mathcal{K}$. But this is immediate from (ii). As regards conclusion (iv), pick a compact interval J containing λ in its interior such that $E(\bullet)$ is concentrated on J. If $\{E(\lambda) - E(\lambda^-)\}x = x$, then it is easy to see that

$$Ax = \left\{ \int_J^{\oplus} s \, dE(s) \right\} \{E(\lambda) - E(\lambda^-)\} \, x = \lambda x.$$

Thus

$$\{E(\lambda) - E(\lambda^-)\}X \subseteq \{x \in X : Ax = \lambda x\}.$$

To obtain the reverse inclusion, choose a sequence $\{p_n\}$ of polynomials such that $\{p_n\}$ is uniformly bounded in $BV(J)$ and tends pointwise on J to the characteristic function of the singleton set $\{\lambda\}$. If $Ay = \lambda y$, we have, with the aid of Proposition 1.9,

$$p_n(\lambda)y = p_n(A)y = \int_J^{\oplus} p_n(s) \, dE(s)y \to \{E(\lambda) - E(\lambda^-)\}y.$$

Since $\{p_n(\lambda)\} \to 1$, the demonstration of Proposition 1.22 is complete.

Although we shall not make use of it, we close this section with the following analogue for well-bounded operators of the Hille-Sz. Nagy theorem for semigroups of self-adjoint operators [2, Corollary 4.14].

Proposition 1.23. *Let X be a reflexive Banach space, and let $\{T_t : t \geq 0\}$ be a strongly continuous one-parameter semigroup of well-bounded operators on X. For each $t \geq 0$, let $F_t(\bullet)$ denote the spectral family of T_t. Then:*

(i) *for each* $t \geq 0$, $\sigma(T_t) \geq 0$ *and* T_t *is one-to-one;*

(ii) $\sup\{\|F_t(\lambda)\|: t \geq 0,\ \lambda \in \mathbf{R}\} < \infty$;

(iii) *there is a unique spectral family* $P(\cdot)$ *such that for some constant* $b \in \mathbf{R}$, $P(\lambda) = I$ *for* $\lambda \geq b$, *and*

$$T_t x = \lim_{a \to +\infty} \int_{-a}^{a} e^{\lambda t}\, dP(\lambda)x\,, \qquad t \geq 0\,,\ x \in X\,.$$

2. Trigonometrically Well-bounded Operators: General Theory

The theory of well-bounded operators originated in 1960. Its success depended on its suitability for applications to classical operators and contexts. Despite a slow start in this regard, the theory began to gather some momentum in 1972, when Dowson and Spain produced the first solid example of a classical well-bounded operator [14]. The setting was the space $\ell^p(\mathbf{Z})$, $1 < p < \infty$ ($\mathbf{Z}$ being the additive group of integers). One of the most celebrated operators on this space is the discrete Hilbert transform H, which is the operator on $\ell^p(\mathbf{Z})$ given by convolution with the discrete Hilbert kernel h, where $h(n) = n^{-1}$, for $n \neq 0$, $h(0) = 0$. Dowson and Spain showed that $\pi I + iH$ is well-bounded (and scalar-type spectral only when $p = 2$). The question then remained of whether their result was an isolated curiosity or a glimpse at yet uncovered structure. A clue pointing toward greater generality is provided by the observation that H is the Fourier multiplier transform on $\ell^p(\mathbf{Z})$ corresponding to the function f on $\mathbf{T}$ defined by $f(e^{it}) \equiv -i(\pi - t)$ for $0 < t < 2\pi$, $f(1) = 0$. However, the right shift, R_{-1} (= the operator of translation by -1 on $\ell^p(\mathbf{Z})$, is the Fourier multiplier transform corresponding to complex conjugation on $\mathbf{T}$. Straightforward considerations with Fourier transforms now show that $R_{-1} = \exp\{i(\pi I + iH)\}$. Thus, the translation operator R_{-1} on $\ell^p(\mathbf{Z})$ can be expressed in the form e^{iA}, where A is well-bounded. This fact is capable of sweeping generalization. Specifically, using Fourier multiplier theory for abelian groups, Gillespie showed in [20] that if G is any locally compact abelian group, $x \in G$, and $1 < p < \infty$, then $R_x^{(p)}$, the operator of translation by x on $L^p(G)$, has the form e^{iA}, where A is a well-bounded operator on $L^p(G)$. This prompts the following definition.

Definition 2.1. Let X be a reflexive Banach space. An operator $U \in \mathcal{B}(X)$ will be called *trigonometrically well-bounded* provided $U = e^{iA}$ for some well-bounded operator $A \in \mathcal{B}(X)$.

Trigonometrically well-bounded operators were introduced in [4] (after some preliminary consideration of the abstract notion in [2]). In this section we shall be concerned with the general theory of trigonometrically well-bounded operators. In the next section we shall treat the abstract Fourier series machinery of the trigonometrically well-bounded operators U which are power-bounded (that is, $\sup\{\|U^n\|: n \in \mathbf{Z}\} < \infty$). Thereafter we shall consider applications of trigonometrically well-bounded operators to general analysis. In particular, it will then be seen that Gillespie's theorem asserting that translation operators are trigonometrically well-bounded has a far-reaching generalization.

We shall denote by $BV(\mathbf{T})$ the algebra of all complex-valued functions on $\mathbf{T}$ having bounded variation. $BV(\mathbf{T})$ becomes a Banach algebra when endowed with the norm $\|\cdot\|_{\mathbf{T}}$ defined by

$$\|f\|_{\mathbf{T}} = |f(1)| + \operatorname{var}(f, \mathbf{T}).$$

We shall denote by $AC(\mathbf{T})$ the closed subalgebra of $BV(\mathbf{T})$ consisting of the absolutely continuous functions on $\mathbf{T}$. By the term *trigonometric polynomial* will be meant a function $Q: \mathbf{T} \to \mathbf{C}$ having the form

$$Q(z) \equiv \sum_{n=-N}^{N} a_n z^n,$$

where N is a positive integer and the a_n's are complex constants. It is an elementary fact that the trigonometric polynomials are dense in $AC(\mathbf{T})$. We first take up a fundamental canonical relationship between well-bounded operators and trigonometrically well-bounded operators [4, Proposition 3.1].

Proposition 2.2. *Let U be a trigonometrically well-bounded operator on a reflexive Banach space X. Then there is a unique well-bounded operator $A \in \mathcal{B}(X)$ (denoted $\arg U$) such that:*

(i) $U = e^{iA}$;

(ii) $\sigma(A) \subseteq [0, 2\pi]$; and

(iii) (2π) is not an eigenvalue of A.

The operators U and $\arg U$ have the same commutants.

Proof: The existence of a well-bounded operator A having the properties (i), (ii), (iii) is a technical adjustment of Definition 2.1 that is easily established by the method of proof for [2, Proposition 3.11]. For such a well-bounded operator A we denote its spectral family by $E(\bullet)$. By Proposition 1.22 $E(\cdot)$ is concentrated on $[0, 2\pi]$,

$$A = \int_{[0,2\pi]}^{\oplus} \lambda \, dE(\lambda), \qquad E((2\pi)^-) = I.$$

Thus,

$$U = \int_{[0,2\pi]}^{\oplus} e^{i\lambda} \, dE(\lambda).$$

Suppose $0 < \alpha < \beta < 2\pi$. Let $\{Q_n\}$ be a sequence of trigonometric polynomials that is uniformly bounded in $AC(\mathbf{T})$ and has the property that $Q_n(e^{it}) \to \chi_{[\alpha,\beta]}(t)$ for $t \in [0, 2\pi]$, where "χ" denotes characteristic function. By Proposition 1.9,

2.3 $$Q_n(U) \to \{E(\beta) - E(\alpha^-)\}$$

in the strong operator topology. This shows that if A_j, $j = 1, 2$, are well-bounded operators satisfying (i), (ii), (iii), and having respective spectral families $E_j(\bullet)$, then $E_1(\beta) - E_1(\alpha^-) = E_2(\beta) - E_2(\alpha^-)$ for $0 < \alpha < \beta < 2\pi$. Letting $\beta \to (2\pi)^-$, we see that $E_1(\alpha^-) = E_2(\alpha^-)$ for $\alpha \in (0, 2\pi)$. For any $\lambda \in (0, 2\pi)$, we can now let $\alpha \to \lambda^+$ to get $E_1(\lambda) = E_2(\lambda)$. It follows by strong right continuity that $E_1(0) = E_2(0)$. Thus $E_1(\bullet) = E_2(\bullet)$, whence $A_1 = A_2$. This establishes the uniqueness assertion. It follows readily from (2.3) that U and $\arg U$ have the same commutants. This completes the proof of Proposition 2.2. ∎

Corollary 2.4. *Let U be a trigonometrically well-bounded operator on a reflexive Banach space X. Then there is a unique spectral family $E(\cdot)$ such that $E(\cdot)$ is concentrated on $[0, 2\pi]$, $E((2\pi)^-) = I$, and*

$$U = \int_{[0,2\pi]}^{\oplus} e^{i\lambda} \, dE(\lambda).$$

This unique spectral family $E(\cdot)$, called the spectral decomposition of U, coincides with the spectral family of $\arg U$. The commutants of U and $\{E(\lambda): \lambda \in \mathbf{R}\}$ are identical.

Proof: Put $A = \arg U$, and let $E(\cdot)$ be the spectral family of A. It is clear that $E(\cdot)$ has the properties in the statement of the corollary. If $F(\bullet)$ is another such spectral family, then it is easy to see that $\int_{[0,2\pi]}^{\oplus} \lambda \, dF(\lambda)$ satisfies the defining properties of $\arg U$, and hence $F(\bullet)$ is the spectral family of $\arg U$. The assertion about commutants follows from the last conclusion of Proposition 2.2 and Proposition 1.22-(i). This completes the proof of Corollary 2.4. ■

Corollary 2.5. *Let X be a reflexive Banach space, and let $U \in \mathcal{B}(X)$. Then U is trigonometrically well-bounded if and only if there is a spectral family $E_0(\bullet)$ concentrated on $[0, 2\pi]$ such that*

$$U = \int_{[0,2\pi]}^{\oplus} e^{i\lambda}\, dE_0(\lambda).$$

Proof: If there is such a spectral family $E_0(\bullet)$, then $U = e^{iB}$, where

$$B = \int_{[0,2\pi]}^{\oplus} \lambda\, dE_0(\lambda).$$

The converse is immediate from Corollary 2.4. ■

As matters now stand, in order to verify trigonometric well-boundedness and any advantages therefrom, one is required to obtain somehow a logarithm whose product with $(-i)$ is well-bounded–that is, has a suitable functional calculus. This is a cumbersome task at best. For convenience and structural clarity it is desirable to have a more direct and intrinsic characterization of trigonometrically well-bounded operators. The following result is crucial for these purposes.

Proposition 2.6. ([4, Theorem 2.3]). *Let X be a reflexive Banach space, and let $U \in \mathcal{B}(X)$. In order that U be trigonometrically well-bounded, it is necessary and sufficient that U have an $AC(\mathbf{T})$-functional calculus. If this is the case, then the spectral decomposition $E(\cdot)$ of U satisfies*

2.7 $$\sup\{\|E(\lambda)\|: \lambda \in \mathbf{R}\} \leq 3\|\mathbf{\Phi}\|,$$

where Φ is the unique $AC(\mathbf{T})$-functional calculus for U.

Comments on the Proof. If U is trigonometrically well-bounded, then the existence of its $AC(\mathbf{T})$-functional calculus follows directly from Corollary 2.5. The heart of the present proposition is the sufficiency assertion. The latter is shown, along with (2.7), in [4, Theorem 2.3] by directly producing a spectral decomposition for U. The method of proof we used for the *only if* part of Theorem 1.13 above was devised in [4] so that, with suitable technical modifications, it also handles the reasoning required for the heart of the present proposition.

Corollary 2.5 provides us with a slight, but nonetheless suggestive, reformulation of trigonometric well-boundedness: an operator U on a reflexive Banach space is trigonometrically well-bounded if and only if U is invertible and the sequence $\{U^n\}_{n=-\infty}^{\infty}$ is the Fourier-Stieltjes transform of a spectral family concentrated on $[0, 2\pi]$. The latter description has the following analogue for one-parameter groups of trigonometrically well-bounded operators [2, Theorem 4.20].

Theorem 2.8. Generalized Stone's Theorem. *Let $\{U_t : t \in \mathbf{R}\}$ be a strongly continuous one-parameter group of trigonometrically well-bounded operators on a reflexive Banach space X such that*

$$K \equiv \sup\{\|E_t(\lambda)\| : t \in \mathbf{R},\ \lambda \in \mathbf{R}\} < \infty,$$

where $E_t(\bullet)$ is the spectral decomposition of U_t for each $t \in \mathbf{R}$. Then:

(i) there is a unique spectral family $\varepsilon(\bullet)$ in X (called the Stone-type spectral family of $\{U_t\}$) such that

$$U_t x = \lim_{u \to +\infty} \int_{-u}^{u} e^{it\lambda}\, d\varepsilon(\lambda) x, \qquad \text{for } t \in \mathbf{R},\ \mathbf{x} \in \mathbf{X};$$

(ii) $\{U_t : t \in \mathbf{R}\}$ and $\{\varepsilon(\lambda) : \lambda \in \mathbf{R}\}$ have the same commutants;

(iii) the domain $\mathcal{D}(\mathcal{A})$ of the infinitesimal generator $\mathcal{A}$ of $\{U_t\}$ equals

$$\left\{ x \in X :\ \lim_{u \to +\infty} \int_{-u}^{u} \lambda\, d\varepsilon(\lambda) x \quad \text{exists} \right\},$$

and

$$\mathcal{A}(x) = i \lim_{u \to +\infty} \int_{-u}^{u} \lambda\, d\varepsilon(\lambda) x \quad \text{for } x \in \mathcal{D}(\mathcal{A});$$

(iv) $\sup\{\|\varepsilon(\lambda)\| : \lambda \in \mathbf{R}\} \le \mathbf{24K^3}$.

Comments on the Proof. Standard proofs of Stone's theorem for unitary representations in Hilbert space make crucial use of measure theory in order to produce a spectral measure (see, for example, [24, S36E]). In the circumstances at hand, however, spectral measures are neither attainable nor desirable, and we must fashion a non-measure-theoretic proof. As a preliminary step we note that in [29] Sz.-Nagy developed a proof of Stone's theorem for unitary groups without complete reliance on measure-theoretic principles (Sz.-Nagy's proof is reproduced in [25, pp. 381–383]). The key idea of his proof is to *periodize* a given one-parameter unitary group. The proof outlined below uses periodization as its starting point, and then develops techniques suitable to the broad context under consideration. For each $t \in \mathbf{R}$, denote $\arg U_t$ by A_t, and let $V_t = U_t\{\exp(-itA_1)\}$. Thus $\{V_t: t \in \mathbf{R}\}$ is a strongly continuous one-parameter group such that $V_1 = I$. We shall obtain a description of the group $\{V_t\}$ which will enable us to reach the required conclusions concerning the given group $\{U_t\}$. The periodicity of $\{V_t\}$ automatically gives us certain facts about $(C,1)$-summability (compare, for example, [22, proof of Theorem 16.7.2]). Specifically, for each $n \in \mathbf{Z}$, let

2.9 $$P_n = \int_0^1 e^{-2\pi\, int} V_t \, dt ,$$

the integral being taken as the strong limit of Riemann sums. It is straightforward to see that

2.10 $$V_s P_n = e^{2\pi\, ins} P_n , \qquad \text{for } s \in \mathbf{R},\ \mathbf{n} \in \mathbf{Z} .$$

Hence, with the aid of (2.9), each P_n is idempotent, and $P_n P_m = 0$, for $n \neq m$. The periodicity and strong continuity of the group $\{V_t\}$ enable us to apply Fejér's Theorem, thereby inferring that:

2.11 $$V_t = (C,1) \text{ sum of } \sum_{n=-\infty}^{\infty} e^{2\pi\, int} P_n$$

in the strong operator topology, for each $t \in \mathbf{R}$.

The crux of our proof consists of showing that in the present circumstances the qualification of $(C,1)$-summability can be removed from (2.11)–that is, the series on the right will converge in the strong operator topology (in fact, with separate convergence of $\sum_{n=0}^{\infty} e^{2\pi\, int} P_n$ and $\sum_{n=1}^{\infty} e^{-2\pi\, int} P_{-n}$). We shall only indicate how this last assertion can be accomplished, and why it implies the conclusions of the Generalized Stone's Theorem. For complete details we refer the reader to [2]. Let k be a positive integer, and put

$$\widetilde{A}_k = \sum_{j=0}^{k-1} (kA_{1/k} - 2\pi j) \left\{ E_{1/k} \left[(2\pi(j+1)k^{-1})^- \right] - E_{1/k} \left[(2\pi j k^{-1})^- \right] \right\} .$$

Then,

$$\exp(i\widetilde{A}_k) = \exp(ikA_{1/k}) = U_{1/k}^k = U_1 .$$

It is not difficult to check that $\widetilde{A}_k$ possesses the remaining properties in the definition of $\arg(U_1)$. Thus, in our present notation, $\widetilde{A}_k = A_1$. Substituting in the equation above which defined $\widetilde{A}_k$, we see that

$$(2.12)\quad A_{1/k} - k^{-1}A_1 = \sum_{j=1}^{k-1} 2\pi jk^{-1}\left\{E_{1/k}\left[(2\pi(j+1)k^{-1})^-\right] - E_{1/k}\left[(2\pi jk^{-1})^-\right]\right\}$$

for each positive integer k. For notational convenience we shall write $E^{(k)}(\bullet)$ instead of $E_{1/2^k}(\bullet)$. From (2.12) we have

$$2.13\qquad V_{1/2^k} = \sum_{j=0}^{2^k-1} e^{2\pi ij/2^k}\left\{E^{(k)}\left[(2\pi(j+1)2^{-k})^-\right] - E^{(k)}\left[(2\pi j2^{-k})^-\right]\right\}.$$

Now let n be a fixed but arbitrary non-negative integer. Let k be an integer such that $2^k - 1 > n$. Put

$$\mathcal{F}_{n,k} = E^{(k)}\left[(2\pi(n+1)2^{-k})^-\right] - E^{(k)}\left[(2\pi n2^{-k})^-\right].$$

If $x \in \mathcal{F}_{n,k+1}X$, then, by (2.13), $V_{1/2^{k+1}}(x) = e^{2\pi\, in/2^{k+1}}x$. So

$$V_{1/2^k}(x) = \left(V_{1/2^{k+1}}\right)^2 x = e^{2\pi\, in/2^k}x.$$

This shows that

$$2.14\qquad \mathcal{F}_{n,k+1}X \subseteq \mathcal{F}_{n,k}X, \qquad \text{for } 2^k - 1 > n \geq 0.$$

Using (2.10) and (2.13), we see that

$$2.15\qquad P_nX \subseteq \mathcal{F}_{n,k}X, \qquad \text{for } 2^k - 1 > n.$$

Suppose $m \in \mathbf{Z}$ and $m \neq n$. Obviously there is some k_0 such that $2^{k_0} - 1 > n$ and $(m-n) \notin 2^{k_0}\mathbf{Z}$. It follows by (2.10) and (2.13) that $\mathcal{F}_{n,k_0}P_m = 0$, and hence by (2.14) that $\mathcal{F}_{n,k}P_m = 0$ for $k \geq k_0$. Combining this with (2.15) gives us

$$2.16\qquad \mathcal{F}_{n,k}x \to P_nx, \qquad \text{as } k \to +\infty \text{ for } x \in X_0,$$

where X_0 is the linear manifold spanned by the subspaces P_jX for all $j \in \mathbf{Z}$. However, $K \equiv \sup\{\|E_t(\lambda)\|\colon t, \lambda \in \mathbf{R}\} < \infty$ by hypothesis, and so the projections $\mathcal{F}_{n,k}$ for $2^k - 1 > n$ are uniformly bounded. Specializing (2.11) to the case $t = 0$, we see that X_0 is dense in X. It now follows that (2.16) holds for all $x \in X$, and so for any non-negative integer N, we have

$$2.17\qquad \sum_{n=0}^{N} P_n = \lim_{k\to+\infty}\sum_{n=0}^{N}\mathcal{F}_{n,k} \qquad \text{in the strong operator topology}.$$

Recalling the definition of $\mathcal{F}_{n,k}$, we notice that

$$\sum_{n=0}^{N} \mathcal{F}_{n,k} = E^{(k)}\left[(2\pi(N+1)2^{-k})^{-}\right].$$

Applying this to (2.17) gives

2.18 $$\left\| \sum_{n=0}^{N} P_n \right\| \le K, \qquad \text{for } N \ge 0.$$

Since it is obvious that $\sum_{n=0}^{\infty} P_n x$ converges for $x \in X_0$, and X_0 is dense in X, we get from (2.18)

2.19 $$\sum_{n=0}^{\infty} P_n \quad \text{converges in the strong operator topology of } \mathcal{B}(X).$$

The strong convergence of $\sum_{n=1}^{\infty} P_{-n}$ can be demonstrated by first showing that the one-parameter group $\{U_{-t}: t \in \mathbf{R}\}$ satisfies the hypotheses of the Generalized Stone's Theorem, and then applying the foregoing considerations to the latter group. We now apply the $(C,1)$ summability in (2.11) for the special case $t = 0$ to the just established convergence of the two series in the P_n's. This gives

$$I = \sum_{n=0}^{\infty} P_n + \sum_{n=1}^{\infty} P_{-n}.$$

With the aid of (2.10) it follows that

2.20 $$V_t = \sum_{n=0}^{\infty} e^{2\pi\, int} P_n + \sum_{n=1}^{\infty} e^{-2\pi\, int} P_{-n}, \qquad \text{for all } t \in \mathbf{R},$$

the two series on the right converging in the strong operator topology. Applying the definition of the group $\{V_t\}$ to (2.20), we see that

2.21 $$U_t = \sum_{n=0}^{\infty} e^{2\pi\, int} P_n[\exp(itA_1)] + \sum_{n=1}^{\infty} e^{-2\pi\, int} P_{-n}[\exp(itA_1)], \text{ for } t \in \mathbf{R}.$$

Although a number of detailed arguments and calculations are still needed in order to establish the conclusions of the Generalized Stone's Theorem, (2.21) is the essential step. For example, it can be shown with the aid of (2.21) that the Stone-type spectral family $\mathcal{E}(\cdot)$ of $\{U_t\}$ is expressed by

$$\mathcal{E}(2\pi\lambda) = \sum_{n=-\infty}^{[\lambda]-1} P_n + P_{[\lambda]} E_1(2\pi(\lambda - [\lambda])), \qquad \text{for all } \lambda \in \mathbf{R},$$

where $[\lambda]$ denotes the greatest integer not exceeding λ.

An analysis of the method of proof for the Generalized Stone's Theorem provides the following corollary [2, Corollary (4.46)].

Corollary 2.22. *Let $\{U_t\colon t \in \mathbf{R}\}$ be a strongly continuous one-parameter group of trigonometrically well-bounded operators on a reflexive Banach space X such that*

$$K \equiv \sup\{\|E_t(\lambda)\|\colon t \in \mathbf{R},\ \lambda \in \mathbf{R}\} < \infty,$$

where $E_t(\cdot)$ is the spectral decomposition of U_t for each $t \in \mathbf{R}$. If $U_{2\pi} = I$, then there is a unique sequence of projections $\{P_n\}_{n=-\infty}^{\infty} \subseteq \mathcal{B}(X)$ such that:

(i) $P_m P_n = 0$ for $m \neq n$, and

(ii) for each $t \in \mathbf{R}$,

$$U_t = \sum_{n=0}^{\infty} e^{int} P_n + \sum_{n=1}^{\infty} e^{-int} P_{-n},$$

each series converging in the strong operator topology.

This unique sequence $\{P_n\}_{n=-\infty}^{\infty}$ is expressed by

$$P_n x = (2\pi)^{-1} \int_0^{2\pi} e^{-int} U_t x \, dt, \qquad \text{for } x \in X,\ n \in \mathbf{Z}. \tag{2.23}$$

2.24. Remarks. It follows from Gillespie's proof [20, Theorem 1] of trigonometric well-boundedness for a translation operator on $L^p(G)$ (G a locally compact abelian group, $1 < p < \infty$) that a strongly continuous one-parameter group of translation operators on $L^p(G)$ satisfies the hypotheses of the Generalized Stone's Theorem (2.8). Application of (2.8) and (2.22) respectively to the translation groups on $L^p(\mathbf{R})$ and $L^p(\mathbf{T})$ produces the M. Riesz partial sum operators as the respective Stone-type spectral families (see [2, Examples (4.47)-(i), (ii)] for a detailed account). In particular, these remarks realize the *Hilbert space minded* operator-theoretic approach to Fourier inversion in $L^p(\mathbf{T})$ which was mentioned in §I. In §IV we shall see that Gillespie's spectral theorem for $L^p(G)$-translations has a sweeping generalization and that there are extensive applications of (2.8) and (2.22).

3. Trigonometrically Well-bounded Operators Which are Power-bounded

Recall that an invertible operator V on a Banach space Y is said to be *power-bounded* provided

$$\sup\{\|V^n\|\colon n \in \mathbf{Z}\} < \infty.$$

In this section we shall establish a powerful association between power-bounded trigonometrically well-bounded operators and Fourier series for operator-valued functions.

Our purpose is twofold. First, the results illustrate the value of trigonometric well-boundedness as a property. Secondly, they also provide explicit formulas for the arg and spectral decomposition of a power-bounded trigonometrically well-bounded operator directly in terms of the operator, and this sets the stage for the applications in §IV, where the logic is reversed, and the formulas are used to establish trigonometric well-boundedness for all the power-bounded operators on a wide variety of spaces.

Let U be a power-bounded trigonometrically well-bounded operator on a reflexive Banach space X, and let $E(\cdot)$ be the spectral decomposition of U. The map

$$g \in BV(\mathbf{T}) \to g(U) \equiv \int_{[0,2\pi]}^{\oplus} g(e^{i\lambda})\, dE(\lambda)$$

is a $BV(\mathbf{T})$-functional calculus for U. Let $f \in BV(\mathbf{T})$. We denote the Fourier-coefficient sequence of f by $\widehat{f}$. Consider the operator-valued function $f(e^{it}U)$, defined for $t \in \mathbf{R}$. Formal substitution of $e^{it}U$ for $e^{i\vartheta}$ in the Fourier series $\sum_{n=-\infty}^{\infty} \widehat{f}(n)e^{in\vartheta}$ of f leads us to associate with $f(e^{it}U)$ the *purely formal* Fourier series expression $\sum_{n=-\infty}^{\infty} \widehat{f}(n)U^n e^{int}$. Our first object in this section is to show that the latter series converges in the strong operator topology, and to describe (see (3.2)) a suitable sense in which it is the Fourier series of $f(e^{it}U)$. (Here and henceforth the convergence of a bilateral series $\sum_{n=-\infty}^{\infty} x_n$ will mean the convergence of its sequence of balanced partial sums $\left\{\sum_{n=-N}^{N} x_n\right\}$.) For convenience we write $\varphi(t)$ instead of $f(e^{it}U)$. Thus

$$\varphi(t) = \int_{[0,2\pi]}^{\oplus} f(e^{it}e^{i\lambda})\, dE(\lambda) \qquad \text{for all } t \in \mathbf{R},$$

and

$$\|\varphi(t)\| \le 2\|f\|_{\mathbf{T}} \sup\{\|E(\lambda)\| \colon \lambda \in \mathbf{R}\},$$

for $t \in \mathbf{R}$. It can be seen that for each $x \in X$, $\varphi(t)x$ is an X-valued Lebesgue measurable function on $\mathbf{R}$. For each $n \in \mathbf{Z}$, we define $\widehat{\varphi}(n) \in \mathcal{B}(X)$ by setting

$$\text{3.1} \qquad \widehat{\varphi}(n)x = (2\pi)^{-1} \int_0^{2\pi} e^{-int}\varphi(t)x\, dt, \qquad \text{for } x \in X.$$

Theorem 3.2. *For $n \in \mathbf{Z}$,*

$$\widehat{\varphi}(n) = \widehat{f}(n)U^n.$$

Observe that the substitution

$$\varphi(t) = \int_{[0,2\pi]}^{\oplus} f(e^{it}e^{i\lambda})\, dE(\lambda)$$

into the right-hand side of (3.1) followed by a *formal* change in the order of integration yields (3.2). However, this procedure is not allowable since $E(\bullet)$ need not stem from a projection-valued measure. Nevertheless, there are technical means for simulating this procedure with approximating sums so as to rigorize the argument and obtain (3.2). We omit the details, which can be found in [6, Theorem (3.6)] as well as in the research announcement [5, proof of Theorem (2.1)]. Theorem (3.2) allows us to establish the following [6, Theorems (3.10), (3.21)].

Theorem 3.3. *Let U be a power-bounded trigonometrically well-bounded operator on a reflexive Banach space X, and let $E(\bullet)$ be the spectral decomposition of U. For each $f \in BV(T)$, the series $\sum_{n=-\infty}^{\infty} \widehat{f}(n)U^n$ converges in the strong operator topology to $\int_{[0,2\pi]}^{\oplus} 2^{-1}\{F_1(\lambda) + F_2(\lambda)\}\, dE(\lambda)$, where $F_1(\lambda)$ (resp., $F_2(\lambda)$) is defined by $F_1(\lambda) = \lim_{s\to\lambda+} f(e^{is})$ (resp., $F_2(\lambda) = \lim_{s\to\lambda-} f(e^{is})$) for $\lambda \in \mathbf{R}$. There is a constant C_U depending only on U such that*

$$\left\| \sum_{n=-N}^{N} \widehat{f}(n)U^n \right\| \le C_U \|f\|_{\mathbf{T}}, \qquad \text{for } N \ge 0,\ f \in BV(T). \tag{3.4}$$

Proof: Let $x \in X$. Since the function $\Phi(t)x$ (in the notation of Theorem (3.2)) has the form

$$\Phi(t)x = \int_{[0,2\pi]}^{\oplus} f(e^{it}e^{i\lambda})\, dE(\lambda)x\,,$$

we see from Proposition (1.9) that

$$\lim_{t\to 0+} \Phi(t)x = \int_{[0,2\pi]}^{\oplus} F_1(\lambda)\, dE(\lambda)x\,,$$

$$\lim_{t\to 0-} \Phi(t)x = \int_{[0,2\pi]}^{\oplus} F_2(\lambda)\, dE(\lambda)x\,.$$

We can now apply the analogue for vector-valued functions of Fejér's Theorem [23, Theorem 1.3.1]) to the function $\Phi(t)x$ (with Fourier series $\sum_{n=-\infty}^{\infty} e^{in\vartheta}\widehat{f}(n)U^n x$) to deduce that $\sum_{n=-\infty}^{\infty} \widehat{f}(n)U^n x$ converges $(C,1)$ to $\int_{[0,2\pi]}^{\oplus} 2^{-1}\{F_1(\lambda) + F_2(\lambda)\}\, dE(\lambda)x$. However, $|\widehat{f}(n)| \le (2\pi|n|)^{-1}\mathrm{var}(f,\Pi)$ for $n \in \mathbf{Z}\setminus\{0\}$, and so the nth Fourier coefficient of $\Phi(t)x$, $\widehat{\Phi}(n)x$, is $O(|n|^{-1})$ in view of Theorem (3.2) and the power-boundedness of U. A Tauberian theorem of G. H. Hardy [23, Theorem II.2.2] now enables us to replace the $(C,1)$ convergence of $\sum_{n=-\infty}^{\infty} \widehat{f}(n)U^n x$ by actual convergence of the balanced partial sums $\sum_{n=-N}^{N} \widehat{f}(n)U^n x$. This establishes the first conclusion of the theorem. We now take up the proof of (3.4). Let $\{K_N\}$ denote the Fejér kernel. We have

3.5
$$\sum_{n=-N}^{N}\left[1-\frac{|n|}{N+1}\right]\widehat{f}(n)U^n x=(2\pi)^{-1}\int_0^{2\pi}K_N(t)\Phi(t)x\,dt\,.$$

However,

$$\left\|(2\pi)^{-1}\int_0^{2\pi}K_N(t)\Phi(t)x\,dt\right\|\le 2\|f\|_T\,\|x\|\,\sup\{\|E(\lambda)\|\colon\lambda\in\mathbf{R}\}\,,$$

and

$$\left\|\sum_{n=-N}^{N}\frac{|n|}{N+1}\,\widehat{f}(n)U^n x\right\|\le\pi^{-1}\mathrm{var}(f,T)c\|x\|\,,$$

where $c=\sup\{\|U^n\|\colon n\in\mathbf{Z}\}$. Use of these estimates in (3.5) gives

$$\left\|\sum_{n=-N}^{N}\widehat{f}(n)U^n\right\|\le 2\|f\|_{\mathbf{T}}\,\sup\{\|E(\lambda)\|\colon\lambda\in\mathbf{R}\}+\pi^{-1}\mathbf{c}\ \mathrm{var}(\mathbf{f},\mathbf{T})\,.$$

This completes the proof of the theorem. ■

We are now in a position to state explicit descriptions for the arg and spectral family of U ([6, Theorems (3.18) and (3.20)]).

Theorem 3.6. *Let U be a power-bounded trigonometrically well-bounded operator on a reflexive Banach space X. Then*

$$\arg U=\pi I-\pi\lim_{n\to\infty}n^{-1}\left\{\sum_{k=0}^{n-1}U^k\right\}+i\lim_{n\to\infty}\left\{\sum_{k=-n}^{n}{}'\,k^{-1}U^k\right\},$$

where the limits are taken in the strong operator topology, and the prime superscript denotes (here and henceforth) omission of 0 as a summation index.

Proof: Now and henceforth let $g_0\in BV(T)$ be the function defined by $g_0(e^{it})=i(\pi-t)$ for $0<t<2\pi$, $g_0(1)=0$. It is elementary that $\widehat{g}_0(0)=0$ and $\widehat{g}_0(n)=n^{-1}$ for $n\ne 0$. Denoting the spectral decomposition of U by $E(\cdot)$ and applying the first conclusion of Theorem (3.3) to U and g_0, we see that

3.7
$$\sum_{n=-\infty}^{\infty}{}'\,n^{-1}U^n=\int_{[0,2\pi]}^{\oplus}g_0(e^{i\lambda})\,dE(\lambda)\,.$$

With the aid of the fact that $E((2\pi)^-) = I$, the integral on the right of (3.7) is easily calculated to get

3.8 $$\sum_{n=-\infty}^{\infty}{}' \, n^{-1}U^n = i\pi\{I - E(0)\} - i(\arg U).$$

The proof of Theorem (3.6) is easily completed by substituting for $E(0)$ in (3.8) according to the following lemma. ■

Lemma 3.9. *Let U be a power-bounded trigonometrically well-bounded operator on a reflexive Banach space X, and let $E(\bullet)$ be the spectral decomposition of U. Then for $0 \le \lambda < 2\pi$,*

$$n^{-1}\left\{\sum_{k=0}^{n-1} e^{-ik\lambda}U^k\right\} \longrightarrow E(\lambda) - E(\lambda^-) \qquad \text{as } n \to \infty$$

in the strong operator topology.

Proof: Calculations based on the $BV(T)$-functional calculus of U (see [6, Theorem (3.14)]) show that for $0 \le \lambda < 2\pi$:

3.10 $$\{E(\lambda) - E(\lambda^-)\}X = \{x \in X: Ux = e^{i\lambda}x\};$$

3.11 $$[I - \{E(\lambda) - E(\lambda^-)\}]X = \overline{(e^{i\lambda} - U)X},$$

where the bar superscript denotes closure.

However, by [16, Corollaries VIII.5.2 and VIII.5.4], the power-boundedness of U and the reflexivity of X are alone sufficient to ensure that the discrete averages

$$n^{-1}\left[\sum_{k=0}^{n-1} e^{-ik\lambda}U^k\right]$$

converge in the strong operator topology to a projection having the same range and null space as those exhibited for $\{E(\lambda) - E(\lambda^-)\}$ in (3.10) and (3.11).

Theorem 3.12. *Let U be a power-bounded trigonometrically well-bounded operator on a reflexive Banach space X, and let $E(\bullet)$ be the spectral decomposition of U. For each $t \in [0,2\pi)$ let P_t be the strong limit of the sequence*

$$\left\{n^{-1}\sum_{k=0}^{n-1} e^{-ikt}U^k\right\}.$$

Then for $t \in [0,2\pi)$,

$$E(t) = (2\pi i)^{-1}\left\{itI - \sum_{n=-\infty}^{\infty}{}' n^{-1}e^{-int}U^n + \sum_{n=-\infty}^{\infty}{}' n^{-1}U^n\right\} + 2^{-1}(P_t + P_0),$$

both series converging in the strong operator topology.

Proof: For $0 \le t < 2\pi$, let $G_t \in BV(T)$ be the characteristic function of $\{e^{i\lambda}\colon 0 \le \lambda \le t\}$. Application of Theorem (3.3) to G_t gives us:

$$\sum_{k=-\infty}^{\infty} \widehat{G}_t(k)U^k = 2^{-1}\{E(t^-) + E(t) - E(0)\}.$$

Use of Lemma (3.9) and direct calculation of the Fourier coefficients $\widehat{G}_t(k)$ complete the proof. ■

The series convergence occurring in Theorems (3.3), (3.6), and (3.12) bears a certain formal resemblance to the Coifman and Weiss notion of transference. In essence, for appropriate settings of group representations, the latter notion describes a method for *transferring* convolution operators and their associated bounds to the L^p-spaces of measures. Here is a particular version of it.

Theorem 3.13. General Transference ([12, Theorem 2.4]). *Suppose G is a locally compact abelian group, M is a σ-finite measure space, $1 < p < \infty$, and S is a closed subspace of $L^p(M)$. Let $u \to R_u$ be a strongly continuous representation of G by operators on S such that $c \equiv \sup\{\|R_u\|\colon u \in G\} < \infty$. Suppose that $\Psi \in L^1(G)$ has compact support, and denote by C_Ψ the operator of convolution by Ψ on $L^p(G)$. Let us put*

$$(\mathcal{H}_\Psi F)(x) = \int_G \Psi(u)(R_{u^{-1}}F)(x)\,du, \qquad \text{for } F \in S,\ x \in M.$$

Then $\mathcal{H}_\Psi$ is a bounded linear mapping of S into S such that

$$\|\mathcal{H}_\Psi\| \le c^2\|\mathcal{C}_\Psi\|.$$

Steckin's Theorem for $\mathbf{Z}$ (see [13, Theorem 20.7]) states that for $f \in BV(T)$ and $1 < p < \infty$, convolution by the Fourier transform $\widehat{f}$ defines a bounded operator on $\ell^p(\mathbf{Z})$. In the setting of Theorem (3.3) the additive group $\mathbf{Z}$ has a uniformly bounded representation $n \to U^n$ in the reflexive Banach space X. From the viewpoint of this representation, the series convergence in the conclusion of Theorem (3.3) states *in a purely formal way* that for $f \in BV(T)$, convolution by the Fourier transform of f transfers to X. Similarly, the series convergence in the conclusion of Theorem (3.12) states that formally the unimodular multiples of U transfer the discrete Hilbert transform to X. There are of course significant differences between the context of the formal transferences and the setting of the General Transference Theorem (3.13). For one thing, the convolution kernels $\hat{f}$ for $f \in BV(T)$ do not generally have compact support. Nevertheless, formal transference analogies provide us with a suggestive heuristic language which will gain impetus as we proceed. In particular, for power-bounded operators on reflexive spaces, the ability to transfer the discrete Hilbert transform actually characterizes trigonometric well-boundedness, a fact which we now make precise ([8, Theorems (3.1) and (3.43)]).

Theorem 3.14. *Let U be a power-bounded operator on the reflexive Banach space X. The following assertions are equivalent.*

(i) U is trigonometrically well-bounded;

(ii) for each $t \in [0,2\pi)$,

$$\sum_{n=-\infty}^{\infty}{}' n^{-1} e^{-int} U^n$$

converges in the strong operator topology to an operator B_t, and

$$\sup\{\|B_t\| : t \in [0,2\pi)\} < \infty ;$$

(iii)
$$\sup\left\{\left\|\sum_{n=-N}^{N}{}' n^{-1} e^{-int} U^n\right\| : N \geq 1,\ t \in [0,2\pi)\right\} < \infty .$$

Comments on the proof. That (i) implies (ii) and (iii) follows directly by applying Theorem (3.3) to the translates of the function g_0 (defined in the proof of Theorem (3.6)). The implication (iii) implies (ii) can be seen by decomposing X into the direct sum of the kernel of $(e^{it} - U)$ and $\overline{(e^{it} - U)X}$. It is then not difficult to see that the series occurring in (ii) converges pointwise on each of these direct summands. The main feature of the present theorem is the implication (ii) $\Rightarrow$ (i). This is demonstrated in [8] in the following manner. All the terms on the right-hand side of the equation in the statement of Theorem (3.12) are meaningful in the present circumstances–the projections P_t by power-boundedness and reflexivity, and the series by virtue of the hypotheses in (ii).

It is then shown that the operator-valued function defined by the right-hand side of the equation in Theorem (3.12) actually serves as a spectral decomposition for U. The actual details are lengthy and arduous, although including some items of independent interest. Essentially, we shall only require the implication (iii) $\Rightarrow$ (i), and that has the following short proof.

Theorem 3.15. *Let U be an invertible operator on a reflexive Banach space X such that*

$$M \equiv \sup \left\{ \left\| \sum_{n=-N}^{N}{}' n^{-1} e^{-int} U^n \right\| : N \geq 1, t \in [0, 2\pi) \right\} < \infty.$$

Then U is trigonometrically well-bounded, and the spectral decomposition $E(\cdot)$ of U satisfies

3.16 $$\sup\{\|E(\lambda)\| : \lambda \in \mathbf{R}\} \leq \mathbf{3}[\mathbf{1} + (\mathbf{2}\pi)^{-1}\mathbf{M}].$$

Proof: (N. B. : The present theorem does not require U to be power-bounded.) In view of Proposition (2.6) it suffices to show that for every trigonometric polynomial Q

3.17 $$\|Q(U)\| \leq \left[1 + (2\pi)^{-1} M\right] \|Q\|_{\mathbf{T}}.$$

Let $Q(z) \equiv \sum_{n=-K}^{K} \widehat{Q}(n) z^n$ $(z \in T)$. Thus,

$$\begin{aligned} Q(U) &= \sum_{n=-K}^{K} (2\pi)^{-1} \int_0^{2\pi} Q(e^{it}) e^{-int} U^n \, dt \\ &= \widehat{Q}(0) I + (2\pi)^{-1} \int_0^{2\pi} Q(e^{it}) \left[\sum_{n=-K}^{K}{}' e^{-int} U^n \right] dt \\ &= \widehat{Q}(0) I + (2\pi)^{-1} \int_0^{2\pi} Q(e^{it}) \, d \left[\sum_{n=-K}^{K}{}' (-in)^{-1} e^{-int} U^n \right]. \end{aligned}$$

Integration by parts in the last integral gives us

$$Q(U) = \widehat{Q}(0) I + (2\pi)^{-1} \int_0^{2\pi} \left[\sum_{n=-K}^{K}{}' (in)^{-1} e^{-int} U^n \right] dQ(e^{it}).$$

An easy estimate from this equation now shows (3.17). This completes the proof of Theorem (3.15). ■

4. Applications of Trigonometric Well-boundedness in Modern Analysis

A highly fruitful combination occurs when the foregoing considerations are joined to the class of UMD spaces. A Banach space X_0 is said to possess the unconditionality property for martingale differences (written $X_0 \in$ UMD) provided that for $1 < p < \infty$, X_0-valued martingale difference sequences are unconditional in $L^p([0,1],X_0)$ with a uniform unconditionality constant $K_p(X_0)$ (see [11] for a more detailed account of this definition). UMD spaces have recently been characterized by Bourgain and Burkholder ([10], [11]) as those Banach spaces X_0 such that the Hilbert kernel of $\mathbf{R}$ defines a bounded convolution operator on $L^p(\mathbf{R},\mathbf{X_0})$ for some value of p in the range $1 < p < \infty$, and hence for all values of p in this range. This characterization of UMD spaces is the one we shall be concerned with. We remark in passing that it remains a valid characterization of UMD spaces when $\mathbf{R}$ is replaced by the circle group. It is well-known that UMD spaces are super-reflexive, that the UMD property is inherited by subspaces, dual spaces, and quotient spaces, and that the class UMD contains many of the classical spaces. In particular, for $1 < p < \infty$, the von Neumann-Schatten p-class $\mathcal{C}_p$ and $L^p(\mu)$ (μ an arbitrary measure) are UMD spaces. The fact that $\mathcal{C}_p$ is a UMD space, rests on a well-known argument of Gohberg and Krein which employs M. Cotlar's *bootstrap* method for the Hilbert transform. This argument is improved in [8, S6] to show that the non-commutative L^p-spaces associated with a semifinite von Neumann algebra belong to the class UMD for $1 < p < \infty$. Thus the results of this section concerning operator theory in UMD spaces will apply simultaneously both to commutative and non-commutative analysis.

In view of the importance of the discrete Hilbert kernel in the formulas of Theorem (3.14)-(ii), (iii), it is desirable to characterize the class UMD in terms of the discrete Hilbert transform. More precisely, for a Banach space X, let $D_{p,X}$ denote convolution by the discrete Hilbert kernel h on $L^p(\mathbf{Z},\mathbf{X})$, and let $D_{N,p,X}$ denote convolution by h_N, where, for $n \in \mathbf{Z}$, $h_N(n) = h(n)$ if $|n| \leq N$, and $h_N(n) = 0$ otherwise. The following theorem was shown in [8, Theorem (2.8) and Corollary (2.18)].

Theorem 4.1. *Suppose that X is a Banach space and $1 < p < \infty$. The following statements are equivalent:*

(i) $X \in$ UMD;

(ii) $D_{p,X} \in \mathcal{B}(L^p(\mathbf{Z},\mathbf{X}))$;

(iii) $\sup\{\|D_{N,p,X}\|: N \geq 1\} < \infty$.

Comments on the proof. The classical proofs of Marcel Riesz proceeding from the Hilbert kernel of $\mathbf{R}$ show that (i) implies (ii) and (iii). (These implications are all we shall require below.) It is elementary that (iii) implies (ii). The implication (ii) implies (i) is the most difficult, and appears to be a new result in the generality stated here. Its proof is reduced by invoking the continuity and commutativity with dilations of the classical Hilbert transform on $L^p(\mathbf{R})$. Thereby it suffices to show that convolution by the Hilbert kernel of $\mathbf{R}$ is bounded on the X-valued step functions taking constant values on the intervals $(n,n+1)$, $n \in \mathbf{Z}$. We omit the remaining technical details. ■

The link between trigonometric well-boundedness and the UMD property is provided by the methodology of transference. In the special case of $G = \mathbf{Z}$, we adapt the proof of the Coifman and Weiss General Transference Theorem (Theorem (3.13)) so as to obtain a suitable vector-valued variant ([8, Theorem (4.1)]).

Theorem 4.2. *Let Y be a Banach space, and suppose V is a power-bounded operator on Y. For each trigonometric polynomial f and $1 \le p < \infty$, let $\Omega_{f,p,Y}$ be the bounded operator on $L^p(\mathbf{Z},\mathbf{Y})$ defined by convolution with $\widehat{f}$. Then*

$$\|f(V^{-1})\| \le c^2 \|\Omega_{f,p,Y}\|,$$

where $c = \sup\{\|V^n\|: n \in \mathbf{Z}\}$.

Proof: Put $a_n = \widehat{f}(n)$ for $n \in \mathbf{Z}$, and pick N a positive integer so that $a_n = 0$ for $|n| > N$. Suppose $y \in Y$. Let ν be an arbitrary positive integer. Since

$$\|f(V^{-1})y\|_Y^p \le c^p \|V^k f(V^{-1})y\|_Y^p, \qquad \text{for all } k \in \mathbf{Z},$$

we obviously have

$$\|f(V^{-1})y\|_Y^p \le c^p (2\nu+1)^{-1} \sum_{k=-\nu}^{\nu} \|V^k f(V^{-1})y\|_Y^p.$$

Expressing $f(V^{-1})$ as $\sum_{n=-N}^{N} a_n V^{-n}$ in the sum on k, we obtain

$$\|f(V^{-1})y\|_Y^p \le c^p (2\nu+1)^{-1} \sum_{k=-\nu}^{\nu} \left\| \sum_{n=-N}^{N} a_n V^{k-n} y \right\|_Y^p.$$

Let ξ denote the characteristic function, relative to $\mathbf{Z}$, of $\{j \in \mathbf{Z}: |\mathrm{j}| \le \nu + \mathrm{N}\}$, and definte $x \in L^p(\mathbf{Z},\mathbf{Y})$ by setting $x_j = \xi_j V^j y$ for $j \in \mathbf{Z}$. The last inequality can now be rewritten

$$\|f(V^{-1})y\|_Y^p \le c^p (2\nu+1)^{-1} \sum_{k=-\nu}^{\nu} \|(\widehat{f} \star x)(k)\|_Y^p,$$

which gives

$$\|f(V^{-1})y\|_Y^p \leq c^p(2\nu+1)^{-1}\ \|\Omega_{f,p,Y}\|^p \sum_{|j|\leq\nu+N} \|V^j y\|_Y^p .$$

Finally, then,

$$\|f(V^{-1})y\|_Y^p \leq c^{2p}\|\Omega_{f,p,Y}\|^p\ (2\nu+1)^{-1}(2\nu+2N+1)\|y\|_Y^p .$$

Letting $\nu \to +\infty$ now gives the desired conclusion.

We come now to the central theorem of this section for single operator theory [8, Theorem (4.5)].

Theorem 4.3. *Let U be a power-bounded operator on a UMD space X, Then U is trigonometrically well-bounded, and there is a constant K_X depending only on X such that the spectral decomposition $E(\cdot)$ of U satisfies*

$$\sup\{\|E(\lambda)\|:\ \lambda \in \mathbf{R}\} \leq \mathbf{c^2 K_X},$$

where $c = \sup\{\|U^n\|:\ n \in \mathbf{Z}\}$.

Proof: Fix p in the range $1 < p < \infty$, and let f be the trigonometric polynomial

$$f(e^{is}) \equiv \sum_{n=-N}^{N}{}' \ n^{-1}e^{ins} .$$

By Theorem (4.2) we have

$$\left\| \sum_{n=-N}^{N}{}' \ n^{-1}U^n \right\| \leq c^2\|D_{N,p,X}\|.$$

Now replace U by $e^{-it}U$ and apply Theorems (4.1) and (3.15) to complete the proof. ■

Corollary 4.4. *Let X be a UMD space. Then:*

(i) every power-bounded operator on X has a logarithm in $\mathcal{B}(X)$;
(ii) every $A \in \mathcal{B}(X)$ such that $\sup\{\|e^{itA}\|:\ t \in \mathbf{R}\} < \infty$ is well-bounded.

The next theorem [8, S5] provides the analogue of Theorem (4.3) for one-parameter groups.

Theorem 4.5. *Let $\{U_t\colon t \in \mathbf{R}\}$ be a strongly continuous one-parameter group of operators on a UMD space X such that $\sup\{\|U_t\|\colon t \in \mathbf{R}\} < \infty$. Then the group $\{U_t\}$ satisfies the hypotheses of the Generalized Stone's Theorem (2.8). If $\varepsilon(\cdot)$ denotes the Stone-type spectral family of $\{U_t\}$, then:*

(i) *for each $s \in \mathbf{R}$,*

$$(\pi i)^{-1} \int_{\delta<|t|<1/\delta} t^{-1} e^{ist} U_{-t}\, dt$$

converges in the strong operator topology, as $\delta \to 0^+$, to an operator J_s;

(ii) *$\varepsilon(s) = I + 2^{-1}(J_s - J_s^2)$, for all $s \in \mathbf{R}$.*

Proof: The group $\{U_t\}$ satisfies the hypotheses of the Generalized Stone's Theorem by virtue of Theorem (4.3). The remaining conclusions (i), (ii) are shown in [8, Theorem (5.16)]. We shall only indicate what is involved. A vector-valued transference argument analogous to that used above in the proof of Theorem (4.2) can be developed to show that the integrals in (i) are uniformly bounded as s and δ vary. This uniform boundedness eventually yields (i) and (ii), but not until after a number of details have been attended to by using the Stone-type spectral family of $\{U_t\}$ to decompose the space appropriately.

Theorem (4.5) can be applied to holomorphic semigroup extensions for uniformly bounded one-parameter groups on UMD spaces [9, S4]. In effect, the theory of holomorphic extensions for such groups is completely analogous to that of unitary groups in Hilbert space, even though in the UMD setting the Stone-type spectral family need not stem from a spectral measure. The following theorem describes the specifics.

Theorem 4.6. *Let $\{U_t\}$, $t \in \mathbf{R}$, be a uniformly bounded, strongly continuous one-parameter group of operators on a UMD space X, and let $\varepsilon(\cdot)$ be the Stone-type spectral family of $\{U_t\}$, $t \in \mathbf{R}$. Then $\{U_t\}$, $t \in \mathbf{R}$, can be extended to a strongly continuous semigroup $\{U_z\}$, $\Im m\, z \geq 0$, such that $\{U_z\}$, $\Im m\, z > 0$, is holomorphic if and only if there is a real number b such that the infinitesimal generator of $\{U_t\}$, $t \in \mathbf{R}$, has its spectrum contained in $\{i\lambda\colon \lambda \geq b\}$. If this is the case, then the semigroup $\{U_z\}$, $Im\, z \geq 0$, is uniquely determined, and is given by*

$$U_z \alpha = \lim_{u\to+\infty} \int_{-u}^{u} e^{i\lambda z}\, d\varepsilon(\lambda)\alpha, \qquad \text{for } \alpha \in X,\ \Im m\, z \geq 0.$$

Moreover, in this case $\sup\{\|U_z\|\colon \Im m\, z \geq 0\} < \infty$ if and only if the infinitesimal generator of $\{U_t\}$, $t \in \mathbf{R}$, has its spectrum contained in $\{i\lambda\colon \lambda \geq 0\}$.

In addition to the foregoing broad applications, we have already indicated in (2.24) how Theorem (4.5) can be used to recover the partial sum operators in $L^p(T)$ and $L^p(\mathbf{R})$, $1 < p < \infty$. We shall devote the remainder of this section to illustrating further applications by considering some specific topics.

4.7. Ergodic Flows. In [18] D. Fife considered one-parameter isometric groups in $L^p(\mu)$, $1 < p < \infty$, induced by a one-parameter group of measure-preserving transformations in the underlying measure space satisfying appropriate measurability and continuity conditions (he also assumed that μ is σ-finite). Fife showed that such a group has a spectral decomposition $\varepsilon(\cdot)$ which satisfies conclusions (i) and (ii) of Theorem (4.5). Thus Theorem (4.5) broadly extends Fife's results. Of course Theorem (4.5) applies to ergodic flows in non-commutative L^p-spaces, and this aspect and its ramifications are currently under investigation in joint work of the author with T. A. Gillespie and P. S. Muhly.

4.8. Helson's Theory of Invariant Subspaces and Generalized Analyticity. Let Γ be a dense subgroup of the additive group $\mathbf{R}$. Endow Γ with the discrete topology, and let K be the dual group of Γ. For $\lambda \in \Gamma$ let χ_λ denote the corresponding character of K. Helson defines a closed subspace M of $L^2(K)$ to be *simply invariant* provided $\chi_\lambda M \subseteq M$ for all $\lambda > 0$, but for some $\alpha < 0$, $\chi_\alpha M$ is not a subset of M. A simply invariant subspace M is said to be *normalized* provided $M = \cap\{\chi_\lambda M\colon \lambda \in \Gamma,\ \lambda < 0\}$. (For a comprehensive account of Helson's theory see [21]). It is elementary that the real line forms a dense subgroup in K. A cocycle on K is a Borel measurable function $A\colon \mathbf{R}\times\mathbf{K} \to \mathbf{T}$ such that the equation $U_t f = A(t,\cdot)R_t f$ for $f \in L^2(K)$, $t \in \mathbf{R}$, defines a family of operators U_t satisfying the group property (here R_t denotes translation by t). More precisely, the Borel measurable unimodular function $A(\cdot,\cdot)$ is required to satisfy

$$A(t+u,x) = A(t,x)A(u,x+t) \qquad \text{for } t \in \mathbf{R},\ \mathbf{u} \in \mathbf{R},\ \mathbf{x} \in \mathbf{K}.$$

Helson's classic theory assigns to each cocycle the range of $\{I - \varepsilon(O^-)\}$, where $\varepsilon(\cdot)$ is, in our parlance, the Stone-type spectral family of the one-parameter group $\{U_t\}$ just described. This correspondence is a one-to-one map of the class of all cocycles onto the class of all normalized simply invariant subspaces in $L^2(K)$. The surjectivity of this mapping relies on the fact that each closed subspace of $L^2(K)$ is the range of an orthogonal projection. In [9] we have shown, with the aid of Theorem (4.5), that Helson's theory extends to $L^p(K)$ $(1 < p < \infty)$ in complete analogy with the L^2-case. In particular, the analogous mapping of cocycles to subspaces via Stone-type spectral families is a one-to-one map of the cocycles onto the class of all normalized simply invariant subspaces in $L^p(K)$. This extension of Helson's theory requires more than Theorem (4.5)–in particular, the surjectivity of the mapping of cocycles requires special attention. Other aspects of Helson's L^2-theory involving generalized analyticity make use of the spectral *measure* in Stone's theorem for unitary groups. These aspects are handled in the $L^p(K)$-setting with the aid of Theorem (4.6).

4.9. Abstract Conjugate Function Operators. In a recent paper [7, S4] T. A. Gillespie and I illustrated Theorem (4.5) with the following special case.

Theorem 4.10. *Let $\{U_t\}$, $t \in \mathbf{R}$, be a strongly continuous one-parameter group of operators on a UMD space X such that $U_{2\pi} = I$. Then there is a unique sequence of projections $\{P_n\}_{n=-\infty}^{\infty} \subseteq \mathcal{B}(X)$ such that:*

(i) $P_n P_m = 0$ for $m \neq n$;

(ii) for each $t \in \mathbf{R}$,

$$U_t = \sum_{n=0}^{\infty} e^{int} P_n + \sum_{n=1}^{\infty} e^{-int} P_{-n} ,$$

where each series on the right converges in the strong operator topology.

This unique sequence $\{P_n\}_{n=-\infty}^{\infty}$ has the representation

$$P_n x = (2\pi)^{-1} \int_0^{2\pi} e^{-int} U_t x \, dt , \qquad \text{for } n \in \mathbf{Z}, \ \mathbf{x} \in \mathbf{X}.$$

By the Principle of Uniform Boundedness, a periodic group is automatically uniformly bounded. Hence Theorem (4.10) follows from Theorem (4.5) and Corollary (2.22). If the UMD space X in Theorem (4.10) is replaced by an arbitrary Banach space, then as a consequence of Fejér's Theorem the conclusions of Theorem (4.10) remain valid except that on the right of (4.10)-(ii) we must substitute the $(C,1)$-sum in the strong operator topology of $\sum_{n=-\infty}^{\infty} e^{int} P_n$. Thus the effect of the UMD hypothesis in Theorem (4.10) is to replace $(C,1)$-summability by the separate convergence of the two series in (4.10)-(ii). This distinction allows us to associate an abstract *harmonic conjugation* operator $C_{\mathcal{U}}$ with the group $\mathcal{U} \equiv \{U_t\}$ in Theorem (4.10).

Definition. Let $\mathcal{U} \equiv \{U_t\}$ and X be as in the hypotheses of Theorem (4.10). We define the operator $C_{\mathcal{U}} \in \mathcal{B}(X)$ by setting

$$C_{\mathcal{U}} = -i \sum_{n=1}^{\infty} P_n + i \sum_{n=1}^{\infty} P_{-n} ,$$

where each series on the right converges in the strong operator topology.

The operator $C_{\mathcal{U}}$ provides an abstract unified approach to harmonic conjugation. In particular, if we take $X = L_h^p$, $1 < p < \infty$, the Bergman space of all complex-valued harmonic functions in $L^p(\mathbf{D})$, and the group $\mathcal{U} = \{U_t\}$ to be the translation group on L_h^p, then the operator $C_{\mathcal{U}}$ shows directly that harmonic conjugation is continuous in the $L^p(\mathbf{D})$-norm on the real functions in L_h^p.

In closing, we remark that the results in SIV for operators on UMD spaces do not remain valid if the UMD hypothesis is removed (see the examples in [6, S5]). For example, there are a reflexive Banach space X and a surjective isometry U_0 of X such that U_0 has no logarithm in $\mathcal{B}(X)$.

References

1. Barry, J. Y., On the convergence of ordered sets of projections, Proc. Amer. Math. Soc. **5** (1954), 313–314.
2. Benzinger, H, E. Berkson, and T. A. Gillespie, Spectral families of projections, semigroups, and differential operators, Trans. Amer. Math. Soc. **275** (1983), 431–475.
3. Berkson, E. and H. R. Dowson, On uniquely decomposable well-bounded operators, Proc. London Math. Soc. (3) **22** (1971), 339–358.
4. Berkson, E. and T. A. Gillespie, *AC* functions on the circle and spectral families, J. Operator Theory **13** (1985), 33–47.
5. ———, The existence of spectral decompositions in L^p-subspaces, Proc. Japan Academy **61A** (1985), 172–175.
6. ———, Steckin's Theorem, transference, and spectral decompositions, J. Functional Analysis **70** (1987), 140–170.
7. ———, Fourier series criteria for operator decomposability, Integral Equataions and Operator Theory **9** (1986), 767–789.
8. Berkson, E., T. A. Gillespie, and P. S. Muhly, Abstract spectral decompositions guaranteed by the Hilbert transform, Proc. London Math. Soc. (3) **53** (1986), 489–517.
9. ———, Analyticity and spectral decompositions of L^p for compact abelian groups, Pacific J. Math. **127** (1987), 247–260.
10. Bourgain, J., Some remarks on Banach spaces in which martingale difference sequences are unconditional, Arkiv för Matematik **21** (1983), 163–168.

11. Burkholder, D. L., A geometric condition that implies the existence of certain singular integrals of Banach-space-valued functions, Proc. of Conference on Harmonic Analysis in Honor of Antoni Zygmund (Chicago, Illinois, 1981), Wadsworth Publishers, Belmont, California, 1983.
12. Coifman, R. R. and G. Weiss, Transference methods in analysis, CBMS Regional Conference Series in Math **31** Amer. Math. Soc., Providence, 1977.
13. Dowson, H. R., *Spectral Theory of Linear Operators*, London Math. Soc., Monographs **12** Academic Press, New York, 1978.
14. Dowson and P. G. Spain, An example in the theory of well-bounded operators, Proc. Amer. Math. Soc., **32** (1972), 205–208.
15. Dunford, N. Spectral operators, Pacific J. Math., **4** (1954), 321–354.
16. Dunford, N. and J. T. Schwartz, Linear operators. Part I: General theory, *Pure and Applied Math.*, Vol. 7, Interscience, New York, 1958.
17. Edwards, R. E. and G. I. Gaudry, Littlewood-Paley and multiplier theory, *Ergebnisse der Math. und ihrer Grenzgebiete*, Vol. 90, Springer-Verlag, New York, 1977.
18. Fife, D., Spectral decomposition of ergodic flows on L^p, Bull. Amer. Math. Soc., **76** (1970), 138–141.
19. Fong, C. K. and L. Lam, On spectral theory and convexity, Trans. Amer. Math. Soc., **264** (1981), 59–75.
20. Gillespie, T. A., A spectral theorem for L^p translations, J. London Math. Soc., (2)**11** (1975), 499–508.
21. Helson, H., Analyticity on compact abelian groups, *Algebras in Analysis*, Proc. of 1973 Birmingham Conference–NATO Advanced Study Institute, Academic Press, London, 1975, 1–62.
22. Hille, E. and R. S. Phillips, Functional analysis and semi-groups, *Amer. Math. Soc. Colloq. Publ.*, Vol. 31, Amer. Math. Soc., Providence, 1957.
23. Katznelson, Y., *An Introduction to Harmonic Analysis*, Dover, New York, 1976.
24. Loomis, L. H., *An Introduction to Abstract Harmonic Analysis*, Van Nostrand, New York, 1953.
25. Riesz, F. and B. Sz-Nagy, *Functional Analysis*, Ungar, New York, 1955.
26. Ringrose, J. R., On well-bounded operators, J. Australian Math. Soc., **1** (1960), 334–343.
27. ———, On well-bounded operators II, Proc. London Math. Soc., (3) **13** (1963), 613–638.

28. Smart, D. R., Conditionally convergent spectral expansions, J. Australian Math. Soc., **1** (1960), 319–333.
29. Sz-Nagy, B., Über messbare Darstellungen Liescher Gruppen, Math. Ann., **112** (1936), 286–296.

Earl Berkson
Department of Mathematics
University of Illinois at Urbana-Champaign
Urbana, Illinois 61801
U.S.A.

The work of the author described in these lectures was supported by the National Science Foundation.

Some Remarks on Principal Currents and Index Theory for Single and Several Commuting Operators

by

Richard W. Carey

1. Introduction

Let $M_1, \ldots, M_n$ be n finite matrices and let

$$\mathbf{Tr}(M_1 \cdots M_n)$$

be the trace of the product. Since $\mathbf{Tr}\, AB = \mathbf{Tr}\, BA$, it follows that this product is invariant under cyclic permutations, e.g.,

$$\mathbf{Tr}(M_1 \cdots M_n) = \mathbf{Tr}(M_n M_1 \cdots M_{n-1}) = \mathbf{Tr}(M_{n-1} M_n M_1 \cdots), \qquad \text{etc.}$$

However, this phenomenon fails for infinite matrices as it can happen that $M_1 M_2 - M_2 M_1$ lies in the trace class ideal and has a non-zero trace.

The lack of a cyclic invariant trace for the class of infinite matrices is related to the nonvanishing of the Noether index,

$$\dim\ \ker T - \dim \operatorname{Coker} T = \mathbf{Tr}(TR - RT),$$

where R is an approximate inverse of T. In this note we shall discuss how traces associated with single and several commuting Fredholm operators lead to a natural concept of index for the lattice of operator ranges and for the maximal ideals of commuting subalgebras of $L(\mathbf{H})$, the bounded operators on a complex separable Hilbert space. In particular, we shall discuss a surprising integrality result for subnormal operators.

Motivation for this study comes in part from an extension problem in cyclic homology and perhaps is best viewed in terms of a reciprocity law which holds for tuples of commuting Fredholm operators [13].

During the course of this paper, we give some highlights of the interplay between geometric measure theory and operator theory. The style stresses motivation. A number of open problems are included.

2. Cyclic Homology of Algebras

Let A be an associative $\mathbf{C}$-algebra with unity. The Hochschild homology of A with coefficients in itself $H_n(A,A)$ is the homology of the complex

$$\dots A\otimes A\otimes A \xrightarrow{b_2} A\otimes A \xrightarrow{b_1} A \longrightarrow 0,$$

where the boundary operator b is given by

$$b(a_0,\dots,a_n)=\sum_{i=0}^{n-1}(-1)^i(a_0,\dots,a_ia_{i+1},\dots,a_n)+(-1)^n(a_na_0,\dots,a_{n-1}).$$

Then $b_{n-1}b_n = 0$ and $H_n(A,A) = \ker b_n/\operatorname{Im} b_{n+1}$. In particular, $b_1(a_0,a_1) = a_0a_1 - a_1a_0$ the commutator of a_0 and a_1. If A is commutative, then $H_0(A,A) = A$ and $(H_1(A,A) = \Omega^1_{A/\mathbf{C}}$ the module of Kähler differentials, generated by symbols dx for x in A with the relations $d(x+y) = dx + dy$, $d(xy) = xdy + ydx$ and $d(\mathbf{C}) = \mathbf{0}$.

Let $t : \mathbf{A}^{\otimes n+1} \to \mathbf{A}^{\otimes n+1}$ be the cyclic operator

$$t(a_0,a_1,\dots,a_n)=(-1)^n(a_n,a_0,\dots,a_{n-1}).$$

We can divide $\mathbf{A}^{\otimes n+1}$ by the action of t to get $C_n(A) = \mathbf{A}^{\otimes n+1}/(1-t)$ so that b is still well-defined and therefore $(C_n(A),b)$ is a chain complex.

The *cyclic homology* of A is $HC_n(A) = H_n(C_n(A),b)$. For instance, if A is commutative, $HC_0(A) \simeq A$ and $HC_1(A) \simeq \Omega^1_{A/\mathbf{C}}/dA$, where $(a_0,a_1) \simeq a_0da_1$,. Let $\Omega^n_{A/\mathbf{C}}$ denote the n-th exterior power. The deRham complex

$$A=\Omega^0_A \to \Omega^1_A \to \dots \to \Omega^n_A \xrightarrow{d} \Omega^{n+1}_A \to \dots$$

where $d(a_0da_1 \wedge da_n) = da_0 \wedge da_1 \wedge\cdots\wedge da_n$ gives rise to deRham cohomology $H^\star_{DR}(A)$.

Theorem 2.1 (Loday-Quillen [31]). *If A is commutative and smooth, then*

$$HC_n(A)=\frac{\Omega^n_A}{d\Omega^{n-1}_A}\oplus H^{n-2}_{DR}(A)\oplus H^{n-4}_{DR}(A)\oplus\cdots.$$

Cyclic homology was introduced by A. Connes, *Non-commutative differential geometry*, Inst. des Hautes Études Sci. Publ., Math. No. **62** (1985), 41–144. A basic theorem says that there is a long exact sequence

$$\cdots\to H_n(A,A)\to HC_n(A)\xrightarrow{S} HC_{n-2}(A)\to H_{n-1}(A,A)\to\cdots$$

where S is a periodicity operator on the cyclic homology. In this note we shall consider the interest of this homology for the case of planar function algebras, that is for $HC_1(A)$.

3. Examples (Subnormal Operators)

Let T be a subnormal operator acting on a Hilbert space H. Let N be the normal extension on a space H_ℓ and let P be the orthogonal projection from H_ℓ to H so that

$$T = PN|H.$$

If T is finitely-rationally cyclic, a theorem due to Berger-Shaw [2] says that the commutator $PN - NP = PN(1-P)$ belongs to the Schatten ideal $\mathcal{L}^p(\mathbf{H}_\ell)$, $p = 2$, that is to say the set of operators A on $\mathbf{H}_\ell$ such that $\mathbf{Tr}(|A|^p) < \infty$, $(|A| = (A^\star A)^{1/2})$. Let M^p denote the set of 2×2 matrices over the ring $\mathcal{L}(\mathbf{H})$ formed by matrices $a = (a_{ij})$ such that $a_{ij} \in \mathcal{L}^{p+1}(H)$ if $i \neq j$. If we identify $\mathbf{H}$ with its orthogonal complement, it follows that

$$N = \begin{pmatrix} T & PN(1-P) \\ (1-P)NP & (1-P)N(1-P) \end{pmatrix}$$

belongs to M^1.

Now $HC_1(M^1)$ comes equipped with a naturally defined cocycle τ (see [14]) given by

$$\begin{aligned} \tau(a^0, a^1) &= \mathbf{Tr} \begin{pmatrix} 1 & 0 \\ 0 & -1 \end{pmatrix} \begin{pmatrix} 0 & a_{12}^0 \\ a_{21}^0 & 0 \end{pmatrix} \begin{pmatrix} 0 & a_{12}^1 \\ a_{21}^1 & 0 \end{pmatrix} \\ &= \mathbf{Tr} \left(a_{12}^0 a_{21}^1 - a_{21}^0 a_{12}^1 \right) . \end{aligned}$$

If we put $a^0 = N^\star$ and $a^1 = N$, then

$$\begin{aligned} \tau(N^\star, N) &= \mathbf{Tr}[PN^\star(1-P)NP - (1-P)N^\star PN(1-P)] \\ &= -\mathbf{Tr}[(1-P)N^\star PN(1-P)] \\ &= \mathbf{Tr}[TT^\star - T^\star T]. \end{aligned}$$

The inclusion $PN - NP \in \mathcal{L}^2(\mathbf{H}_\ell)$ is stable under C^∞-functions, so if we let $A = C^\infty(N)$, then τ restricts to a functional on $HC_1(A) = \Omega'_{A/\mathbf{C}}/dA$, i.e., to one-forms modulo exact forms. Note that τ is closed since $\partial\tau(f) = \tau(1 \cdot df) = 0$.

A closely related example is obtained by taking $\mathcal{A}$ commutative and a linear representation $\pi : \mathcal{A} \to \mathcal{L}(\mathbf{H})$ so that the composition of π with the quotient map $\mathcal{L}(\mathbf{H}) \to \mathcal{L}(\mathbf{H})/\mathcal{L}^1(\mathbf{H})$ is multiplicative.

Such operators can essentially be represented as one-dimensional singular integral operators, i.e., operators with Cauchy-type singularities. This gives a functional on $HC_1(\mathcal{A})$ by

$$\tau(a^0, a^1) = -\mathbf{Tr} b_1 \pi(a^0, a^1).$$

4. An Extension Problem

The trace invariant τ is clearly a measure of the nonnormality of T and so it is fundamental and natural to compute it in terms of the *spectral* theory of the operator N or the algebra $\mathcal{A}$. However, a nice formula in this generality seems rather hopeless, so instead we smooth τ by integration; that is, we look for a functional (current) $\mathcal{T}$ on $\Omega^2_{\mathcal{A}}$ which makes the diagram commute:

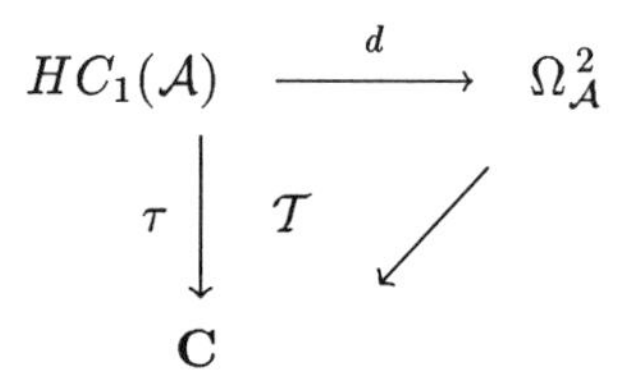

the image of d is onto (modulo constants, i.e., using cohomology with compact supports, $H^1_c(\mathbf{C};\mathbf{C}) \simeq \mathbf{C}$) and the relation $\mathcal{T}(dw) = \tau(w)$ determines $\mathcal{T}$ up to a constant (in the category of flat currents).

Theorem 4.1. *Let $T \in \mathcal{L}(\mathbf{H})$ and suppose $T^\star T - TT^\star$ is in $\mathcal{L}^1(\mathbf{H})$. Then there exists a unique Lebesgue integrable function $g : \mathbf{C} \to \mathbb{R}$ (the principal function) such that if $\eta \in \Omega^1_{\mathcal{A}/\mathbf{C}}$ represents a cycle w in $HC_1(\mathcal{A})$,*

$$\tau(w) = \mathcal{T}(d\eta) = \frac{i}{2\pi}\int g\, d\eta\,. \tag{4.2}$$

Note that with $w \simeq f\,dh$, (4.2) can be written

$$\begin{aligned}\mathbf{Tr}[f(T),h(T)] &= \frac{i}{2\pi}\int g\, df \wedge dh \\ &= \frac{i}{2\pi}\int g(f_z h_{\bar z} - h_z f_{\bar z})\, dz \wedge d\,\bar z\ .\end{aligned}$$

Note also that the signum of g (i.e., the orientation of $\mathbf{C}$) is determined by viewing $\mathcal{A} \otimes \mathcal{A}$ as an $\mathcal{A}$ module via $a \mapsto a \otimes 1$. Using $a \to 1 \otimes a$ would alter the sign.

The current $\mathcal{T}$ has the following properties:

(i) $\tau(df/f) = \partial\mathcal{T}(df/f) = \mathrm{ind} f(T)$ if $f \neq 0$ on $\sigma_{\mathrm{ess}}(T)$, the essential spectrum of T;

(ii) $\partial\mathcal{T}(RdS) = \mathbf{Tr}[R(T),S(T)]$;

(iii) $|g(\zeta)| \leq \mathrm{Rank}[T^\star,T]$;

(iv) $\mathrm{Mass}(\mathcal{T}) = \frac{1}{\pi}\int |g| \leq \|[T^\star,T]\|_{\mathcal{L}^1(\mathbf{H})}$;

(v) if F is a smooth map $\mathbf{C} \to \mathbf{C}$, then $\mathcal{T}_F$, the principal current for the "operator $F(T)$" is the push forward $F_\sharp(\mathcal{T})$. More accurately, $F_\sharp(\mathcal{T})$ is the principal current for the operator class $F(T)$ in $\mathcal{L}(\mathbf{H})/\mathcal{L}^1(\mathbf{H})$, since the map F only determines an operator modulo $\mathcal{L}^1(\mathbf{H})$.

It is known that for any prescribed g there exists an operator class for which (4.2) is valid. Moreover, there are representatives T in $\mathcal{L}(\mathbf{H})$ with $\text{Mass}(T) = \|[T^\star,T]\|_{\mathcal{L}^1(\mathbf{H})}$, i.e., there exist operators with minimal mass. For example, all hyponormal operators do.

Question: Does every class contain representatives with minimal mass?

In the parlance of geometric measure theory, we deduce (from Federer's characterization of flat currents) that a 1-cocycle in the plane is associated to an operator in $\mathcal{L}(\mathbf{H})$ if and only if it is an exact flat chain [20; S4.12–S4.20]. Note that τ is the boundary of a unique flat chain having compact support. Adding a constant to g also gives a solution to (4.20) and therefore we might suspect that the local variation of g is related to the action of T on $\mathbf{H}$.

Suppose T has a rank one self-commutator. Then the scalar function g is a complete unitary invariant for the nonnormal component of T, and we have the following theorems:

Theorem 4.3. *Let T be pure hyponormal with* $\text{Rank}[T^\star,T] = 1$. *Then $\sigma_{ess}(T)$ consists of those points z for which $-g(\zeta)$ is not (exclusively) equal to zero or to one almost everywhere as ζ ranges over a neighborhood of ζ.*

When T is the unilateral shift, $-g$ is just the characteristic function of the open unit disc and we have the well-known result that $\sigma_{\text{ess}}(T) = \{z : |z| = 1\}$.

Theorem 4.4. *Under the hypothesis of (4.3), a complex number z is an eigenvalue (necessarily of multiplicity one) for $T^\star$ iff there exists a neighborhood B of z such that*

$$\int_B \frac{1+g(\zeta)}{|\zeta - z|^2}\,\frac{d\zeta \wedge d\bar{\zeta}}{2i} < \infty .$$

In particular, z is a Lebesgue point for g with value -1.

A description of the eigenspace is given in [34].

Let $T = WP$ be the polar decomposition. A subspace invariant for both W and P is said to be polar invariant for T. If T satisfies (4.3), there is the following criterion for polar invariant subspace for $T^\star$.

Theorem 4.5. *$T^\star$ has a nontrivial polar invariant subspace if there is a set of positive measure $\Sigma \subset [0,\infty)$ such that*

$$\int_\Sigma dr \int_0^{2\pi} \log \sin |2\pi g(\mathrm{re}^{i\vartheta})|\, d\vartheta > -\infty .$$

These theorems are proved through the analysis of a canonical model constructed from the principal function together with some dilation theory for partial isometries which depend upon a parameter [9], [7].

For operators with higher rank self-commutators, such precise results are no longer valid. However, there are some counterparts. If T is pure hyponormal, then $\sigma(T)$ coincides with the support of $\mathcal{T}$ [33], while $g(\zeta) = \mathrm{ind}(T - \zeta)$ whenever $\zeta \notin \sigma_{\mathrm{ess}}(T)$ for general T [7].

In view of our earlier remarks that any g can occur as principal function, our next result is quite striking.

5. Integrality for Subnormal Operators

Theorem 5.1. *The principal function of a subnormal operator with trace class self-commutator assumes almost everywhere nonpositive integer values.*

This result had first been established for Swiss cheese examples and subnormal operators T with $[P,N] \in \mathcal{L}^1(\mathbf{H}_\ell)$. In each of these cases, τ is representable by integration over the *measure* boundary of $\sigma(T)$. It is an interesting fact that the integrality is a geometric property of subnormal operators and is not associated with any smoothness or thinness of the essential spectrum. In particular, if T is rationally cyclic, a pointwise estimate of Berger says that $-1 \leq g(\zeta) \leq 0$, [3] and therefore $-g$ equals the characteristic function of a set E. In other words, the cocycle τ associated to a cyclic subnormal operator is the boundary of some measurable set whose closure equals $\sigma(T)$.

The integrality property provides the following criteria concerning the existence of *cyclic* vectors.

Theorem 5.2. *Let T be a subnormal operator having a cyclic vector and principal function $-\chi_\Omega$. If ζ is a point of density for Ω, then $(T-\zeta)^2$ cannot have a cyclic vector. More generally, if f is a complex-valued function analytic in a neighborhood of $\sigma(T)$ such that $\#\{f^{-1}(\{z\}) \cap \Omega\} \geq 2$ on a set of positive area, then $f(T)$ fails to have a cyclic vector.*

Proof: For the first part of the theorem we note that it suffices to consider $\zeta = 0$. Now $\mathcal{L}^2(\Omega \cap -\Omega) > 0$, for otherwise ζ would be a point of rarefaction for $-\Omega$ which is impossible. Now $g_{T^2}(z) = -\sharp\{\omega \in \Omega : \omega^2 = z\} = -2$ if $z \in [\Omega \cap -\Omega]^2$. Thus on the set $[\Omega \cap -\Omega]^2$, which is of positive mmeasure, g_{T^2} is equal to -2. On the other hand, Berger's theorem requires that g_{T^2} be bounded below by -1, if indeed T^2 had a cyclic vector.

The second statement of the theorem follows from the hypothesis, by combining Berger's estimate and integrality with the transformation law $g_{f(T)}(z) = \sum\{g_T(\omega) : f(\omega) = z\}$.

Problem 5.3. Describe the kinds of sets E which occur for rationally (polynomially) cyclic operators.

In this connection we mention the following results proved in [8].

Parts decomposition of g. Let X be a compact subset of the complex plane and let $R(X)$ denote the closure of the rational functions with poles off X in the sense of the sup norm. Then X has a partition into Gleason parts $\{P_n\}_{n=1}^{\infty} \cup P_0$ with P_0 taken as the collection of peak points. Let ν be a Radon measure on X and let $R^2(X,\nu)$ denote the closure in $L^2(X,\nu)$ of $R(X)$. Following Foias and Suciu [25], ν is called a Szegö measure relative to $R(X)$, if ν is not a zero measure and the following implication holds: If B is a Borel subset of X and $\chi_B L^2(X,\nu) \subset R^2(X,\nu)$, then $\nu(\Omega) = 0$.

Now the canonical representation

$$\pi : R(X) \longrightarrow \text{multiplication operators on } R^2(X,\nu),$$

splits into a direct sum decomposition for the subnormal multiplication operator

$$f(z) \xrightarrow{T} zf(z).$$

The nonnormal component of T corresponds to the maximal Szegö component of ν. Henceforth, we shall assume that the given measure ν is itself a Szegö measure, so T is pure subnormal.

The following theorem appears in [26].

Theorem 5.4. *There are disjoint Borel sets* $F_1, F_2, \ldots$ *such that for* $n = 1,2,\ldots$

(i) $P_n \subset F_n \subset$ closure P_n.
(ii) *Every representing measure for points in* P_n *is carried on* F_n.

(iii) $R^2(X,\nu) = \sum_n \oplus R^2(F_n,\nu)$.

Thus T decomposes into a sum $\oplus_n T_n$. Let g_n denote the principal function of T_n so that $g_n = -\chi_{E_n}$ for some $E_n \subset$ closure P_n. In spite of the fact that (closure $P_n)\backslash P_n$ may have positive area, we have:

Theorem 5.5. *$g_n = g\bullet\chi_{P_n}$ for $n = 1,2,\ldots$.*

Corollary 5.6. *$E_n = E \cap P_n$. In particular, $E \cap P_0$ has measure zero.*

Although $[T^\star,T]$ may have rank > 1, we still have the following result:

Proposition 5.7. *If z is a bounded point evaluation for T, then*

$$\int_B \frac{1-\chi_E(\zeta)}{|z-\zeta|^2} < \infty,$$

where B is some neighborhood of z.

Corollary 5.8. *If P_n is a nontrivial Gleason part of X and $z \in P_n$, then*

$$\sum_{m=1}^{\infty} 2^{2m} \operatorname{Area}(A_m\backslash P_n) \leq \sum_{m=1}^{\infty} 2^{2m} \operatorname{Area}(A_m\backslash E_n) < \infty$$

for $A_m = \{\zeta : 2^{-m-1} \leq |z-\zeta| \leq 2^{-m}$.

(Cf. Browder [6] and Melnikov [32] concerning the fact that P_n has metric density one at all of its points.)

We shall now sketch a proof of Theorem 5.1. The reader is referred to [10] for full details.

Let $T = \frac{(T+T^\star)}{2} + i\,\frac{(T-T^\star)}{2} \equiv U+iV$, denote the Cartesian decomposition. Clearly, we may assume that T is pure. A computation of g in terms of spectral invariant proceeds by *slicing* (with respect to U) the $C^\star$-algebra generated by T.

By scattering theory one knows that there are $C^\star$–algebra homomorphisms

$$S_\pm(U;\bullet) : \text{Norm closure } \{A \in \mathcal{L}(\mathbf{H}) : [\mathbf{U},\mathbf{A}] \in \mathcal{L}^1(\mathbf{H})\}$$
$$\longrightarrow \text{Commutant of } U$$

defined by taking strong limits

$$S_\pm(U;A) = s-\lim_{t\to\pm\infty} e^{itU} A e^{-itU} P_a(U).$$

Note that the compact operators are contained in the kernel of both homomorphisms $S_{\pm}(U;\bullet)$.

Here $P_a(U)$ denotes the projection onto the absolutely continuous subspace for U. However, a result due to C. Putnam [35] says that $P_a(U) = 1$ if T is pure hyponormal. Therefore, in a direct integral space corresponding to U, $\mathbf{H}$ has the form

$$\int \oplus \mathbf{H}_x \, dx$$

and the operators $S_{\pm}(U;V)$ are decomposable

$$S_{\pm}(U;V) = \int \oplus S_{\pm}(U;V)(x) \, dx \, .$$

Moreover, for almost all x,

$$S_{+}(U;V)(x) - S_{-}(U;V)(x) = \sum_{\alpha} \lambda_\alpha \phi_\alpha(x) \otimes \phi_\alpha(x) \in \mathcal{L}^1(\mathbf{H}_x) \, .$$

Here $[T^*,T] = \sum \lambda_\alpha \phi_\alpha \otimes \phi_\alpha$.

The set of pairs (A,B) in $\mathcal{L}(\mathbf{H}_x)$ such that $A - B \in \mathcal{L}^1(\mathbf{H}_x)$ is an algebra M_x^0 and the map $(A,B) \xrightarrow{\beta} \det(A^{-1}B)$ defines a multiplicative character on the Whitehead group $K_1(M_x^0)$ defined by Bass [1]. A basic fact is that

$$\beta(S_{+}(U;V-z)(x), S_{-}(U;V-z)(x)) = \exp\left\{ \int g(x+iy) \, \frac{dy}{y-z} \right\} , \qquad \operatorname{Im} z \neq 0 \, .$$

Consequently, our analysis of g will rest upon a study of the symbols $S_{\pm}(U;V)(x)$ as x varies over $\sigma(U)$.

So far this approach works for most any operator T having a trace class self-commutator. Our proof of integrality will depend on special information contained in the symbols $S_{\pm}(U;V)$ when T is subnormal. In particular, the existence of a zero order functional calculus for T, combined with the fact that projections have integer traces, is essential.

First of all, the fibre H_x, which is the smallest closed subspace reducing $S_{\pm}(U;V)$ (x) is actually spanned by the vectors $\{\phi_\alpha(x)\}$. Consequently, $\dim \mathbf{H}_x \leq \operatorname{Rank}[P,N]$.

Second, and most important, is the fact that the symbols $S_{\pm}(U;V)(x)$ must have singular spectra, e.g., the spectral measures of the self-adjoint operators $S_{\pm}(U;V)(x)$ are singular relative to one-dimensional Lebesgue measure. From this fact it is not difficult to then show that g is integer-valued. So let us focus on the problem of showing that $S_{\pm}(U;V)(x)$ have singular spectra.

Suppose A, B are self-adjoint and $A - B \in \mathcal{L}^1(\mathbf{H})$. If $\dim \mathbf{H} < \infty$, then for arbitrary functions f, it follows that $f(A) - f(B) \in \mathcal{L}^1(\mathbf{H})$. Similarly, if A,B are compact operators, we may choose any f which is smooth near the origin. From this we may guess that: If A, B have singular spectra, there should be a collection of functions f having many singularities so that $f(A) - f(B) \in \mathcal{L}^1(\mathbf{H})$ remains valid. In the back of our minds is the idea that this cannot happen in the presence of absolutely continuous spectra. In contrast, we note there is an interesting example of a *Lipschitz* function f and a diagonal operator A so that $B - A \in \mathcal{L}^1(\mathbf{H})$ but $f(B) - f(A) \notin \mathcal{L}^1(\mathbf{H})$ [19].

Our proof of integrality ultimately depends on exhibiting a continuous function $f : \mathbb{R} \to \mathbb{R}$ such that for almost all x, the following two seemingly contradictory assertions hold:

5.9
$$(f[S_+(U;V)(x)], f[S_-(U;V)(x)]) \in M_x^0$$

and

5.10
$$\text{ap} - \lim_{y \to x} \left| \frac{f(x) - f(y)}{x - y} \right| = +\infty,$$

where $\text{ap} - \lim_{y \to x}$ refers to the *Denjoy* approximate derivative.

In view of the fact that for arbitrary functions F the set x at which the usual limit

$$\lim_{h \to 0} \left| \frac{F(x+h) - F(x)}{h} \right| = +\infty$$

is of measure zero, it is surprising that there are any continuous functions which satisfy (5.9) and (5.10). Nevertheless, the existence of continuous functions satisfying (5.10) has been demonstrated by A. Khintchine [28] and V. Jarnik [27].

In order to produce singular functions which satisfy (5.9), we consider the spectral representation of the normal extension N. Recall that

$$[T^*, T] = [P, N][P, N^*] \in \mathcal{L}^1(\mathbf{H}).$$

Consequently, $K \equiv [P, N] \in \mathcal{L}^2(\mathbf{H}_\ell)$, and there is a matrix kernel function $K(z, \zeta)$ so that

$$Kh(z) = \int_{\sigma(N)} K(z, \zeta) h(\zeta) \, d\nu(\zeta),$$

where ν is the spectral measure for N and $K(z, \zeta)$ is square summable with respect to the product measure $\nu \times \nu$. Now, for f in $L^\infty(\sigma(N), \nu)$ with

5.11
$$\iint \left| \frac{f(z) - f(\zeta)}{z - \zeta} \right|^2 |K(z, \zeta)|^2 \, d\nu(z) \, d\nu(\zeta) < \infty,$$

we get

$$K_f \equiv [P, f(N)] \in \mathcal{L}^2(\mathbf{H}_\ell)$$

and

$$[U, Pf(N)P] \in \mathcal{L}^1(\mathbf{H}).$$

In addition, if f is continuous, $[P, f(N)]$ is compact. Consequently, $S_\pm(U; Pf(N)P) = f(S_\pm(U;T))$ and

$$(f[S_+(U;T)(x)], f[S_-(U;T)(x)]) \in M_x^0$$

for almost all x.

The proof is completed by showing that the Jarnik construction [27] can be applied to yield functions which satisfy (5.11). ■

Since kernel $S_\pm(U; \cdot)$ contains the compact operators, it is not difficult to prove that

$$\sigma(S_\pm(U;V)(x)) \subset \{y : x + iy \in \sigma_{\text{ess}}(T)\}.$$

In order to probe more deeply into Problem 5.3, we ask the following:

Problem 5.12. Can the symbols $S_\pm(U;V)(x)$ have continuous spectra?

Problem 5.13. Find a proof of integrality using complex function theory.

Problem 5.14. If T is hyponormal and has a cyclic vector, is $g = -\chi_E$?

6. $\mathcal{L}^2(\mathbf{H})$-Invariance of the Principal Function

Let $\widetilde{\mathcal{L}}(\mathbf{H})$ denote the set of operators T in $\mathcal{L}(\mathbf{H})$ with $[T^*, T] \in \mathcal{L}^1(\mathbf{H})$. Since $\mathbf{Tr}[A, B] = 0$ if $A \in \mathcal{L}^1(\mathbf{H})$, the map, $\widetilde{\mathcal{L}}(\mathbf{H}) \to$ Currents, factors through $\widetilde{\mathcal{L}}(\mathbf{H})/\mathcal{L}^1(\mathbf{H})_{\text{iso}}$, the set of isomorphism classes (under unitary equivalence). On a more subtle level, we may consider what happens if T is perturbed by a compact operator K so chosen that T and $T+K$ both belong to $\widetilde{\mathcal{L}}(\mathbf{H})$. One reason to consider compact operators is due to the fact that since they leave the index unchanged, $\tau_T = \tau_{T+K}$ whenever $\sigma_{ess}(T)$ has zero area. On the other hand, if $\sigma_{ess}(T)$ has positive area, for arbitrary $\varepsilon > 0$, it is usually possible to find elements K in $\mathcal{L}^{2+\varepsilon}(\mathbf{H})$ with $\tau_T \neq \tau_{T+K}$. As far as I know, the case of $\mathcal{L}^2(\mathbf{H})$ invariance is unsettled although Voiculescu has shown in [38] that $\tau_T = \tau_{T+K}$, whenever T has finite multiplicity. If T is subnormal, the multiplicity restriction can be dropped. Voiculescu's proof of this is based upon his result that normal operators can be diagonalized by adding an operator in $\mathcal{L}^2(\mathbf{H})$. We would like to give here an alternate, more direct proof based upon the Besicovitch differentiation of measures.

Theorem 6.1. *Let T be subnormal and $[T^\star,T] \in \mathcal{L}^1(\mathbf{H})$. Suppose $K \in \mathcal{L}^2(\mathbf{H})$ and $[(T+K)^\star,T+K] \in \mathcal{L}^1(\mathbf{H})$. Then $\tau_T = \tau_{T+K}$.*

Proof: By polarization, it suffices to show that

$$\mathbf{Tr}[f(T^\star + K^\star, T+K)^\star, f(T^\star + K^\star, T+K)] = \mathbf{Tr}[f(T^\star,T)^\star, f(T^\star,T)]$$

when f is a polynomial. We can write

$$\begin{aligned}\mathbf{Tr}[f(T^\star + K^\star, T+K)^\star, f(T^\star + K^\star, T+K)] &= \mathbf{Tr}[f(T^\star,T)^\star, f(T^\star,T)] \\ &\quad + \mathbf{Tr}[f(T^\star,T)^\star, L] + [L^\star, f(T^\star,T)]\end{aligned}$$

since $\mathbf{Tr}[L^\star,L] = 0$ because $L \equiv f(T^\star + K^\star, T+K) - f(T^\star,T) \in \mathcal{L}^2(\mathbf{H})$. Consequently, it suffices to check that

$$\mathbf{Tr}\big([f(T^\star,T)^\star, L] + [L^\star, f(T^\star,T)]\big) = 0\,.$$

Put $f \equiv f(N^\star,N)$ and extend L to $\mathbf{H}_\ell \ominus \mathbf{H}$ by the zero operator. Noting that $[P,f] \in \mathcal{L}^2(\mathbf{H}_\ell)$, we have

$$\begin{aligned}\mathbf{Tr}([f(T^\star,T)^\star, L] + [L^\star, f(T^\star,T)]) &= \mathbf{Tr}(Pf^\star L - Lf^\star + L^\star fP - fL^\star) \\ &= \mathbf{Tr}([f^\star,L] + [L^\star,f])\end{aligned}$$

since $\mathbf{Tr}\big((1-P)f^\star L\big) = \mathbf{Tr}(L^\star f(1-P)) = 0$.

Suppose $\mathbf{H}_\ell = \int \bigoplus \mathbf{H}_\mathbf{z}\, \mathbf{d}\nu(\mathbf{z})$ and L has the kernel $L(z,\zeta)$. Then

$$M \equiv [f^\star,L] + [L^\star, f]$$

has the kernel

$$M(z,\zeta) = [\bar{f}\,(\bar{z}\,,z) - \bar{f}\,(\bar{\zeta}\,,\zeta)]L(z,\zeta) + [f(\bar{\zeta}\,,\zeta) - f(\bar{z}\,,z)]L^\star(z,\zeta)$$

with pointwise equality almost everywhere with respect to the product measure $\nu \times \nu$. As $M \in \mathcal{L}^1(\mathbf{H}_\ell)$, we note that for ν-almost all ζ, the map $z \to M(z,\zeta)$ is approximately continuous at $z = \zeta$ [cf. 10; pp. 37–38]. Moreover,

$$\mathbf{Tr}M = \int \mathbf{Tr}M(z,z)\, d\nu(z)\,.$$

Consequently, the proof comes down to showing that $M(z,z) = 0$ for ν-almost all z. This is a consequence of the following lemma.

Lemma 6.2. *Let μ be a Borel regular measure in* $\mathbf{C}$ *and suppose that L is square summable with respect to the product measure $\mu \times \mu$. Then, for μ-almost all z, the map $z \to (z-\zeta)L(z,\zeta)$ has Lebesgue value zero at $z = \zeta$.*

Proof: Since area measure and μ are Borel regular and every bounded set has finite μ-outer measure, the set Γ of those points z where the derivate $(B_delta(z) = \{\zeta : |z-\zeta| < \delta\})$

$$\frac{\lim_{\delta\to 0} \text{Area}[B_\delta(z)]}{\mu[B_\delta(z)]}$$

exists and is finite, is a set of full μ-measure [20; Theorem 2.9.5]. For μ-almost all z, the function $L(z,\bullet)$ is μ-measurable and square integrable. Let $\delta > 0$ and consider

$$\begin{aligned} F_\delta(z) &= \frac{1}{\mu[B_\delta(z)]} \int_{B_\delta(z)} |z-\zeta|\,|L(z,\zeta)|\,d\mu(\zeta) \\ &\le \frac{1}{\mu[B_\delta(z)]} \left(\int_{B_\delta(z)} |z-\zeta|^2\,d\mu(\zeta) \right)^{1/2} \left(\int_{B_\delta(z)} |L(z,\zeta)|^2\,d\mu(\zeta) \right)^{1/2} \\ &\le \frac{1}{\sqrt{\pi}} \left(\frac{\text{Area}[B_\delta(z)]}{\mu[B_\delta(z)]} \right)^{1/2} \left(\int_{B_\delta(z)} |L(z,\zeta)|^2\,d\mu(\zeta) \right)^{1/2} . \end{aligned}$$

It follows that for μ-almost all z,

$$\lim_{\delta\to 0} F_\delta(z) = 0 ,$$

and therefore z belongs to the Lebesgue set of $(z-\zeta)L(z,\zeta)$ and has the value zero. This completes the proof of Lemma 6.2. ■

7. Index Theory for Operator Ranges

Despite the fact that the usual Noether index is always an integer, the existence of an operator T for arbitrary g gives reason to ask if there is a notion of index which generalizes the ordinary one and satisfies the following analogue of (4.2):

$$\tau(w) = \frac{i}{2\pi} \int \text{ind}(T-\zeta)\,d\eta,\ w = d\eta.$$

A real-valued index theory for II_∞ von-Neumann algebras was discovered by Breuer [4], [5]. However, the commutators which occur here are compact and so the II_∞ theory does not apply. Recently we have considered this question for operators having finite rank self-commutators and found that an appropriate answer relies on the extension of *index to the category of operator ranges* $\mathcal{L}$.

The space $\mathcal{L}$ was first systematically studied in a series of papers by Dixmier in the late 1940's [16], [17], [18]. Subsequent study was taken up by Foias [23], [24], and Fillmore-Williams [22]. Our approach is to blend these algebraic considerations with geometric measure theory [12].

We now discuss the concepts and results which are needed to handle operators with finite rank self-commutators. A full understanding for general trace class commutators is still open for investigation.

Definition 7.1. Let f,g,h belong to $L^\infty([0,\infty))$, $h \geq 0$, with f,g,h having compact support in the closed interval $[0,\infty)$. The function f is said to be h-related to g at the origin, provided

$$\left|\int \frac{f(x)-g(x)}{x+\ell}\,dx\right| \leq \int \frac{h(x)}{x+\ell}\,dx \qquad \text{for } \ell > 0\,.$$

Denote by $f \underset{h}{\sim} g$ the property f is h-related to g at the origin. If $f \underset{h}{\sim} g$ for some h which is approximately continuous with value 0 at the origin, we shall write $f \simeq g$. Note that $\simeq$ is an equivalence relation and the family of equivalence classes $|f|$ forms an abelian group $K(\mathcal{L})$ under the natural identification $[f \pm f'] \equiv [f] \pm [f']$. The group $K(\mathcal{L})$ is viewed as an index group for the lattice of operator ranges.

Proposition 7.2. *Suppose g and g' are representatives of $[f]$ with g and g' approximately continuous at the origin. Then $g(0) = g'(0)$. In other words, there is a natural injective homomorphism of the additive group of real numbers $\mathbb{R}$ into $K(\mathcal{L})$.*

It is not hard to show that this map fails to be surjective, (see [12]).

Definition 7.3. Let $f \in L^\infty([0,\infty))$. The origin is said to be in the extended Lebesgue domain of $[f]$ if there is a $g \in [f]$ which is approximately continuous at $x = 0$. We shall denote this value by $f^*(0)$.

Note that the image of $\mathbb{R}$ in $K(L)$ consists of those classes $[f]$ for which Definition 7.3 holds.

We next describe the idea of congruence with respect to an ideal in $\mathcal{L}$.

Definition 7.4. Suppose R and S are elements of the lattice $\mathcal{L}$, and suppose I is an ideal in $\mathcal{L}$. Then R and S are said to be congruent modulo I, provided $R + d = S + d'$ for some elements d,d' in I.

Congruence is an equivalence relation which we denote by $R \equiv S(I)$.

Suppose F is the invariant ideal of $\mathcal{L}$ consisting of the finite-dimensional subspaces. If $R \equiv S(F)$, then there are positive operators $\{A,B,C,D\}$ with

$$\begin{aligned}
&\mathrm{Ran}(\sqrt{A}) = R \\
&\mathrm{Ran}(\sqrt{B}) = S \\
&\mathrm{Ran}(\sqrt{A+C}) = \mathrm{Ran}(\sqrt{B+D}) \\
&\mathrm{Ran}(\sqrt{C}),\ \mathrm{Ran}(\sqrt{D}) \in F\,.
\end{aligned}$$

It is known from the work of M. G. Krein [29] that for $X,Y \in L(\mathbf{H})$ with $X - Y \in L^1(\mathbf{H})$ and X,Y self-adjoint, there is a unique summable function ξ having compact support such that

$$\mathbf{Tr}[f(X) - f(Y)] = \int f'(\lambda)\xi(\lambda)\,d\lambda$$

for smooth functions f. Moreover, if $X - Y$ has p negative eigenvalues and q positive eigenvalues (counting multiplicity), then $-p \le \xi(\lambda) \le q$. The function ξ is called the phase shift for the perturbation $X \to Y$. We shall write this as

$$X \xrightarrow{\ \xi\ } Y\,.$$

The following result is basic to achieving our goal of an index for operator ranges, [12].

Theorem 7.5. *Suppose* $R \equiv S(F)$. *Let* A,B,C,D *be as above, and let*

$$A \xrightarrow{\ \xi\ } A + C \qquad \textit{and} \qquad B \xrightarrow{\ \eta\ } B + D$$

be the associated phase shifts. Then the equivalence class $[\xi - \eta] = [\xi] - [\eta]$ *in* $K(\mathcal{L})$ *is independent of the choice of* A,B,C,D, *and therefore depends only on the subspaces* R *and* S. *In other words, if* $\{V,W,X,Y\}$ *is another choice of positive operators with* $\mathrm{Ran}(\sqrt{V}) = R$, $\mathrm{Ran}(\sqrt{W}) = S$, $\mathrm{Ran}(\sqrt{V+X}) = \mathrm{Ran}(\sqrt{W+Y})$ *and* $\mathrm{Ran}(\sqrt{X})$, $\mathrm{Ran}(\sqrt{Y}) \in F$, *with*

$$V \xrightarrow{\ \alpha\ } V + X \qquad \textit{and} \qquad W \xrightarrow{\ \beta\ } W + Y\,,$$

then $[\alpha] - [\beta] = [\xi] - [\eta]$.

On the basis of this result we can define

$$\mathrm{ind}(R,S) = [\xi] - [\eta]\,.$$

When the origin belongs to the extended Lebesgue domain of $[\xi - \eta]$, then the number ind(R,S) measures the relative density of R and S.

Clearly, $\mathrm{ind}(R,S) = -\mathrm{ind}(S,R)$. It is also easy to check that for T Fredholm, $\mathrm{ind}(\mathrm{Ran}\, T^*,\mathrm{Ran}\, T)$ agrees with the usual Noether index. Moreover, whenever $\mathrm{Ran}\, T \equiv \mathrm{Ran}\, T^*(F)$ and K is finite rank, it follows that

$$\mathrm{ind}(\mathrm{Ran}\, T^*,\mathrm{Ran}\, T) = \mathrm{ind}(\mathrm{Ran}(T^* + K^*),\mathrm{Ran}(T + K)).$$

On the other hand, it can happen that $R \supset S$, $R \neq S$, and yet $\mathrm{ind}(R,S) = 0$.

The following theorem solves our problem in the finite rank case.

Theorem 7.6. *Suppose* $\mathrm{Ran}[T,T^*] \in F$. *Then for almost all* ζ *in* $\mathbf{C}$,

7.7 $$g(\zeta) = \mathrm{ind}(\mathrm{Ran}(T^* - \zeta^*),\mathrm{Ran}(T - \zeta)).$$

If g is of bounded variation (e.g., τ is represented by integration), then by [21; Property VIII, p. 306] for almost all points ζ in $\mathbf{C}$ (with respect to one-dimensional Hausdorff measure), $g(\zeta)$ can be redefined as $\frac{1}{2}[\lambda + \mu] \equiv g^*(\zeta)$, the average of its upper- and lower-approximate limits. Under these circumstances, (7.7) holds for all such ζ with $g(\zeta)$ replaced by $g^*(\zeta)$. As a simple example, take T to be the unilateral shift and ζ a complex number with $|\zeta| = 1$. Then, $R(T^* - \overline{\zeta}) \supset R(T - \zeta)$ and both ranges are nonclosed dense subspaces of $\mathbf{H}$. The quotient space $\mathrm{Ran}(T^* - \overline{\zeta})/\mathrm{Ran}(T - \zeta)$ is one-dimensional so the relative index is defined. But $-g$ is the characteristic function of the unit disc and therefore has bounded variation. Clearly $g^*(\zeta) = -1/2$. Thus, the correct assignment of the relative dimension of $\mathrm{Ran}(T^* - \overline{\zeta})/\mathrm{Ran}(T - \zeta)$ is not 1 but $1/2$.

Based on these remarks, together with the integrality result for subnormal operators, we see that it is certainly of clear geometrical significance to try and understand the implications of the new index which can measure the relationship of two possibly dense nonclosed subspaces of $\mathbf{H}$.

In this connection we mention the following problem. Suppose T is subnormal and has a cyclic vector. Then $[T^*,T]$ is trace class and the principal function is $-\chi_E$ for some measurable set E. Thus, for almost all ζ, we expect $\mathrm{ind}(\mathrm{Ran}(T - \zeta)^*,\mathrm{Ran}(T - \zeta)) = -\chi_E(\zeta)$. Now $\mathrm{Ran}(T - \zeta)^* \supset \mathrm{Ran}(T - \zeta)$, but the trouble is we don't know if the index is defined, since $\mathrm{Ran}[T^*,T] = \infty$, generically.

Question 7.8. If T is cyclic hyponormal, is the usual vector space

$$\dim \mathrm{Ran}(T - \zeta)^*/\mathrm{Ran}(T - \zeta) \leq 1.$$

The answer is clearly yes for $\zeta \notin \sigma_{\text{ess}}(T)$. So what happens if $\zeta \in \sigma_{\text{ess}}(T)$?

8. Index Theory for Several Operators

Motivated by developments in the case of a single operator, we would like to apply the methods of geometric measure theory to the study of operator algebras which require several generators. However, at the present time, the lack of a good notion of spectrum (i.e. spanning the essential spectrum) precludes working in this generality. So, as a natural first step, we confine our considerations to algebras which are generated by several commuting operators. In this case, there is a nice notion of spectrum and essential spectrum due to J. Taylor [37] with which one can hope to build a good spectral type index theory from both the local and global points of view.

To begin with, suppose $T = (T_1, T_2)$ is a pair of commuting operators acting on a Hilbert space $\mathbf{H}$. Let $\mathcal{A}$ denote the Calkin algebra and let $\text{Sp}(\underline{T}; \mathbf{H})$ and $\text{Sp}(\underline{T}; \mathcal{A})$ denote the Taylor spectrum and essential spectrum of the pair $\underline{T}$. For example, take $\underline{T}$ to be the 2-tuple of Toeplitz operators on $H^2(\partial B_2)$ with symbols z_1, z_2. Here B_2 is the unit ball in $\mathbf{C}^2$. Then $\text{Sp}(\underline{T}; \mathbf{H}) = \mathbf{B_2}$ and $\text{Sp}(\underline{T}; \mathcal{A}) = \partial B_2$. For a point $z \notin \partial B_2$, the index of $\underline{T} - z \cdot 1$ is defined as the Euler-Poincaré characteristic of the complex $K(\underline{T} - z) \otimes \mathbf{H}$ (here $K(\underline{T} - z)$ is the Koszul complex of the pair $\underline{T} - z$) and $\text{ind}(\underline{T} - z) = -\chi_{B_2}(z)$ [15]. However, it is perhaps more interesting to consider examples where $\text{Sp}(\underline{T}; \mathcal{A})$ is an oriented system of smooth real curves (perhaps with multiplicities, i.e., a one-current τ in $\mathbf{C}^2$) with $\text{Sp}(\underline{T}; \mathbf{H})$ as its polynomial convex hull. By results of Wermer [39], $V = \text{Sp}(\underline{T}; \mathbf{H}) \setminus \text{Sp}(\underline{\mathbf{T}}; \mathcal{A})$ is a one-dimensional complex-analytic subvariety of $\mathbf{C}^2 \setminus \text{Sp}(\underline{T}; \mathcal{A})$ and therefore has a decomposition into a locally finite union $\cup_\alpha V_\alpha$ of irreducible subvarieties V_α. From the viewpoint of index theory, we should expect to have integers g_α attached to each V_α in a manner which is consistent with the multiplicities on the boundary; thus, $\tau = \partial \sum_\alpha g_\alpha [V_\alpha]$. At this point recall that if V^* denotes the set of smooth points of V, then V is irreducible if and only if V^* is connected, so that if $V = V_1 \cup V_2$, then $V^* = (V_1^* \setminus V_1 \cap V_2) \cup V_2^* \setminus (V_1 \cap V_2)$ gives a disconnection of V^* and we can think of $V_1^* \setminus (V_1 \cap V_2)$ and $V_2^* \setminus (V_1 \cap V_2)$ as different components of the Fredholm spectrum of $\underline{T}$ in which the local indices g_α are constant. *However, as we pass through the singular locus, the index will jump, and therefore represents a new phenomenon not present in single operators.*

Density, Lelong number, multiplicity local index. Suppose that $\mathcal{L}$ is a current in $\mathbf{C}^n$ of dimension p (i.e., a linear functional on p-forms with compact support). Let $c(p)$ denote the volume of the unit ball in $\mathbb{R}^p$. If $\mathcal{L}$ has measure coefficients (i.e., is of order zero), then the density $\Theta(\mathcal{L}, z)$ of L at x is defined to be

$$\lim_{r \to 0+} \frac{M_{B_r(z)}(\mathcal{L})}{c(p) r^p} .$$

The density is a measure of the mass (volume) of $\mathcal{L}$ near z compared to the volume of a p-plane through z.

For $\mathcal{L} = \sum_\alpha n_\alpha [V_\alpha]$, we define the local degree of $\mathcal{L}$ at z as

$$\sum_\alpha n_\alpha \Theta([V_\alpha], z) .$$

Recall that $[V_\alpha]$ denotes the current of integration over the regular points of the (irreducible) variety V_α with the induced orientation of $\mathbf{C}^n$ [30].

Now the density for a variety V has several meanings:

(algebraic)	(1)	The multiplicity of the local ring of V at z.
(geometric)	(2)	The minimal sheeting number of V near z of a complex p-plane through z.
(topological)	(3)	The degree of the tangent cone $C_z(V)$ to V at z considered as a subvariety of P^{n-1}.
(analytic)	(4)	Density or Lelong number of $[V]$ at z.

Recall that $\Theta([V_\alpha], z) = 1$ if and only if z is a smooth point. Also, z may belong to several of the components V_α.

For a single operator T, V = union of the components V_α of the complement of $\sigma_{\text{ess}}(T)$ and $g_\alpha = \text{ind}(\underline{T} - z)$ for z in V_α and all points are smooth.

Theorem 8.1. *Suppose $\underline{T} = (T_1, \ldots, T_s)$ is a tuple of commuting operators on a Hilbert space $\mathbf{H}$ and $\sigma_{\text{ess}}(\underline{T})$ is a system of smooth curves. Then there is a principal current $\mathcal{T} = \sum_\alpha g_\alpha [V_\alpha]$ whose local degree at $z \notin \sigma_{\text{ess}}(T)$ equals*

$$\text{ind}(\underline{T} - z) \equiv \lim_{n \to \infty} \frac{1}{n} \dim \frac{\mathbf{H}}{(\underline{T} - z)^n \mathbf{H}} - \lim_{n \to \infty} \frac{1}{n} \dim \frac{\mathbf{H}}{(\underline{T}^* - \overline{z})^n \mathbf{H}} .$$

Here $(\underline{T} - z)^n$ (resp. $(\underline{T}^* - \overline{z})^n$) refers to the nth-power of the ideal generated by $(\underline{T} - z)$ (resp. $(\underline{T}^* - \overline{z})$). Note that each of the limits above exists. The choice of large n is to eliminate the effect of the nilpotent elements, (see [13]).

Corollary 8.2. *In addition to the hypothesis above,suppose the $\star$-algebra generated by $\underline{T}$ is commutative modulo $L^1(\mathbf{H})$ and τ is the induced cocycle. Then*

$$\tau = \partial\left(\sum_\alpha g_\alpha[V_\alpha]\right).$$

In other words,

$$\mathbf{Tr}[R(\underline{T}),S(\underline{T})] = \sum_\alpha g_\alpha[V_\alpha](dR \wedge dS).$$

An alternate description of $\operatorname{ind}(\underline{T}-z)$ can be given in terms of root subspaces [11], [13].

If $\pi^\perp$ is a projection $\mathbf{C}^n \to \mathbf{C}^1$ so that

(i) $\operatorname{kernel}(\pi - \pi(z)) \cap C_z(V) = \{0\}$.
(ii) $\pi^\perp(T - z)$ is a Fredholm operator, then

$$\begin{aligned}\operatorname{ind}(\underline{T}-z) = \dim\ \text{Root space}\ \pi(\underline{T}-z)\Big|\ker\pi^\perp(T-z) \\ - \dim\ \text{Root space}\ \pi(\underline{T}-z)\|\operatorname{Coker}\pi^\perp(T-z).\end{aligned}$$

Here the expression Root space refers to the set of vectors which are annihilated by some power of the operator tuple $(\underline{T}-z)$.

From this formula, we see that if $\operatorname{ind}(\underline{T}-z) \neq 0$, then there are vectors in $\mathbf{H}$ which are simultaneous eigenvectors for all operators in the n-tuple $(\underline{T}-z)$ or $(\underline{T}^*-\overline{z})$.

Finally, we comment that in the case of several commuting operators, the construction of a current T which spans the cocycle τ is based upon an operator theory version of L'Hopital's rule: see [13] and the forthcoming paper *On Local Index and the Cocycle Property of Lefschetz Numbers*, to appear in Integral Eq. and Operator Theory, special edition dedicated to M. S. Livsic. The starting point is to consider A and B as a pair of commuting Fredholm operators acting on a Hilbert space $\mathbf{H}$. The vector spaces $\operatorname{Ker} A$ and $\operatorname{Coker} A$ are finite dimensional, and B induces a linear map on both spaces. The multiplicative Lefschetz number of B relative to A is defined as a ratio of determinants

$$\psi(B;A) = \frac{\det(B|\operatorname{Ker} A)}{\det(B|\operatorname{Coker} A)},$$

provided that $\det(B|\operatorname{Coker} A) \neq 0$. Similarly, by exchanging A and B, one can form $\psi(A;B)$ as well as their ratio

$$\Psi(B;A) = \frac{\psi(B;A)}{\psi(A;B)}.$$

Moreover, we can insert parameters z,w so that $\Psi(B-z;A-w)$ is indeed well defined whenever (w,z) belongs to the complement of the Taylor spectrum $\mathrm{Sp}(A,B;\mathbf{H})$. The remarkable fact is that locally, $\Psi(B-z;A-w)$ has a non-vanishing holomorphic extension through $\mathrm{Sp}(A,B;\mathbf{H})$ so that $\Psi(B,A)$ has *an intrinsic meaning even if some of the determinants vanish*. It is this reciprocity property which leads to the construction of a holomorphic line bundle and meromorphic section whose divisor gives the spanning current $\mathcal{T}$. For a proof of this reciprocity result and its connection with algebraic K-theory, the reader should consult [13].

References

1. Bass, H., *Algebraic K-Theory*, Benjamin, New York, 1968.
2. Berger, C. A. and B. L. Shaw, Self commutators of multicyclic hyponormal operators are always trace class, Bull. Amer. Math. Soc. **79** (1973), 1193–1199.
3. ———, Sufficiently high powers of hyponormal operators have rationally invariant subspaces, Integal Eq. and Operator Theory **1**, (1978), 444–447.
4. Breuer, M., Fredholm theories in von Neumann algebras II, Math. Ann. **178** (1968), 243–254.
5. ———, Fredholm theories in von Neumann algebras II, Math. Ann. **180** (1969), 313–325.
6. Browder, A., Point derivations on function algebras, J. Funct. Anal. **1** (2967), 22–27.
7. Carey, R. W. and J. D. Pincus, An invariant for certain operator algebras, Proc. Nat. Acad. Sci. U.S.A. **71** (1974), 1952–1956.
8. ———, Principal functions, index theory, geometric measure theory and function algebras, Integral Eq. and Operator Theory **2** (1979), 441–483.
9. ———, The structure of intertwining partial isometries II, (preprint), 1975.
10. ———, An integrality theorem for subnormal operators, Integral Eq. and Operator Theory **4** (1981), 10–44.
11. ———, Principal currents, Integral Eq. and Operator Theory **8** (1985), 614–640.
12. ———. Index theory for operator ranges and geometric measure theory, Proc. Symp. Pure Math. **44** (1986), 149–161.
13. ———, Reciprocity for Fredholm operators, Integral Eq. and Operator Theory **9** (1986), 469–501.

14. Connes, A. and M. Karoubi, Caractére multiplicatif d'un module de Fredholm, C. R. Acad. Sc. Paris **229** (1984), 963–968.
15. Curto, R. E., Fredholm and invertible n-tuples of operators. The deformation problem. Trans. Amer. Math. Soc. **226** (1981), 129– 159.
16. Dixmier, J., Etude sur les variétés et les operateurs de Julia, Bull. Soc. Math. France **77** (1949), 11–101.
17. ———, Sur les variétés J. d'un espace de Hilbert, J. Math. Pures Appl. **28** (1949), 321–358.
18. ———, Les ideaux dans l'ensemble des variétés d'un espace hilbertien, Ann. Fac. Sci. Toulouse Math. **10** (1949), 91–114.
19. Farforoskaya, Ju. B., Example of a Lipschitzian function of self- adjoint operators that gives a nonnuclear increment under a nuclear perturbation, (Russian) Investigations of linear operators and the theory of functions III, Zap. Naucn. Sem. Leningrad Otdel Math. Inst. Steklov (LOMI) **30** (1972), 146–153.
20. Federer, H., *Geometric Measure Theory*, Springer–Verlag, New York, 1969.
21. ———, Federer, H. Colloquium lectures on geometric measure theory, Bull. Amer. Math. Soc. **84** (1974), 291–338.
22. Fillmore, P. A. and J. P. Williams, On operator ranges, Adv. in Math. **7** (1971), 897–900.
23. Fois, C., Invariant para–closed subspaces, Indiana Univ. Math. J. **20** (1971), 254-281.
24. ———, Invariant para–closed subspaces, Indiana Univ. Math. J. **21** (1971), 887–906.
25. ——— and I. Suciu, Szegö measures and spectral theory in Hilbert spaces, Rev. Roumaine Math. Pures Appl. **11, 2** (1966), 147–159.
26. Gamelin, T. W., *Uniform Algebras*, Englewood Cliffs, New Jersey, Prentice–Hall, 1969.
27. Jarnik, V., Sur les nombres dérivés approximatifs, Fundam. Math. **22** (1934), 4–16.
28. Khintchine, A., Recherches sur la structure des fonctions measurables, Rec. Math. Soc. Math. Moscow **31** (1924), 261–285 and 377–433.
29. Krein, M. G., Perturbation determinants and a formula for the trace of unitary and self–adjoint operators, Soviet Math. Dokl. **3** (1962), 707–710.
30. Lelong, P., Intégration sur un ensemble analytique complexe, Bull. Soc. Math. France **85** (1957), 239-262.

31. Loday, J.-L. and D. Quillen, Cyclic homology and the Lie algebra homology of matrices, Comment. Math. Helvetici **59** (1984), 565–591.
32. Melnikov, M. S., The structure of the Gleason part of the algebra $R(E)$, Funct. Anal. Appl. **1** (1967), 84–86.
33. Pincus, J. D., The spectrum of seminormal operators, Proc. Nat. Acad. Sci. U.S.A. **68** (1971), 1684–1685.
34. ———, D. Xia, and J. Xia, The analytic model of a hyponormal operator with rank one self–commutator, Integral Eq. and Operator Theory **7** (1984), 516–535.
35. Putnam, C. R., An inequality for the area of hyponormal spectra, Math. Z. **116** (1970), 323–330.
36. Saks, S., *Theory of the Integral*, Dover, New York, 1964.
37. Taylor, J. L., A joint spectrum for several commuting operators, J. Funct. Anal. **6** (1970), 172–191.
38. Voiculescu, D., Remarks on Hilbert–Schmidt perturbations of almost–normal operators, Topics in Modern Operator Theory, (Timisoara/Herculane, 1980) 311-318, Operator Theory: Avd. Appl. **2** Birkhäuser, Basel-Boston, Massachusetts, 1981.
39. Wermer, J. The hull of a curve in C^n, Ann. of Math. (2) **68** (1958), 550–561.

Richard Carey
Department of Mathematics
University of Kentucky
Lexington, Kentucky 40506
U.S.A.

Hyponormal and Subnormal Toeplitz Operators

by

Carl C. Cowen

This paper is my view of the past, present, and future of Problem 5 of Halmos's 1970 lectures "Ten Problems in Hilbert Space " [12] (see also [13]):

Is every subnormal Toeplitz operator either normal or analytic?

We recall that for $\varphi \in L^\infty$, the *Toeplitz operator* T_φ is the operator on the Hardy space H^2 of the unit disk D, given by $T_\varphi h = P\varphi h$ where h is in H^2 and P is the orthogonal projection of $L^2(\partial D)$ onto H^2. An operator S on a Hilbert space $\mathcal{H}$ is *subnormal* if there is a normal operator N on $\mathcal{K} \supset \mathcal{H}$ such that $\mathcal{H}$ is invariant for N and $N|_{\mathcal{H}} = S$.

The question is natural because the two classes, the normal and analytic Toeplitz operators, are fairly well understood and are obviously subnormal. The normal Toeplitz operators were characterized by Brown and Halmos in 1964.

Theorem 1. ([4], page 98) *The Toeplitz operator T_φ is normal if and only if $\varphi = \alpha + \beta\rho$ where α and β are complex numbers and ρ, in L^∞, is real valued.*

A Toeplitz operator T_φ is called *analytic* if φ is in H^∞, that is, φ is a bounded analytic function on D. These are easily seen to be subnormal: $T_\varphi h = P\varphi h = \varphi h = L_\varphi h$ for h in H^2, where L_φ is the normal operator of multiplication by φ on $L^2(\partial D)$.

All progress on this question has begun with the study of the self-commutator of T_φ. A subnormal operator S is *hyponormal*, that is, its self-commutator, $S^*S - SS^*$, is positive. It is not difficult to show that the range of the self-commutator of a subnormal operator is an invariant subspace of S^* ([20], Theorem 5).

Halmos almost certainly believed the answer to his question would be yes, and his intuition was soon bolstered as several results appeared that showed the answer is yes for certain classes. I'll prove the first of these, a 1972 theorem of Ito and Wong, because its proof is typical in its use of the self-commutator. An *inner function* is a function in H^∞ that has modulus 1 almost everywhere on the unit circle.

Theorem 1. ([14], Theorem 1). *If φ is a polynomial in χ and $\bar{\chi}$, where χ is an inner function, then T_φ is subnormal if and only if it is normal or analytic.*

Proof: Since $T_{f\circ\chi}$ is unitarily equivalent to an inflation of T_f for any f in L^∞, it is sufficient to prove the result for $\chi(z) = z$ ([8], Theorem 1).

Suppose T_φ is subnormal but not analytic, say

$$\varphi = a_{-n}\bar{z}^n + \cdots + a_0 + \cdots + a_m z^m$$

where $n > 0$ and $a_{-n} \neq 0$. Let $C = T_\varphi^* T_\varphi - T_\varphi T_\varphi^*$, the self-commutator of T_φ.

If $j \geq k = \max\{m, n\}$, then, $z^j\varphi$ and $z^j\bar{\varphi}$ are both analytic so $z^k H^2 \subset$ kernel C. That is, range C is contained in span $\{1, z, \ldots, z^{k-1}\}$.

If C were not 0, then we could choose f of maximal degree in range C, say, $f = b_0 + \cdots + b_l z^l$, where $b_l \neq 0$. But since T_φ is subnormal, range C is invariant for T_φ^* and $T_\varphi^* f = \cdots + \bar{a}_{-n} b_l z^{n+l}$ which would have higher degree than f. Thus $C = 0$ and T is normal. ■

Ito and Wong were apparently the first to consider hyponormal Toeplitz operators. In the same paper ([14], page 158), they gave the interesting example $T_{z+\frac{1}{2}\bar{z}}$. This was the first *easy* example of a hyponormal operator that is not subnormal. Their example generalizes.

Proposition 3. *Let A be hyponormal. Then $A + \lambda A^*$ is hyponormal if and only if $|\lambda| \leq 1$.*

Proof:

$$(A + \lambda A^*)^*(A + \lambda A^*) - (A + \lambda A^*)(A + \lambda A^*)^* = (1 - |\lambda|^2)(A^*A - AA^*). \quad ■$$

Later, Amemiya, Ito, and Wong showed by a similar argument that the answer to Halmos's question is yes if T_φ is quasinormal.

Theorem 4. [2] *If T_φ commutes with $T_\varphi^* T_\varphi$ then either T_φ is normal or φ is analytic and $\varphi = \lambda\chi$, where χ is an inner function.*

Corollary. *If T_φ is subnormal with rank 1 self-commutator, then $\varphi = \alpha + \beta\chi$ where α and β are numbers and χ is an inner linear fractional transformation.*

(This corollary also follows from a more general theorem of B. Morrel [18], page 508.)

The deepest work in this direction is that of Abrahamse, for which we need the following definition.

Definition. ([16], page 187) A function φ in L^∞ is of *bounded type* (or in the *Nevanlinna class*) if there are functions ψ_1, ψ_2 in $H^\infty(D)$ such that

$$\varphi(e^{i\vartheta}) = \frac{\psi_1(e^{i\vartheta})}{\psi_2(e^{i\vartheta})}$$

for almost all ϑ in ∂D.

Clearly, rational functions in $L^\infty(\partial D)$ are of bounded type: they are quotients of analytic polynomials. A polynomial p in z and $\bar{z}$ is of bounded type because on the unit circle, $p(z,\bar{z}) = p(z, z^{-1})$ which is a rational function.

Theorem 5. [1] *If*

(1) T_φ is hyponormal,

(2) φ or $\bar{\varphi}$ is of bounded type, and

(3) $\ker(T_\varphi^* T_\varphi - T_\varphi T_\varphi^*)$ is invariant for T_φ, then T_φ is normal or analytic.

Since (3) holds for any subnormal operator ([20], Theorem 5), Abrahamse obtains the conclusion for subnormals.

Corollary. *If T_φ is subnormal and φ or $\bar{\varphi}$ is of bounded type, then T_φ is normal or analytic.*

Abrahamse concluded his paper by asking several questions, including the following:

> Which hyponormal weighted shifts are unitarily equivalent to Toeplitz operators?
>
> Is the Bergman shift unitarily equivalent to a Toeplitz operator?

Recall that an operator W is a (unilateral) *weighted shift* if there is an orthonormal basis $e_0, e_1, \ldots$ and weights $w_n > 0$ such that $We_j = w_j e_{j+1}$ for $j = 0, 1, 2, \ldots$. An easy calculation shows W is hyponormal if and only if $w_0 \le w_1 \le w_2 \le \cdots$. It is more difficult to show that W is subnormal if and only if

$$(w_0 w_1 \cdots w_{k-1})^2 = \int t^{2k} d\nu(t)$$

for some probability measure ν with support in $[0, \|W\|]$ (due independently to Berger and to Gellar and Wallen [7], page 159). The *Bergman shift* is the subnormal shift with weights $w_n^2 = (n+1)(n+2)^{-1}$ for $n = 0, 1, 2, \cdots$.

Sun Shunhua proved the following remarkable theorem by carefully examining the action of the self-commutator on e_0 in the shifted basis.

Theorem 6. [22] *If T_φ is a hyponormal weighted shift, then there is a number α, $0 \le \alpha \le 1$ so that the weights are $w_n^2 = (1-\alpha^{2n+2})\|T_\varphi\|^2$.*

The case $\alpha = 0$ is T_z. Sun Shunhua left open the question of existence for the cases $\alpha > 0$, but this did answer Abrahamse's second question.

Corollary. *The Bergman shift is not unitarily equivalent to a Toeplitz operator.*

Proof:

$$\frac{n+1}{n+2} \neq 1 - \alpha^{2n+2} \text{ for any } \alpha > 0 \quad \blacksquare$$

Examination of the proof of the theorem reveals that such a φ must have $\psi = \varphi - \alpha\bar{\varphi}$ in H^∞. Moreover, the matrix for T_ψ in the shifted basis is a compact perturbation of

$$\begin{pmatrix} 0 & -\alpha & 0 & 0 & \\ 1 & 0 & -\alpha & 0 & \\ 0 & 1 & 0 & -\alpha & \\ 0 & 0 & 1 & 0 & \\ & & & & \ddots \end{pmatrix}$$

which is the matrix for $T_{z-\alpha\bar{z}}$ in the usual basis. It follows that the Fredholm indices and the essential spectra of these Toeplitz operators must be the same. Since T_ψ is an analytic Toeplitz operator, ψ must be a conformal mapping of D onto the interior of the ellipse with vertices $\pm(1+\alpha)i$ and passing through $\pm(1-\alpha)$, and $\varphi = (1-\alpha^2)^{-1}(\psi + \alpha\bar{\psi})$. Knowing that this would have to be the symbol made it possible to verify that this works.

Theorem 7. [9] *Let $0 < \alpha < 1$ be given and let ψ be a conformal map of the disk onto the interior of the ellipse with vertices $\pm(1+\alpha)i$ and passing through $\pm(1-\alpha)$. If $\varphi = (1-\alpha^2)^{-1}(\psi+\alpha\bar{\psi})$, then T_φ is a weighted shift with weight sequence $w_n^2 = 1-\alpha^{2n+2}$ and is subnormal but neither normal nor analytic.*

The next step forward was Sun Shunhua's characterization of those Toeplitz operators such that $T_\varphi + T_\varphi^*$ and $T_\varphi T_\varphi^*$ commute, what Campbell called the Θ-class. Surprisingly, the answer again involved these ellipse maps: $\varphi = \psi + \beta\bar{\psi}$ gives rise to a Θ-class Toeplitz operator for a certain choice of β. Moreover, since $|\beta| < 1$, the operator T_φ is hyponormal and a theorem of Campbell, [5], implies T_φ is subnormal.

Thus, at this point, we know that $T_{\psi+0\bar{\psi}}$ (which is analytic), $T_{\psi+\alpha\bar{\psi}}$, $T_{\psi+\beta\bar{\psi}}$, and $T_{\psi+1\bar{\psi}}$ (which is normal) are all subnormal, and from the generalization of Ito and

Wong's observation, $T_{\psi+\lambda\bar{\psi}}$ is hyponormal for $|\lambda| \leq 1$. This strongly suggests the question:

> For which λ is $T_{\psi+\lambda\bar{\psi}}$ subnormal? The answer to this question is provided by the following recent theorem.

Theorem 8. ([10], Theorem 2.4) *Let λ be a complex number, let $0 < \alpha < 1$, and let ψ be the conformal map of the disk onto the interior of the ellipse with vertices $\pm i(1+\alpha)$ passing through $\pm(1-\alpha)$ where $0 < \alpha < 1$. For $\varphi = \psi + \lambda\bar{\psi}$, the Toeplitz operator T_φ is subnormal if and only if $\lambda = \alpha$ or $\lambda = (\alpha^k e^{i\vartheta} + \alpha)(1 + \alpha^{k+1}e^{i\vartheta})^{-1}$ for some $k = 0, 1, 2, \ldots$ and $0 \leq \vartheta < 2\pi$.*

Note that $k = 0$ in the theorem means $|\lambda| = 1$, that is, $T_{\psi+\lambda\bar{\psi}}$ is normal; $\lambda = \alpha$ (which corresponds to $k = \infty$) is the weighted shift case; $k = 1, \vartheta = \pi$ is the analytic case $T_{\psi+0\bar{\psi}}$; and the Θ-class case of Sun Shunhua corresponds to $k = 2$.

This theorem follows by simple algebra from the Cowen-Long result and the following result on the special weighted shifts that are multiples of the Toeplitz operators in Theorem 7.

Proposition 9. [10] *Let T be the weighted shift with weights*

$$w_n^2 = \sum_{j=0}^{n} \alpha^{2j}.$$

Then $T + \mu T^*$ is subnormal if and only if $\mu = 0$ or $|\mu| = \alpha^k$ for $k = 0, 1, 2, \ldots$.

The proof of the proposition is a modification of the matrix constructions of Ando [3] and Stampfli [21] for the minimal normal extension of a subnormal operator. The heart of the matter is: If S is subnormal and

$$\begin{pmatrix} S & B \\ 0 & A \end{pmatrix}$$

is its minimal normal extension, what choice do we have for B? The answer: We may replace B by $\widetilde{B}$ as long as $BB^* = \widetilde{B}\widetilde{B}^*$ and their kernels have the same dimension. In particular, by polar factorization, we may replace B by a non-negative operator.

Sketch of proof. Let D be the diagonal operator whose k-th diagonal entry is α^k. Then, D is positive and $D^2 = T^*T - TT^*$. Using rotations and the fact that T is subnormal and $T + T^*$ is normal, it is sufficient to give the proof for $\mu = s$, where $0 < s < 1$. Let $A_0 = T + sT^*$.

Suppose A_0 is subnormal, and

$$\begin{pmatrix} A_0 & X_1 \\ 0 & Y_1 \end{pmatrix}$$

is its minimal normal extension. The normality of this matrix implies $X_1 X_1^* = A_0^* A_0 - A_0 A_0^* = (1 - s^2)D^2$. A theorem of Olin ([17], page 228) implies that either X_1 has kernel (0) or X_1 has infinite-dimensional kernel. Letting $B_1 = r_1 D$ where $r_1^2 = (1 - s^2)$, it follows that the minimal normal extension has a representation as either

$$\begin{pmatrix} A_0 & B_1 \\ 0 & A_1 \end{pmatrix} \quad \textit{or} \quad \begin{pmatrix} A_0 & B_1 & 0 \\ 0 & A_1 & X_2 \\ 0 & 0 & Y_2 \end{pmatrix}.$$

In either case, normality implies that $A_1 = \alpha T + s\alpha^{-1}T^*$. In the former case, we also get $A_1^* A_1 - A_1 A_1^* = -B_1 B_1^*$ which means $\alpha = s$. In the latter case, using a positivity condition that is a consequence of the normality, we find that $\alpha < s < 1$ is impossible.

After $n-1$ such steps, if $s \neq \alpha^k$ for $k = 1, 2, \ldots, n-1$, we have shown that $s < \alpha^{n-1}$ and that, for $A_k = \alpha^k T + s\alpha^{-k}T^*$ and $B_k = r_k D$, the minimal normal extension is unitarily equivalent either to

$$\begin{pmatrix} A_0 & B_1 & \cdots & 0 \\ 0 & A_1 & \cdots & 0 \\ \vdots & \vdots & \ddots & \vdots \\ 0 & 0 & \cdots & A_n \end{pmatrix}$$

in which case normality implies $s = \alpha^n$ or to

$$\begin{pmatrix} A_0 & B_1 & \cdots & 0 & 0 \\ 0 & A_1 & \cdots & 0 & 0 \\ \vdots & \vdots & \ddots & \vdots & \vdots \\ 0 & 0 & \cdots & A_n & X_{n+1} \\ 0 & 0 & \cdots & 0 & Y_{n+1} \end{pmatrix}$$

and normality implies that $\alpha^n < s < \alpha^{n-1}$ is impossible. ■

A miracle has occurred: all the entries in the matrix model were explicitly computable for this weighted shift!

Standard facts about unitary equivalence of shifts and Theorem 7 give some non-obvious unitary equivalences of Toeplitz operators.

Corollary. [10] *The analytic Toeplitz operator T_ψ is unitarily equivalent to each of the non-analytic Toeplitz operators T_φ with*

$$\varphi = ie^{-i\vartheta/2}(1-\alpha^2)^{-1}\left[(1+\alpha^2 e^{i\vartheta})\psi + \alpha(1+e^{i\vartheta})\bar{\psi}\right]$$

for $-\pi < \vartheta < \pi$.

Proof: Since T is a weighted shift, $T + sT^*$ is unitarily equivalent to $T + \lambda T^*$ whenever $|\lambda| = s$. Rewriting this unitary equivalence in terms of the Toeplitz operator T_ψ gives the result. ■

Now Theorem 7 is interesting and surprising, but it is not very encouraging. It says that the subnormality of these Toeplitz operators depends on some combinatorial coincidences and suggests that subnormality of Toeplitz operators may be the wrong question to be studying. Perhaps more progress can be made studying the hyponormality of Toeplitz operators.

Very little work has gone into discovering which symbols in L^∞ give hyponormal Toeplitz operators and into developing an adequate theory for them. Much of what is known is in the form of *folk theorems*. I will give some elementary observations and suggest some questions for further study. The intuition is that T_φ is hyponormal if the analytic part of φ dominates the conjugate analytic part. Some of the results below may be interpreted as making a precise statement supporting this intuition. For example, in the following unpublished proposition of Wogen, we feel $\chi\varphi$ is the same size as φ but is *more analytic*.

Proposition 10. [10] *If χ is inner and φ in L^∞ is such that T_φ is hyponormal, then $T_{\chi\varphi}$ is also hyponormal.*

Proof: For h in H^2 since T_χ is an isometry and T_φ is hyponormal, we have

$$\|T_{\chi\varphi}h\| \geq \|T_{\bar{\chi}}T_{\chi\varphi}h\| = \|T_\varphi h\| \geq \|T_\varphi^* h\| = \|T_{\bar{\varphi}}h\| \geq \|T_{\bar{\chi}}T_{\bar{\varphi}}h\| = \|T_{\chi\varphi}^* h\|.$$

Thus, $T_{\chi\varphi}$ is also hyponormal. ■

We recall the definition of Hankel operators (see for example [19]). As usual, P is the projection from L^2 onto H^2.

Definition. For φ in L^∞, the Hankel operator $H_\varphi : H^2 \to (H^2)^\perp$ is given by $H_\varphi h = (I-P)\varphi h$, for h in H^2.

The following proposition is not completely general because not all φ can be split as in the hypothesis. With care, regarding the Toeplitz operators as unbounded operators, it can be improved to general L^∞ functions.

Proposition 11. *Suppose f and g are in H^∞ and suppose $\varphi = f + \bar{g}$. The following are equivalent.*

(1) *T_φ is hyponormal.*
(2) *For every h in H^2, $\|fh\|^2 - \|P\overline{f}h\|^2 \geq \|gh\|^2 - \|P\overline{g}h\|^2$.*
(3) *$T_f^*T_f - T_fT_f^* \geq T_g^*T_g - T_gT_g^*$.*
(4) *$H_{\overline{f}}^*H_{\overline{f}} \geq H_{\overline{g}}^*H_{\overline{g}}$.*
(5) *For every h in H^2, we have $\|H_{\overline{f}}h\| \geq \|H_{\overline{g}}h\|$.*

Proof: Compute. ■

Corollary. *Let χ be inner, and let F and G be in H^∞ with $f = \chi F$ and $g = \chi G$. If $T_{f+\bar{g}}$ is hyponormal, then $T_{F+\bar{G}}$ is also hyponormal.*

Proof: Using (2) above, we have

$$\begin{aligned}\|Fh\|^2 - \|P\bar{F}h\|^2 &= \|f(\chi h)\|^2 - \|P\overline{f}(\chi h)\|^2 \geq \|g(\chi h)\|^2 - \|P\overline{g}(\chi h)\|^2 \\ &= \|Gh\|^2 - \|P\bar{G}h\|^2.\end{aligned}$$ ■

Using (4) and the fact that for χ inner, $H_{\bar{\chi}}^*H_{\bar{\chi}}$ is the projection of H^2 onto $(\chi H^2)^\perp$, we get the following corollary.

Corollary. *Let χ be inner and g be in H^∞. Then $T_{\chi+\bar{g}}$ is hyponormal if and only if g is in $(z\chi H^2)^\perp$ and $\|H_{\overline{g}}\| \leq 1$.*

If χ is a finite Blaschke product, then $(z\chi H^2)^\perp$ is finite dimensional and for each g in $(z\chi H^2)^\perp$, $H_{\overline{g}}$ is finite rank, so finding the set of g such that $T_{\chi+\bar{g}}$ is hyponormal is a computation in finite dimensional linear algebra. For example, it is tedious but not difficult to verify that $T_{z^2+\bar{g}}$ is hyponormal if and only if $g = \alpha + \beta z + \gamma z^2$ where $|\beta| + |\gamma|^2 \leq 1$.

We are interested more generally in the set of g such that $T_{f+\bar{g}}$ is hyponormal for a given f. This motivates the following definition. To avoid difficulties with splitting functions in L^∞, we formulate the definition for f and g in H^2 without regard to whether they are the analytic parts of functions in L^∞.

Definition. Let $\mathcal{H} = \{h \in H^\infty : h(0) = 0 \text{ and } \|h\|_2 \le 1\}$. For f in H^2, let G_f denote the set of g in H^2 such that for every h in H^2,

$$\sup_{h_0 \in \mathcal{H}} | < hh_0, f > | \ge \sup_{h_0 \in \mathcal{H}} | < hh_0, g > |.$$

Note that in the definition of $\mathcal{H}$, we have used the H^2 norm of the H^∞ function h, and that $\mathcal{H}$ is dense in the unit ball of zH^2. To see how this definition is relevant to our work, suppose p is in H^∞ and h is in H^2. Then we have

$$\sup_{h_0 \in \mathcal{H}} | < hh_0, p > | = \sup_{h_0 \in \mathcal{H}} | < \bar{p}h, \bar{h_0} > | = \sup_{h_0 \in \mathcal{H}} | < (I - P)\bar{p}h, \bar{h_0} > | = \|H_{\bar{p}}h\|.$$

Thus, when f and g are both in H^∞, this means, by (5) of Proposition 11, that $T_{f+\bar{g}}$ is hyponormal if and only if g is in G_f. The following theorem gives some elementary properties of G_f.

Theorem 12. *For f in H^2, the following hold.*

(0) $G_f = G_{f+\lambda}$ for all complex numbers λ.
(1) f is in G_f.
(2) If g is in G_f, then $g + \lambda$ is in G_f for all complex numbers λ.
(3) G_f is balanced and convex; that is, if g_1 and g_2 are in G_f and $|s| + |t| \le 1$, then $sg_1 + tg_2$ is also in G_f.
(4) G_f is weakly closed.
(5) If χ is inner and χg is in G_f, then g is in G_f.

Proof: Properties (0), (1), and (2) are obvious.

(3) For h in H^2, we have

$$\begin{aligned} \sup_{h_0 \in \mathcal{H}} | < hh_0, f > | &\ge |s| \sup_{h_0 \in \mathcal{H}} | < hh_0, f > | + |t| \sup_{h_0 \in \mathcal{H}} | < hh_0, f > | \\ &\ge |s| \sup_{h_0 \in \mathcal{H}} | < hh_0, g_1 > | + |t| \sup_{h_0 \in \mathcal{H}} | < hh_0, g_2 > | \\ &\ge \sup_{h_0 \in \mathcal{H}} | < hh_0, sg_1 + tg_2 > |. \end{aligned}$$

(4) Let h be in H^2. Suppose g_α is in G_f for each α in a directed set and suppose $g_\alpha \to g$ weakly. Fix h_1 in $\mathcal{H}$. Thus,

$$| < hh_1, g > | = \lim_\alpha | < hh_1, g_\alpha > | \le | < hh_1, f > | \le \sup_{h_0 \in \mathcal{H}} | < hh_0, f > |.$$

Since this is true for all such h_1, we have g in G_f.

(5) Let h be in H^2. Since $h_0 \in \mathcal{H}$ implies $\chi h_0 \in \mathcal{H}$, for χg in G_f, we have

$$\sup_{h_0 \in \mathcal{H}} |< hh_0, f >| \geq \sup_{h_0 \in \mathcal{H}} |< hh_0, \chi g >| \geq \sup_{h_0 \in \mathcal{H}} |< h(\chi h_0), \chi g >|$$
$$= \sup_{h_0 \in \mathcal{H}} |< hh_0, g >|,$$

so g is in G_f. ■

Corollary. $T_z^* G_f \subset G_f$.

Proof: By (2) g in G_f implies $g - g(0)$ is in G_f and (5) implies $T_z^* g = (g - g(0))/z$ is in G_f. ■

I am proposing the study of the hyponormal Toeplitz operators; the study of the properties of G_f seems like a reasonable place to start. Several questions seem interesting.

Question 1. *Can G_f be characterized? In particular, do (1) to (5) of Theorem [12] characterize G_f?*

For $f = z^2$, these properties determine G_f, and the computations are easier than the norm computations referred to earlier: By (1) and (2), $z^2 - a^2 \in G_{z^2}$ for all a in the unit disk, so (5) implies

$$-\bar{a}z^2 + (1 - |a|^2)z + a = \left(\frac{z - a}{1 - \bar{a}z}\right)^{-1} (z^2 - a^2) \text{ is in } G_{z^2}.$$

Using (2) and (3) we see that $G_{z^2} \supset \{g(z) = \beta z + \gamma z^2 : |\beta| + |\gamma|^2 = 1\}$, and finally that $G_{z^2} \supset \{g(z) = \alpha + \beta z + \gamma z^2 : |\beta| + |\gamma|^2 \leq 1\}$.

Translation by scalars was used to advantage in the above computation, but it may not always be helpful. It may be more convenient to work with the set

$$G'_f = \{g \in G_f : g(0) = 0\}$$

since if g is in G'_f (taking $h \equiv 1$ in the definition of G_f) we have

$$\|g\|_2 = \sup_{h_0 \in \mathcal{H}} |< h_0, g >| \leq \sup_{h_0 \in \mathcal{H}} |< h_0, f >| = \|f\|_2.$$

This means that G'_f is convex and weakly compact.

Question 2. *What are the extreme points of G'_f? In particular, if $g \in G'_f$ but $\lambda g \notin G'_f$ for $|\lambda| > 1$, is g an extreme point of G'_f?*

A function g is an extreme point of G'_{z^2} if and only if $g(z) = \beta z + \gamma z^2$ where $|\beta| + |\gamma|^2 = 1$, but of course we want a description of the extreme points of G'_f that can be used to compute G'_f, not vice versa. The set of extreme points came up naturally in the computation of G_{z^2} using the properties (1) to (5) above; perhaps it is always so. For the special subset in the question to be the set of extreme points, would require a certain rotundity of the set G'_f. The corollary of Theorem 11 characterizes this subset for f inner, and for $f = z^2$, all are extreme points.

In the study of subnormal Toeplitz operators, these results suggest two more questions.

Question 3. *For which f in H^∞ is there λ, $0 < \lambda < 1$ with $T_{f+\lambda\bar{f}}$ subnormal?*

Question 4. *Suppose ψ is as in Theorem 7. Are there g in G_ψ, $g \neq \lambda\psi + c$, such that $T_{\psi+\bar{g}}$ is subnormal?*

Note that Abrahamse's work [1] relates directly to Questions 3 and 4. If $\varphi = f + \bar{g}$ with f, g, in H^∞, then φ is of bounded type if and only if $\bar{g}$ is of bounded type. Thus T_φ subnormal, neither normal nor analytic, implies that $\bar{f}$ and $\bar{g}$ are not of bounded type.

Although isolated results and examples have appeared in the literature, hyponormality of Toeplitz operators has not been systematically studied. I believe that the study of hyponormal Toeplitz operators is of more importance to the understanding of Toeplitz operators than is the study of subnormality, and that in any case, more progress needs to be made on the hyponormality questions before substantial progress on the subnormality questions can be made. I hope I have given you some hints on where to start.

Acknowledgments. I would like to thank Tom Kriete, Bernie Morrel, Sun Shunhua, Ken Stephenson, and Warren Wogen for helpful discussions on this material and I would like to thank John Conway for his invitation to speak at the conference and publish these notes. During this work, I was supported in part by National Science Foundation Grant #8300883.

Added in Proof: The author has proved (preprint) that T_φ is hyponormal if and only if $g = c + T_{\bar{h}} f$ for some constant c and some function h in H^∞ with $\|h\|_\infty \leq 1$. In consequence, the answer to Question 1 is "yes".

References

1. Abrahamse, M. B., Subnormal Toeplitz opertors and functions of bounded type, Duke Math. J. **43** (1976), 597–604.
2. Amemiya, I., T. Ito, and T. K. Wong, On quasinormal Toeplitz operators, Proc. Amer. Math. Soc. **50** (1975), 254–258.
3. Ando, T., Matrices of normal extensions of subnormal operators, Acta Sci. Mth. (Szeged) **24** (1963), 91–96.
4. Brown, A. and P. R. Halmos, Algebraic properties of Toeplitz operators, J. reine angew. Math. **213** (1963-64), 89-102.
5. Campbell, S. L., Linear operators for which T^*T and $T^* + T$ commute, III, Pacific J. Math. **76** (1978), 17-19.
6. Clancey, K. F. and B. B. Morrel, The essential spectra of some Toeplitz operators, Proc. Amer. Math. Soc. **44** (1974), 129-134.
7. Conway, J. B., *Subnormal Operators*, Pitman, Boston, 1981.
8. Cowen, C. C., On equivalence of Toeplitz operators, Journal of Operator Theory **7** (1982), 167-172.
9. Cowen, C. C. and J. J. Long, Some subnormal Toeplitz operators, J. Reine angew. Math. **351** (1984), 216–220.
10. Cowen, C. C., More subnormal Toeplitz operators, J. reine angew. Math. **367** (1986), 215–219..
11. Douglas, R. G., *Banach Algebra Techniques in Operator Theory*, Academic Press, New York, 1972.
12. Halmos, P. R., Ten problems in Hilbert space, Bull. Amer. Math. Soc. **76** (1970), 887-933.
13. ———, Ten years in Hilbert space, Integral Equations and Operator Theory **2** (1979), 529-564.
14. Ito, T. and T. K. Wong, Subnormality and quasinormality of Toeplitz operators, Proc. Amer. Math. Soc. **34** (1972), 157–164.
15. Long, J. J., Hyponormal Toeplitz Operators and Weighted Shifts, Thesis, Michigan State University, 1984.
16. Morrel, B. B., A decomposition for some operators, Indiana Univ. Math. J. **23** (1973), 497–511.
17. Nevanlinna, R., *Analytic Functions*, translated by P. Emig, Springer Verlag, Berlin, 1970.

18. Olin, R. F., Functional relationships between a subnormal operator and its minimal normal extension, Pacific J. Math. **63** (1976), 221–229.

19. Power, S. C., *Hankel Operators on Hilbert Space*, Pitman, Boston, 1982.

20. Stampfli, J. G., Hyponormal operators and spectral density, Trans. Amer. Math. Soc. **117** (1965), 469–476.

21. ———, Which weighted shifts are subnormal?, Pacific J. Math. **17** (1966), 367–379.

22. Shunhua, Sun, Bergman shift is not unitarily equivalent to a Toeplitz operator, Kexue Tongbao **28** (1983), 1027–1030.

23. ———, On Toeplitz operators in the Θ–class, Scientia Sinica (Series A) **28** (1985), 235–241.

24. ———, On the unitary equivalence of Toeplitz operators, preprint.

25. Wogen, W. R., Unpublished manuscript, 1975.

Carl C. Cowen
Department of Mathematics
Purdue University
West Lafayette, Indiana 47907
U.S.A.

Closures of Similarity Orbits of Hilbert Space Operators

by

Lawrence A. Fialkow

Preface

The title of these lectures refers to a recent theorem of C. Apostol, D. Herrero, and D. Voiculescu [10,5] that characterizes the closure of the similarity orbit of a Hilbert space operator in terms of spectral, Fredholm, and algebraic data. This theorem subsumes many known results in operator approximation theory, particularly those which provide *spectral* characterizations of the norm closure of similarity-invariant sets of operators. For example, the closure of the set of all nilpotent operators and the closure of the set of all operators with spectra contained in a fixed subset of the plane may both be characterized by means of the Similarity Orbit Theorem of [10,5]. This theorem thus provides a powerful tool for the study of operator approximation.

The proof of the orbit theorem is very long and difficult, and is thus beyond the scope of these lectures; rather, our purpose here is to provide an introductory survey (suitable for students or non-specialists) of some highlights of *approximation of operators* as this subject has evolved towards the Similarity Orbit Theorem. Operator approximation and the orbit theorem have received a comprehensive exposition in the recent monographs of D. Herrero [63] and C. Apostol et alia [5]; we acknowledge that our presentation borrows freely from these sources. The subjects of these lectures are as follows:

(1) Introduction; similarity and approximation.

(2) Closure of nilpotent operators.

(3) Closed similarity orbits.

(4) Universal quasinilpotent operators.

(5) The Similarity Orbit Theorem.

The Introduction surveys some of the origins of the recent development of operator approximation theory: P. R. Halmos' *Ten Problems in Hilbert Space*; the Weyl-von Neumann-Berg Theorem; the Brown-Douglas-Fillmore Theorem, the Apostol-Foias-Voiculescu spectral characterization of quasitriangular operators; Voiculescu's Theorem and D. W. Hadwin's work on closures of unitary orbits of operators. These results both motivate and facilitate the study of closures of similarity orbits. We also record for ease of reference some basic tools of operator approximation related to similarity. In Lecture 2 we present the theorem of C. Apostol, C. Foias, and D. Voiculescu [8] characterizing the norm closure of the set of nilpotent operators; the proof illustrates several important techniques of approximation. In Lecture 3, as a further application of these techniques, we derive the theorem of C. Apostol and C. Foias [6] and of D. Herrero [61] describing the operators with closed similarity orbits. By way of contrast, in Lecture 4 we develop the characterization of the *universal* quasinilpotent operators, quasinilpotents for which the closure of the similarity orbit is as large as possible; this is the work of C. Apostol [3] and D. Herrero [60]. Lecture 5 concerns the Similarity Orbit Theorem itself. Here we show how the results of Lectures 2-4 can be recovered from the theorem and we delineate a few recent areas of application of the orbit theorem.

1. Introduction

A principal theme of operator approximation is that the structure of Hilbert space operators may be revealed not only by unitary and affine models for operators, but also by models based on suitable *approximate* versions of unitary equivalence and similarity. Let us discuss several notions of equivalence in the unitary case first. For a unital C^*-algebra $\mathcal{A}$, let $\mathcal{U}(\mathcal{A})$ denote the *unitary group* in $\mathcal{A}, \mathcal{U}(\mathcal{A}) = \{u \in \mathcal{A} : u^*u = uu^* = 1\}$. For $a \in \mathcal{A}$, let $\sigma(a)$ denote the *spectrum* of a and let $\mathcal{U}(a)$ denote the *unitary equivalence class* (*unitary orbit*) of a, i.e., $\mathcal{U}(a) = \{u^*au : u \in \mathcal{U}(\mathcal{A})\}$. An element $b \in \mathcal{A}$ is *unitarily equivalent* to a ($b \approx a$) if $b \in \mathcal{U}(a)$. We say that b is *asymptotically unitarily equivalent* to a if $b \in \mathcal{U}(a)^-$, i.e., $\mathcal{U}(b)^- = \mathcal{U}(a)^-$. ($\mathcal{U}(\cdot)^-$ denotes the *norm closure* of $\mathcal{U}(\cdot)$ in $\mathcal{A}$.)

What are the invariants for this notion of equivalence? Consider first normal operators on Hilbert space. Let $\mathcal{L}(\mathcal{H})$ denote the algebra of all bounded linear operators on a separable infinite dimensional complex Hilbert space $\mathcal{H}$. Let $\mathcal{K}(\mathcal{H})$ denote the ideal of compact operators on $\mathcal{H}$ and let $\mathcal{A}(\mathcal{H}) \equiv \mathcal{L}(\mathcal{H})/\mathcal{K}(\mathcal{H})$ denote the Calkin algebra. For T in $\mathcal{L}(\mathcal{H}), \widetilde{T}$ denotes the image of T in $\mathcal{A}(\mathcal{H})$ and $\sigma_e(T) \equiv \sigma(\widetilde{T})$ denotes the *essential spectrum* of T. If $N \in \mathcal{L}(\mathcal{H})$ is normal, then $\sigma_e(N)$ consists of the topological limit points of $\sigma(N)$ together with the isolated points of $\sigma(N)$ that are eigenvalues of infinite multiplicity; the points of $\sigma(N)\backslash\sigma_e(N)$ are the isolated eigenvalues of finite multiplicity. The spectral theorem and continuity properties of the functional calculus for normal operators imply

$$\mathcal{U}(N)^- = \{M \in \mathcal{L}(\mathcal{H}) : M \text{ is normal}, \ \sigma(M) = \sigma(N), \sigma_e(M) = \sigma_e(N), \tag{1.1}$$
$$\dim \ker M - \lambda = \dim \ker N - \lambda (\lambda \in \sigma(N) \backslash \sigma_e(N))\}.$$

In the Hilbert space setting there are other natural notions of equivalence. For operators $S, T \in \mathcal{L}(\mathcal{H})$, we write $S \approx_K T$ if S and T are *unitarily equivalent up to compact perturbation* (*compalent*), i.e., $S = U^*TU + K$ with U unitary and K compact and K compact; we set $\mathcal{U}_K(T) = \{S \in \mathcal{L}(\mathcal{H}) : S \approx_K T\}$. In the definition of compalence, we might further require that K can be chosen with arbitrarily small norm, in which case we say that S and T are *approximately unitarily equivalent* ($S \approx_a T$). What are the connections between the preceding three notions of equivalence? Recall H. Weyl's classical theorem on compact perturbations of self-adjoint operators [82]: Every self-adjoint operator is the sum of a diagonal self-adjoint operator and a compact operator. J. von Neumann [80] subsequently obtained improvements as to the size of the compact operator in Weyl's Theorem—it can be a Hilbert-Schmidt operator of arbitrarily small Hilbert-Schmidt norm—and von Neumann also showed that the essential spectrum is a complete compalence invariant for self-adjoint operators.

This interplay between compact perturbation and norm approximation is again prominent in P. R. Halmos' work on quasitriangular operators [47] and in Halmos' celebrated *Ten Problems in Hilbert Space* [48]. In [48] Halmos used the *quasidiagonality* of self-adjoint operators to reprove Weyl's Theorem, and in this context asked whether Weyl's Theorem extends to normal operators. The positive answer to this question, achieved by I. D. Berg [19] and W. Sikonia [74], proved to be a key element in the subsequent development of operator approximation:

Theorem 1.1. [19] *Every normal operator is the sum of a diagonal operator and a compact operator of arbitrarily small norm.*

Among the important consequences of this theorem is the following:

Theorem 1.2. (Weyl-von Neumann-Berg Theorem) *If N and M are normal, then $N \approx_K M$ if and only if $\sigma_e(N) = \sigma_e(M)$.*

It is clear from (1.1) and Theorem 1.2 that if N and M are normal and $\sigma(N) = \sigma_e(N) = \sigma_e(M) = \sigma(M)$, then $M \in \mathcal{U}(N)^-$ and $M \approx_K N$; actually, a much stronger conclusion is possible:

Theorem 1.3. (Halmos [50]) *If M and N are normal, $\sigma(N) = \sigma_e(N) = \sigma_e(M) = \sigma(M)$, and $\epsilon > 0$, then there exists a unitary operator U such that $M - U^*NU$ is a compact operator of norm less than ϵ, i.e., $M \approx_a N$.*

By combining (1.1) with Theorem 1.3, and by working a bit with the isolated eigenvalues of finite multiplicity, one may conclude that for normal operators, the notions of asymptotic equivalence and approximate equivalence actually coincide:

Proposition 1.4. *If N is normal and $M \in \mathcal{U}(N)^-$, then $M \approx_a N$; in particular, $\mathcal{U}(N)^- \subset \mathcal{U}_K(N)$.*

How are compalence, asymptotic equivalence, and approximate equivalence related for non-normal operators? An important extension of the Weyl-von Neumann-Berg Theorem to non-normal operators occurs in the Brown-Douglas-Fillmore Theorem [23]. Recall that T is *essentially normal* if $\widetilde{T}$ is normal, i.e., $T^*T - TT^*$ is compact. Recall also that for $T \in \mathcal{L}(\mathcal{H})$ and $\lambda \in \mathbb{C} \backslash \sigma_e(T)$, the *index* of the *Fredholm* operator $T - \lambda$ is defined by $\operatorname{ind}(T-\lambda) = \dim\ker(T-\lambda) - \dim(\mathcal{H}/\operatorname{Ran}(T-\lambda))$; the mapping $\lambda \to \operatorname{ind}(T-\lambda)$ is continuous on the *Fredholm domain* $\rho_F(T) \equiv \mathbb{C} \backslash \sigma_e(T)$, and $\operatorname{ind}(T-\lambda)$ is stable under perturbations that are compact or small in norm (see [64]). The theorem that follows extends the Weyl-von Neumann-Berg Theorem by providing the compalence invariant for essentially normal operators.

Theorem 1.5. (Brown-Douglas-Fillmore [23]) *Let T and S be essentially normal. Then the following are equivalent:*

(1) $T \approx_K S$,
(2) $\widetilde{T} \approx \widetilde{S}$,
(3) $\sigma_e(T) = \sigma_e(S)$ *and* $\operatorname{ind}(T-\lambda) = \operatorname{ind}(S-\lambda)(\lambda \in \rho_F(T))$.

The Brown-Douglas-Fillmore Theorem is a cornerstone of the extension theory of C^*-algebra; from this viewpoint, the Weyl-von Neumann-Berg Theorem provides an identity element in $\operatorname{Ext}(C^*(\widetilde{T}))$ for $\widetilde{T}$ normal. An important operator theoretic ingredient in the Brown–Douglass–Fillmore Theorem is the following very useful result (which can also be proved as an immediate corollary of Theorem 1.5).

Corollary 1.6. (Pulling out normal direct summands) (cf. [23, 66, 67, 73]) *Let T be essentially normal and let M be a normal operator such that $\sigma_e(M) \subset \sigma_e(T)$. Then $T \approx_K T \oplus M$.*

Proof: Use Theorem 1.5 and the calculation $\text{ind}((T \oplus M) - \lambda) = \text{ind}(T - \lambda) + \text{ind}(M - \lambda) = \text{ind}(T - \lambda)(\lambda \in \rho_F(T))$. (Here and in the sequel we employ a slight abuse of notation: if $\longleftrightarrow$ is an equivalence relation on $\mathcal{L}(\mathcal{H}), A \in \mathcal{L}(\mathcal{H})$, and $B \in \mathcal{L}(\mathcal{H}_1)$ for some Hilbert space $\mathcal{H}_1$, the notation $A \longleftrightarrow B$ means that there is a Hilbert space isomorphism $U : \mathcal{H} \to \mathcal{H}_1$ such that $A \longleftrightarrow U^*BU$.)

The next two results together provide an analogue of Proposition 1.4 for essentially normal operators.

Proposition 1.7. *If T is essentially normal, then $\mathcal{U}(T)^- \subset \mathcal{U}_K(T), \mathcal{U}_K(T)$ is closed in $\mathcal{L}(\mathcal{H})$, and $\mathcal{U}(\widetilde{T})$ is closed in $\mathcal{A}(\mathcal{H})$.*

Proof: We will prove that $\mathcal{U}(\widetilde{T})$ is closed in $\mathcal{A}(\mathcal{H})$. Suppose $S \in \mathcal{L}(\mathcal{H}), \{\widetilde{U}_n\} \subset \mathcal{U}(\mathcal{A}(\mathcal{H}))$, and $\lim \|\widetilde{U}_n^* \widetilde{T} \widetilde{U}_n - \widetilde{S}\| = 0$. Since $\mathcal{U}(\widetilde{T})^- = \mathcal{U}(\widetilde{S})^-$, then $\widetilde{S}$ is normal and $\sigma(\widetilde{S}) = \sigma(\widetilde{T})$. Let $\lambda \in \rho_F(T)$. Since

$$\lim \|\widetilde{U}_n^*(\widetilde{T} - \lambda)\widetilde{U}_n - (\widetilde{S} - \lambda)\| = 0,$$

there exists $\{K_n\} \subset \mathcal{K}(\mathcal{H})$ such that

$$\lim \|U_n^*(T - \lambda)U_n + K_n - (S - \lambda)\| = 0,$$

whence

$$\begin{aligned} \text{ind}(S - \lambda) &= \lim\ \text{ind}(U_n^*(T - \lambda)U_n + K_n) \\ &= \lim[\text{ind}(U_n^*) + \text{ind}((T - \lambda)) + \text{ind}(U_n)] \\ &= \text{ind}((T - \lambda)). \end{aligned}$$

Theorem 1.5 now implies that $S \approx_K T$, whence $\widetilde{S} \in U(\widetilde{T})$.

For $T \in \mathcal{L}(\mathcal{H})$ we can write $T = T_{\text{nor}} \oplus T_{\text{abnor}}$, where T_{nor} is normal and T_{abnor} has no normal reducing part. The following extension of Proposition 1.4 is due to D. W. Hadwin [45].

Proposition 1.8. *Let T be essentially normal. For $S \in \mathcal{L}(\mathcal{H})$ the following are equivalent:*

(1) $S \in \mathcal{U}(T)^-$;

(2) $S \approx_a T$;

(3)

$$S_{\text{abnor}} \approx T_{\text{abnor}}, \sigma_e(S) = \sigma_e(T), \sigma(S) = \sigma(T),$$

and

$$\dim\ker(S-\lambda) = \dim\ker(T-\lambda)(\lambda \in \sigma(T)\backslash\sigma e(T)).$$

Thus for essentially normal operators we again see the agreement of asymptotic equivalence and approximate equivalence. In [45] Hadwin also determined $\mathcal{U}(T)^-$ in concrete terms for many other types of operators. The extension of these results to arbitrary operators depends on the powerful theorems of D. Voiculescu [79]. Here we state a simplified version of Voiculescu's Theorem that is suitable for the applications we wish to make.

Let $\mathcal{A}$ denote a separable, unital C^*-algebra and let $\rho_1 : \mathcal{A} \to \mathcal{L}(\mathcal{H}_1)$ and $\rho_2 : \mathcal{A} \to \mathcal{L}(\mathcal{H}_2)$ denote unital *-representations of $\mathcal{A}$; ρ_1 and ρ_2 are *approximately unitarily equivalent*, $\rho_1 \approx_a \rho_2$, if there exists a sequence of unitary operators $\{U_k\}, U_k : \mathcal{H}_1 \to \mathcal{H}_2$, such that

$$\lim \|\rho_1(a) - U_k^*\rho_2(a)U_k\| = 0(a \in \mathcal{A})$$

and $\rho_1(a) - U_k^*\rho_2(a)U_k$ is compact for every $a \in \mathcal{A}, k \geq 1$; in particular $\rho_1(a) \approx_a \rho_2(a)(a \in \mathcal{A})$. Let $\pi : \mathcal{L}(\mathcal{H}) \to \mathcal{A}(\mathcal{H})$ denote the canonical projection.

Theorem 1.9. (Voiculescu's Theorem–Special Case) *If $\mathcal{A}$ is a separable C^*-subalgebra of $\mathcal{L}(\mathcal{H})$ and ρ is a unital *-representation of $\pi(\mathcal{A})$ on a separable Hilbert space, then $1_{\mathcal{A}} \approx_a 1_{\mathcal{A}} \oplus \rho \circ \pi$.*

For $T \in \mathcal{L}(\mathcal{H})$, let $\mathcal{A} = C^*(T)$, the C^*-subalgebra generated by T and 1; since $T \approx_a T \oplus \rho(\widetilde{T})$ (for ρ as above), Voiculescu's Theorem implies that each operator is the norm limit of reducible operators, and it thus answers Halmos' 8th Problem affirmatively. Voiculescu's results also imply the existence of an identity element in $\mathrm{Ext}(C^*(\widetilde{T}))$ and thus plays a role analogous to that of the Weyl-von Neumann-Berg Theorem in the essentially normal case. Voiculescu's results have the following remarkable consequence for operator approximation.

Theorem 1.10. [79] *Let $T \in \mathcal{L}(\mathcal{H})$. If $S \in \mathcal{U}(T)^-$, then $S \approx_a T$; in particular, $S \approx_K T$.*

We thus see that asymptotic equivalence and approximate equivalence agree in all cases. D. Hadwin [46] subsequently proved that $S \approx_a T$ if and only if S and T have equal *reducing operator spectra*, and used this result to present a detailed study of closures of unitary orbits.

In [79] D. Voiculescu also characterized the operators with closed unitary orbits: $\mathcal{U}(T)$ is norm closed if and only if $C^*(T)$ is finite dimensional (equivalently, $T \approx A \oplus B^{(\infty)}$, where A and B act on certain finite dimensional spaces [2]). This result has implications concerning local structures of unitary orbits. An element a of a unital C^*-algebra $\mathcal{A}$ has a *local unitary cross section* if the mapping $\gamma : \mathcal{U}(\mathcal{A}) \to \mathcal{U}(a), \gamma(u) = u^*au$, admits a local norm continuous cross section in an open neighborhood in $\mathcal{U}(a)$; in this case, whenever $\{u_n\} \subset \mathcal{U}(\mathcal{A})$ and $\lim \|u_n^* a u_n - a\| = 0$, there exists $\{v_n\} \subset \mathcal{U}(\mathcal{A}), \lim \|v_n - 1\| = 0$, such that $v_n^* a v_n = u_n^* a u_n$ for each n. There are two very general facts known about local unitary cross sections:

Proposition 1.11.

(1) [5, 33] *If $a \in \mathcal{A}$ has a local unitary cross section, then $\mathcal{U}(a)$ is norm closed in $\mathcal{A}$.*

(2) [31] *If $C^*(a)$ is finite dimensional, then a has a local unitary cross section.*

Combining Proposition 1.11 with Voiculescu's result on operators with closed orbits, we obtain the following characterization of operators with local unitary cross sections:

Proposition 1.12. (D. Deckard and L. Fialkow [31]) *For $T \in \mathcal{L}(\mathcal{H})$, the following are equivalent:*

(1) T has a local unitary cross section;

(2) $\mathcal{U}(T)$ is closed.

The situation in the Calkin algebra is unclear. We have seen that $\mathcal{U}(\widetilde{T})$ is norm closed when $\widetilde{T}$ is normal, and K. Davidson has shown that $\mathcal{U}(\widetilde{T})$ is not closed for certain Kakutani-type weighted shifts [29].

Question 1.13.

(1) Which elements of the Calkin algebra have closed unitary orbits?

(2) Which elements of the Calkin algebra have local unitary cross sections?

Our final application of Voiculescu's Theorem provides an indispensable tool for approximation. Let $\sigma_{le}(T)$ denote the *left essential spectrum* of T,

$$\sigma_{le}(T) = \{\lambda \in \mathbb{C} : \widetilde{T} - \lambda \text{ is not left invertible in } \mathcal{A}(\mathcal{H})\}.$$

The following result summarizes some of the innumerable characterizations of $\sigma_{le}(T)$.

Proposition 1.14. (cf. [43]) *For $\lambda \in \mathbb{C}$ and $T \in \mathcal{L}(\mathcal{H})$, the following are equivalent:*

(1) $\lambda \in \sigma_{le}(T)$;

(2) $\operatorname{Ran}(T-\lambda)$ *is not closed or* $\dim\ker(T-\lambda) = \infty$;

(3) *There exists an infinite rank orthogonal projection P such that $(T-\lambda)P$ is compact;*

(4) [22] *Let $\mathcal{J}$ be a 2-sided ideal of $\mathcal{L}(\mathcal{H})$ property containing the ideal of finite rank operators. Given $\epsilon > 0$, there exist a compact operator $K \in \mathcal{J}, \|K\| < \epsilon$, and an orthogonal decomposition $\mathcal{H} = \mathcal{M} \oplus \mathcal{M}^{\perp}$ (with $\mathcal{M}$ and $\mathcal{M}^{\perp}$ infinite dimensional), relative to which the operator matrix of $T - \lambda - K$ is of the form*

$$\begin{pmatrix} 0 & * \\ 0 & * \end{pmatrix}.$$

(5) *For each finite dimensional subspace $\mathcal{M} \subset \mathcal{H}$ and every $\epsilon > 0$, there exists a unit vector $v \in \mathcal{M}^{\perp}$ such that $\|(T-\lambda)v\| < \epsilon$.*

(6) *There exists an orthonormal sequence $\{e_n\}$ such that $\lim \|(T-\lambda)e_n\| = 0$.*

Successive application of Proposition 1.14 yields the next result:

Lemma 1.15. *Let $\{\lambda_i\}$ be a sequence of distinct points of $\sigma_{le}(T)$. Let $\{\epsilon_i\}$ be a sequence of positive numbers (convergent to 0). Then there exists a sequence $\{P_i\}$ of infinite rank orthogonal projections such that $P_iP_j = 0 (i \neq j)$ and $(T-\lambda_i)P_i$ is a trace class operator with trace norm less than ϵ_i.*

Lemma 1.15 can now be used to derive a basic decomposition for operators due to Apostol, Foias, and Voiculescu [7].

Theorem 1.16. *Let $\sigma \subset \sigma_{le}(T)$ and let $\epsilon > 0$. Then there exist a subspace $\mathcal{H}_\epsilon \subset \mathcal{H}$ and a trace class operator K_ϵ of trace norm less than ϵ, such that $\mathcal{H}_\epsilon$ is invariant for $T - K_\epsilon$, and $T_\epsilon \equiv (T - K_\epsilon)|\mathcal{H}_\epsilon$ is a diagonal operator of uniform infinite multiplicity satisfying $\sigma(T_\epsilon) = \sigma_e(T_\epsilon) = \sigma^{-}$.*

Proof: With the notation of Lemma 1.15, let $\{\lambda_i\}$ be a sequence of distinct points dense in σ, let $\epsilon_i = \epsilon/2^i, \mathcal{H}_\epsilon = \sum \oplus P_i\mathcal{H}$, and $K_\epsilon = \sum (T-\lambda_i)P_i$.

From Theorem 1.16, there exists $\{T_n\} \subset \mathcal{L}(\mathcal{H})$ and for each n, a T_n-invariant subspace $\mathcal{M}_n \subset \mathcal{H}$, such that $T_n|\mathcal{M}_n$ is diagonal, $\sigma(T_n|\mathcal{M}_n) = \sigma_e(T_n|\mathcal{M}_n) = \sigma^{-}, T_n - T \in C_1$ (the trace class), and $\lim \|T_n - T\|_1 = 0$ (trace-norm convergence). If we are willing to settle for small-norm perturbations that are compact, but not necessarily trace class, then we can take each T_n to be unitarily equivalent to a *fixed* operator $S \in \mathcal{U}(T)^-$. This result, one of the key applications of Voiculescu's Theorem, is essentially due to C. Apostol [2].

Proposition 1.17. (Apostol's Lemma) *Let $\sigma \subset \sigma_{le}(T)$. Then there exist $S \in \mathcal{L}(\mathcal{H})$, an S-invariant subspace $\mathcal{M}$, and $\{U_n\} \subset \mathcal{U}(\mathcal{L}(\mathcal{H}))$ such that $S|\mathcal{M}$ is diagonal with uniform infinite multiplicity, $\sigma(S|\mathcal{M}) = \sigma_e(S|\mathcal{M}) = \sigma^-$, and $T_n \equiv U_n^* S U_n$ satisfies $\lim \|T_n - T\| = 0$ and $T_n - T \in \mathcal{K}(\mathcal{H})$.*

Proof: Let $\{\lambda_i\}$ be dense in σ and let $\{P_i\}$ be as in Lemma 1.15. Let

$$\mathcal{A} = C^*(T, 1, P_1, P_2, \ldots),$$

let $\rho : \pi(\mathcal{A}) \to \mathcal{L}(\mathcal{H}_\rho)$ be a unital *-representation on a separable space $\mathcal{H}_\rho$, and let $S' = T \oplus \rho^{(\infty)}(\widetilde{T})$. If $Q_i = 0_{\mathcal{H}} \oplus \rho^{(\infty)}(P_i)$, then $Q_i Q_j = 0$ for $i \neq j$ and $(S' - \lambda_i)Q_i = 0$. Let $\mathcal{M}' = \bigvee Q_i \mathcal{H}_\rho$ (closed span). If $S \in \mathcal{L}(\mathcal{H})$ is unitarily equivalent to S' and $\mathcal{M} \subset \mathcal{H}$ corresponds to $\mathcal{M}'$ under this equivalence, then Voiculescu's Theorem implies that S has all the desired properties.

Similarity and Approximation. Historically, Theorem 1.16 played a significant role in the *spectral* characterization of quasitriangular operators. Recall that $T \in \mathcal{L}(\mathcal{H})$ is *quasitriangular* if there exists an increasing sequence $\{P_n\}$ of finite rank orthogonal projections such that $P_n \to 1$ (strongly) and $\lim \|(1-P_n)TP_n\| = 0$ [47]; (T is *triangular* if $(1 - P_n)TP_n = 0 (n \geq 1)$, for in this case T has an upper triangular matrix relative to a suitable orthonormal basis of $\mathcal{H}$.) The set (QT) of all quasitriangular operators in $\mathcal{L}(\mathcal{H})$ is norm-closed and stable under compact perturbation and similarity; each quasitriangular operator is a small-norm compact perturbation of a triangular operator [33, 47, 48].

Recall also that $T \in \mathcal{L}(\mathcal{H})$ is *semi-Fredholm* if Ran T is closed and either ker T or $\mathcal{H}/\text{Ran } T$ is finite dimensional (equivalently, $\widetilde{T}$ is left or right invertible in the Calkin algebra). The Fredholm index (with its stability properties) extends to the semi-Fredholm domain of $T, \rho_{SF}(T) = \{\lambda \in \mathbb{C} : T - \lambda \text{ is semi-Fredholm}\}$. In [33] R. Douglas and C. Pearcy proved that if $T - \lambda$ is semi-Fredholm with negative index, then T is non-quasitriangular. The following theorem of Apostol, Foias, and Voiculescu gives the complete characterization of quasitriangular operators.

Theorem 1.18. [7] *$T \in \mathcal{L}(\mathcal{H})$ is quasitriangular if and only if* $\text{ind}(T - \lambda) \geq 0 (\lambda \in \rho_{SF}(T))$.

Quite apart from the elegance of the quasitriangularity theorem, its importance stems from the fact that its proof introduced a variety of approximation techniques useful in analyzing other similarity-invariant sets of operators. Two such sets are $\mathcal{N}(\mathcal{H})$ and $\mathcal{Q}(\mathcal{H})$, the sets of all nilpotent and quasinilpotent operators in $\mathcal{L}(\mathcal{H})$. In *Ten Problems in Hilbert Space*, Halmos asked for a description of $\mathcal{N}(\mathcal{H})^-$, and, in particular, whether $\mathcal{N}(\mathcal{H})^- \supset \mathcal{Q}(\mathcal{H})$. The latter question was resolved affirmatively by Apostol and Voiculescu [13]. The complete spectral characterization of $\mathcal{N}(\mathcal{H})^-$ is the theorem of Apostol, Foias, and Voiculescu that we present in Lecture 2.

Perhaps the most natural similarity-invariant set of operators is the *similarity orbit* of an operator T,

$$\mathcal{S}(T) = \{X^{-1}TX : S \text{ is an invertible operator } \in \mathcal{L}(\mathcal{H})\}.$$

These lectures primarily concern invariants for membership in $\mathcal{S}(T)^-$. If $S \in \mathcal{S}(T)^-$ we write $T \underset{\text{sim}}{\longrightarrow} S$ (but we say T *dominates* S); this relation is transitive, but clearly not symmetric. If $T \underset{\text{sim}}{\longrightarrow} S$ and $S \underset{\text{sim}}{\longrightarrow} T$, then $\mathcal{S}(T)^- = \mathcal{S}(S)^-$ and we write $S \# T$ (S and T are *asymptotically similar*).

Up to this point we have made only one brief allusion to operators on finite dimensional spaces, but in fact, finite dimensional results and constructions are quite important in operator approximation; typically, these results involve precise numerical estimates, as in the *orbit exchange* technique of I. D. Berg [20]. Try to guess the $\underset{\text{sim}}{\longrightarrow}$ invariant in the finite dimensional case; then try to prove your answer! We will return to the finite dimensional case in Lecture 5.

We wish to make a few preliminary observations about closures of similarity orbits in the general case, but we first need to recall one of the most fundamental tools of approximation, the Riesz-Dunford analytic functional calculus. Let $\mathcal{A}$ denote a unital Banach algebra (we have in mind $\mathcal{L}(\mathcal{H})$ or $\mathcal{A}(\mathcal{H})$). To each $a \in \mathcal{A}$ and every complex function f analytic in an open neighborhood of $\sigma(a)$, the functional calculus associates an element $f(a) \in \mathcal{A}$. The assignment $f \to f(a)$ is a homomorphism that enjoys the following continuity property.

Proposition 1.19. [63] *Let f be an analytic function defined in a neighborhood of $\sigma(a)$. Given $\epsilon > 0$, there exists $\delta > 0$ such that $f(b)$ is well-defined for all b in $\mathcal{A}$ satisfying $\|a - b\| < \delta$, and in this case, $\|f(a) - f(b)\| < \epsilon$.*

Implicit in this result is *upper-semicontinuity of the spectrum*: If $\mathcal{U}$ is an open neighborhood of $\sigma(a)$, then there exists $\delta > 0$ such that if $\|a - b\| < \delta$, then $\sigma(b) \subset \mathcal{U}$.

Suppose that $a \in \mathcal{A}$ and $\sigma(a)$ is the union of disjoint closed sets σ_1 and σ_2. Let $\mathcal{U}_1$ and $\mathcal{U}_2$ denote open neighborhoods of σ_1 and σ_2 such that $\mathcal{U}_1^- \cap \mathcal{U}_2^- = \phi$. Proposition 1.19 implies that if $\|a - b\|$ is small enough, then

$$\sigma(b) \subset \mathcal{U}_1 \cup \mathcal{U}_2, \sigma(b) \cap \mathcal{U}_1 \neq \phi, \sigma(b) \cap \mathcal{U}_2 \neq \phi;$$

this property is called *upper-semicontinuity of the separate parts of the spectrum.* This result follows by considering the analytic function f defined by $f|\mathcal{U}_1 \equiv 1$ and $f|\mathcal{U}_2 \equiv 0$; the indempotent $f(a)$ is called the Riesz idempotent for a associated with σ_1. In the case $\mathcal{A} = \mathcal{L}(\mathcal{H}), a = A \in \mathcal{L}(\mathcal{H})$, we denote the range of this idempotent by $\mathcal{H}(\sigma_1, A)$. The Riesz Decomposition Theorem [70] states that $\mathcal{H}(\sigma_1, A)$ and $\mathcal{H}(\sigma_2, A)$ are complementary closed A-invariant subspaces and that $\sigma(A|\mathcal{H}(\sigma_i, A)) = \sigma_i (i = 1, 2)$.

Suppose $T \underset{\text{sim}}{\longrightarrow} S$ and let $\{X_n\}$ be a sequence of invertible operators such that $\lim \|X_n^{-1} T X_n - S\| = 0$. Clearly, the spectral properties of T are identical to those of each $X_n^{-1} T X_n$. Since the sets of invertible and Fredholm operators are open in $\mathcal{L}(\mathcal{H})$ and $\lim \|X_n^{-1}(T - \lambda)X_n - (S - \lambda)\| = 0 (\lambda \in \mathbb{C})$, it follows that

1.2 $$\sigma(T) \subset \sigma(S) \text{ and } \sigma_e(T) \subset \sigma_e(S).$$

Moreover, by the upper semicontinuity of the separate parts of the spectrum and essential spectrum

1.3 $$\begin{gathered}\text{every component of } \sigma(S) \text{ intersects } \sigma(T), \text{and} \\ \text{every component of } \sigma_e(S) \text{ intersects } \sigma_e(T).\end{gathered}$$

Since the set of semi-Fredholm operators is open,

1.4 $$\rho_{SF}(S) \subset \rho_{SF}(T),$$

and stability of the index implies

1.5 $$\text{ind}(S - \lambda) = \text{ind}(T - \lambda) \text{ for every } \lambda \in \rho_{SF}(S).$$

We will encounter additional invariants for $\underset{\text{sim}}{\longrightarrow}$ as we progress, but it is worth noting that (1.2)–(1.5) completely determine $\mathcal{S}(T)^-$ in the case when T is normal and $\sigma(T)$ is *perfect* (see Theorem 2.9 below).

We next discuss some useful similarity models for operators. Let $T \in \mathcal{L}(\mathcal{H})$ and suppose $\sigma(T) = \sigma_1 \cup \sigma_2$, where σ_1 and σ_2 are disjoint closed sets. By the Riesz Decomposition Theorem there exist complementary T-invariant subspaces $\mathcal{H}_i = \mathcal{H}(\sigma_i, T) (i = 1, 2)$

such that $\sigma(T|\mathcal{H}_i) = \sigma_i$; moreover, the 2×2 operator matrix (T_{ij}) of T relative to the decomposition $\mathcal{H} = \mathcal{H}_1 \oplus \mathcal{H}_1^\perp$ satisfies $T_{21} = 0$ and $\sigma(T_{ii}) = \sigma_i (i = 1, 2)$. If we set

$$J = \begin{pmatrix} 1 & X \\ 0 & 1 \end{pmatrix},$$

then

$$JTJ^{-1} = \begin{pmatrix} T_{11} & T_{12} - (T_{11}X - XT_{22}) \\ 0 & T_{22} \end{pmatrix}.$$

Since T_{11} and T_{22} have disjoint spectra, the mapping $X \to T_{11}X - XT_{22}$ is invertible (Rosenblum's Theorem [71]), so we may choose X (and hence J) so that $JTJ^{-1} = T_{11} \oplus T_{22}$. By an easy induction argument, we obtain the following similarity model for operators with disconnected spectra.

Proposition 1.20. *Let $\sigma_1, \ldots, \sigma_n$ be mutually disjoint closed subsets of the plane and let $\sigma = \cup_{i=1}^n \sigma_i$. If $T \in \mathcal{L}(\mathcal{H})$ satisfies $\sigma(T) = \sigma$, then there is an orthogonal decomposition $\mathcal{H} = \sum_{i=1}^n \oplus \mathcal{H}_i$ relative to which the matrix of $T, (T_{ij})$, is upper triangular and satisfies $\sigma(T_{ii}) = \sigma_i$; moreover, T is similar to $T_{11} \oplus \ldots \oplus T_{nn}$.*

The elasticity of $\underset{\text{sim}}{\longrightarrow}$ is reflected in the next result.

Lemma 1.21. *If the matrix (T_{ij}) of an operator T, relative to a decomposition $\mathcal{H} = \sum_{i=1}^n \oplus \mathcal{H}_i$, is upper triangular, then $T \underset{\text{sim}}{\longrightarrow} T_{11} \oplus \ldots \oplus T_{nn}$.*

Proof: Consider similarities of T implemented by invertible operators of the form $J = \text{diag}(1, \delta, \delta^2, \ldots, \delta^{n-1})$. (This result makes it apparent why $\underset{\text{sim}}{\longrightarrow}$ is not symmetric: consider the case when each $T_{ii} = 0$.)

A classical result of G. Rota [72] shows that if $\sigma(T)$ is contained in the open unit disk, then T is similar to an invariant part of the backward unilateral shift of infinite multiplicity. We require an analogous model for operators with spectra contained in more general domains.

A nonempty bounded open subset $\mathcal{U}$ of the complex plane is a *Cauchy domain* if $\mathcal{U}$ has finitely many components, the closures of any two of which are disjoint, and if Γ (the boundary) consists of a finite number of pairwise disjoint closed rectifiable Jordan curves; $\mathcal{U}$ is an *analytic* Cauchy domain if, additionally, each boundary curve is a regular analytic Jordan curve.

Let $T \in \mathcal{L}(\mathcal{H})$ and suppose $\sigma(T)$ is contained in the analytic Cauchy domain $\mathcal{U}$. Let $\Gamma = \text{bdry}\mathcal{U}$ and let $M(\Gamma)$ denote the normal operator *multiplication by* λ on the

Hilbert space $L^2(\Gamma)$ (defined with respect to arc-length measure). The subspace $H^2(\Gamma)$ spanned by the rational functions with poles outside $\mathcal{U}^-$ is invariant under $M(\Gamma)$. By $M_+(\Gamma)$ and $M_-(\Gamma)$ we denote the restriction of $M(\Gamma)$ to $H^2(\Gamma)$ and, respectively, the compression of $M(\Gamma)$ to $L^2(\Gamma) \ominus H^2(\Gamma)$; thus $M(\Gamma)$ admits an operator matrix relative to the decomposition $L^2(\Gamma) = H^2(\Gamma) \oplus H^2(\Gamma)^\perp$ of the form

$$M(\Gamma) = \begin{pmatrix} M_+(\Gamma) & Z(\Gamma) \\ 0 & M_-(\Gamma) \end{pmatrix}.$$

Direct sums of the operators $M_+(\Gamma)$ and $M_-(\Gamma)$ were used in [7,33] to model operators with prescribed spectral properties. Among the basic features of these operators are the following:

$$\sigma(M(\Gamma)) = \sigma_e(M(\Gamma)) = \sigma_e(M_+(\Gamma)) = \sigma_e(M_-(\Gamma)) = \Gamma;$$
$$\sigma(M_+(\Gamma)) = \sigma(M_-(\Gamma)) = \mathcal{U}^-;$$

1.6

$M_+(\Gamma)$ is subnormal; $\dim\ker(M_+(\Gamma) - \lambda) = 0$ and

$\operatorname{ind}(M_+(\Gamma) - \lambda) = -1$ for every $\lambda \in \mathcal{U}$.

$M_-(\Gamma)$ is co − subnormal $\dim\ker(M_-(\Gamma) - \lambda)^* = 0$ and

$\operatorname{ind}(M_-(\Gamma) - \lambda) = 1$ for every $\lambda \in \mathcal{U}$.

$M_+(\Gamma)$ and $M_-(\Gamma)$ are essentially normal.

The following similarity model is due to Herrero [58] and Voiculescu [78].

Theorem 1.22. (Generalized Rota Models) *Let $T \in \mathcal{L}(\mathcal{H})$ and suppose $\sigma(T)$ is contained in the analytic Cauchy domain $\mathcal{U}$. Consider $M_+(\Gamma) \otimes 1$ and $1 \otimes T$ acting on the Hilbert space $H^2(\Gamma) \otimes \mathcal{H}$. Let $\mathcal{R} = \operatorname{Ran}(M_+(\Gamma) \otimes 1 - 1 \otimes T)$; then $\mathcal{R}$ is invariant for $M_+(\Gamma) \otimes 1, (M_+(\Gamma) \otimes 1)|\mathcal{R}$ is similar to $M_+(\Gamma) \otimes 1$, and the compression of $M_+(\Gamma) \otimes 1$ to $\mathcal{R}^\perp$ is similar to T.*

In the classical case, $\mathcal{U} = \mathbb{D}$ (the open unit disk), $M(\Gamma)$ is the bilateral shift, $M_+(\Gamma)$ is the unilateral shift, $M_-(\Gamma)$ is the adjoint of the unilateral shift, and Theorem 1.22 yields Rota's Theorem (applied to T^*).

Relative to the decomposition $H^2(\Gamma) \otimes \mathcal{H} = \mathcal{R} \oplus \mathcal{R}^\perp$,

1.7 $$M_+(\Gamma) \otimes 1 = \begin{pmatrix} M_+(\Gamma) \otimes 1|\mathcal{R} & M_{12} \\ 0 & M_{22} \end{pmatrix},$$

where M_{22} is similar to T (notation: $M_{22} \sim T$). If $r(z)$ is a rational function with poles outside $\mathcal{U}^-$, then

$$\|r(M_{22})\| \leq \|r(M_+(\Gamma) \otimes 1)\| = \|r(M_+(\Gamma))\| = \max\{|r(\lambda)| : \lambda \in \Gamma\}.$$

Thus $\mathcal{U}^-$ is a *spectral set* for M_{22}. Since, for every bounded open neighborhood $\mathcal{O}$ of $\sigma(T)$, there exists an analytic Cauchy domain $\mathcal{U}$ such that $\sigma(T) \subset \mathcal{U} \subset \mathcal{U}^- \subset \mathcal{O}$, we have the following result:

Corollary 1.23.

(i) [58] *Given $T \in \mathcal{L}(\mathcal{H})$ and a bounded open neighborhood $\mathcal{O}$ of $\sigma(T), \mathcal{O}^-$ is a spectral set for some operator similar to T.*

(ii) (Rota's Corollary) *If $T \in \mathcal{L}(\mathcal{H})$ is quasinilpotent, then $T \underset{\text{sim}}{\longrightarrow} 0$.*

Corollary 1.24. $M_+(\Gamma) \otimes 1 \underset{\text{sim}}{\longrightarrow} (M_+(\Gamma) \otimes 1) \oplus T$.

Proof: Lemma 1.2 and (1.7).

The next result illustrates how the assumption that an operator has perfect spectrum can be exploited in studying closures of similarity orbits. In the sequel, for a compact subset of the plane σ and $\epsilon > 0, [\sigma]_\epsilon$ denotes the ϵ-neighborhood of σ.

Proposition 1.25. [51] *If M is a normal operator with perfect spectrum, K is compact, and $\sigma(M + K) = \sigma(M)$, then $M \underset{\text{sim}}{\longrightarrow} M + K$.*

Proof: Let N be a diagonal operator of uniform infinite multiplicity such that $\sigma(N) = \sigma(M)$. Suppose for a moment that $N \underset{\text{sim}}{\longrightarrow} N + C$ for every compact operator C such that $\sigma(N + C) = \sigma(N)$. Since $\sigma_e(N) = \sigma(N) = \sigma(M) = \sigma_e(M)$, Theorem 1.3 implies $N \approx_a M$; in particular, there exist a unitary operator U and a compact operator J such that $U^*MU = N + J$. If $K \in \mathcal{K}(\mathcal{H})$ satisfies $\sigma(M + K) = \sigma(M)$, let $C = J + U^*KU$; note that $N + C = U^*MU - J + C = U^*(M + K)U$, whence $\sigma(N + C) = \sigma(M + K) = \sigma(M) = \sigma(N)$. Thus

$$M \approx_a N \underset{\text{sim}}{\longrightarrow} N + C \approx M + K.$$

It thus suffices to prove that $N \underset{\text{sim}}{\longrightarrow} N + C$ whenever $C \in \mathcal{K}(\mathcal{H})$ satisfies $\sigma(N + C) = \sigma(N)$. Suppose C is such an operator. Let $\{e_n\}$ be an orthonormal basis of $\mathcal{H}$ consisting of eigenvectors of N. Let $P_n = P_{<e_1,\ldots,e_n>}$; since C is compact, $\lim \|(N + C) - (N + P_mCP_m)\| = 0$, so it suffices to show that $\lim_{m\to\infty} \operatorname{dist}(\mathcal{S}(N), N + P_mCP_m) = 0$.

Let $\epsilon > 0$; by upper semicontinuity of the spectrum, we can choose m so that $\|(N+C)-(N+P_mCP_m)\| < \epsilon$ and $\sigma(N+P_mCP_m) \subset [\sigma(N+C)]_\epsilon = [\sigma(N)]_\epsilon$; in particular, $T_m \equiv (N+P_mCP_m)|P_m\mathcal{H}$ satisfies $\sigma(T_m) \subset [\sigma(N)]_\epsilon$. Now T_m has an upper triangular matrix relative to a suitable orthonormal basis of $P_m\mathcal{H}$. Let $D_m \in \mathcal{L}(P_m\mathcal{H})$ denote the diagonal matrix corresponding to the diagonal of T_m in this basis; thus $\sigma(D_m) \subset [\sigma(N)]_\epsilon$. Since N is diagonal and $\sigma(N)$ is perfect, there exist m distinct eigenvalues of (N), $z_1,\ldots,z_m$, such that $\|D_m - \text{diag}(z_1,\ldots,z_m)\| < \epsilon$. Proposition 1.19 (and its proof) imply that there is an invertible operator $X_m \in \mathcal{L}(P_m\mathcal{H})$ such that

$$\|X_m^{-1}\text{diag}(z_1,\ldots,z_m)X_m - T_m\| < \epsilon.$$

Now $N \approx \text{diag}(z_1,\ldots,z_m) \oplus N \sim X_m^{-1}\text{diag}(z_1,\ldots,z_m)X_m \oplus N$; the latter operator differs in norm by less than ϵ from $T_m \oplus N$; since $T_m \oplus N \approx P_m(N+C)P_m|P_m\mathcal{H} \oplus N \approx N + P_mCP_m$, the result follows.

The following corollary will be useful in the sequel.

Corollary 1.26. *If $T \epsilon \mathcal{L}(\mathcal{H}), \mathcal{U}$ is an analytic Cauchy domain containing $\sigma(T)$, and M is a normal operator such that $\sigma(M) = \mathcal{U}^-$, then $M \underset{\text{sim}}{\longrightarrow} M \oplus M_+(\Gamma)^{(\infty)} \oplus T$ and $M \underset{\text{sim}}{\longrightarrow} M \oplus M_-(\Gamma)^{(\infty)}$.*

Proof: From (1.6), $M \oplus M_+(\Gamma)$ is essentially normal and $\sigma(M \oplus M_+(\Gamma)) = \sigma_e(M \oplus M_+(\Gamma)) = \mathcal{U}^-(=\sigma_e(M))$. The Brown-Douglas-Fillmore Theorem implies $M \oplus M_+(\Gamma) \approx_K M$. Since $\sigma(M)$ is perfect, Proposition 1.25 implies $M \underset{\text{sim}}{\longrightarrow} M \oplus M_+(\Gamma)$, and thus $M \approx_a M^{(\infty)} \underset{\text{sim}}{\longrightarrow} [M \oplus M_+(\Gamma)]^{(\infty)} \approx M^{(\infty)} \oplus M_+(\Gamma)^{(\infty)} \approx_a M \oplus M_+(\Gamma)^{(\infty)} \underset{\text{sim}}{\longrightarrow} M \oplus M_+(\Gamma)^{(\infty)} \oplus T$ (Corollary 1.24). The proof of the second part is similar.

In addition to Rota-type models, there are important matrix models based on detailed knowledge of the semi-Fredholm behavior of operators. C. Apostol's basic result in [1] associates to each operator a 3×3 upper triangular operator matrix reflecting this behavior. This matrix form is an important tool in studying the effect of compact perturbation on the spectral properties of operators. Here we cite one such result that we need in the sequel.

For $\sigma \subset \mathbb{C}$, let $[\sigma]_{\text{isol}}$ denote the set of isolated points of σ. For $T \in \mathcal{L}(\mathcal{H})$, let $\sigma_0(T) = \{\lambda \in [\sigma(T)]_{\text{isol}} : \dim \mathcal{H}(\{\lambda\}, T) < \infty\} (= [\sigma(T)]_{\text{isol}} \cap \rho_F(T))$. For $\lambda \in \rho_{SF}(T)$, let $\text{min.ind.}(T-\lambda) = \min(\dim\ker(T-\lambda), \dim \mathcal{H}/\text{Ran}(T-\lambda))$. Let $\sigma_{\ell re}(T) = \sigma_{\ell e}(T) \cap \sigma_{re}(T)$.

Theorem 1.27. [1] (cf. [8, Theorem 2.8], [76]) *Let $T\epsilon\mathcal{L}(\mathcal{H})$ and let $\epsilon > 0$; then there exists $K\epsilon\mathcal{K}(\mathcal{H})$ such that:*

$$\|K\| < \epsilon + \max\{\text{dist}[\lambda, \text{bdry } \rho_{SF}(T)] : \lambda \in \sigma_0(T)\}$$

and min.ind.$(T + K - \lambda) = 0$ *for all* $\lambda \in \rho_{SF}(T)$.

Suppose $\sigma_0(T) = \phi$ and $\text{ind}(T - \lambda) = 0(\lambda\epsilon\rho_{SF}(T))$. Let K be as in Theorem 1.27, $\|K\| < \epsilon$. For $\lambda\epsilon\rho_{SF}(T)$, since min.ind.$(T + K - \lambda) = 0$, then $T + K - \lambda$ is invertible. Thus $\sigma(T + K) = \sigma_{\ell re}(T) = \sigma_{\ell re}(T + K)$.

Finally, we note a result (closely related to (1.1)) concerning the distance between unitary orbits of normal operators. In the sequel, $\text{dist}_H(\cdot,\cdot)$ denotes distance in the Hausdorff metric.

Proposition 1.28. *If M and N are normal, $\sigma_e(M) = \sigma(M), \sigma_e(N) = \sigma(N)$, and* $\text{dist}_H(\sigma(M), \sigma(N)) < \epsilon$, *then* $\text{dist}(M, \mathcal{U}(N)) < \epsilon$ *and* $\text{dist}(N, \mathcal{U}(M)) < \epsilon$.

2. Closure of Nilpotent Operators

Our goal in this lecture is to derive the spectral characterization of $\mathcal{N}(\mathcal{H})^-$ due to C. Apostol, C. Foias, and D. Voiculescu [8]. The proof we present, essentially that given by D. Herrero in [63], illustrates several important approximation techniques, including the use of generalized Rota models and Voiculescu's Theorem, and it demonstrates the surprising interplay between normal operators and nilpotent operators in operator approximation. As in [63], this approach permits one to quickly derive D. Herrero's characterization of $\mathcal{S}(N)^-$ for a normal operator N with perfect spectrum [57], and also D. Voiculescu's spectral characterization of the norm closure of algebraic operators [78].

Theorem 2.1. *$\mathcal{N}(\mathcal{H})^-$ coincides with the set of all operators T in $\mathcal{L}(\mathcal{H})$ satisfying the following properties:*

(1) $\sigma(T)$ is connected and contains the origin;
(2) $\sigma_e(T)$ is connected and contains the origin;
(3) $\text{ind}(T - \lambda) = 0$ *for all* $\lambda\epsilon\rho_{SF}(T)$.

(In view of Theorem 1.18, condition (3) is equivalent to the property that T is biquasitriangular.)

The necessity of (1)-(3) follows directly from upper-semicontinuity of the spectrum and essential spectrum, and continuity of the index. To see this, first note that if $T\epsilon\mathcal{L}(\mathcal{H})$, then $\sigma_{\ell e}(T) \neq \phi$ and $\sigma_{re}(T) \neq \phi$ [43]. Thus, if $Q\epsilon\mathcal{N}(\mathcal{H})$, then $\{0\} = \sigma(T) = \sigma_{\ell re}(T)$. Suppose $T\epsilon\mathcal{N}(\mathcal{H})^-$ and let $\{N_k\} \subset \mathcal{N}(\mathcal{H})$ satisfy $\lim \|N_k - T\| = 0$. If T were invertible, then N_k would be invertible for large k; thus $0\epsilon\sigma(T)$. By upper semicontinuity of spectra, if $\sigma(T)$ were disconnected, then $\sigma(N_k)$ would be disconnected for large k; thus $\sigma(T)$ is connected. Similarly, $\sigma_e(T)$ is connected and contains the origin. Moreover, if $T - \lambda$ is semi-Fredholm, then so is $N_k - \lambda$ for large k, so $\lambda \neq 0$, and thus $\text{ind}(T - \lambda) = \lim \text{ind}(N_k - \lambda) = 0$. Thus (1)-(3) hold.

The proof of sufficiency requires several preliminary results that we now develop. An early example of a non-quasinilpotent operator in $\mathcal{N}(\mathcal{H})^-$ is due to Kakutani (see [48]): Let W be the unilateral weighted shift defined on the orthonormal basis $\{e_n\}_{n=1}^{\infty}$ by

$$We_n = \alpha_n e_{n+1} \quad (n \geq 1),$$

where $\{\alpha_n\}$ is given by

$$1, 1/2, 1, 1/4, 1, 1/2, 1, 1/8, 1, 1/2, 1, 1/4, 1, 1/2, 1, 1/16, \ldots .$$

Let W_n be the nilpotent obtained from W be replacing $1/2^n$ with 0. Then $\lim \|W_n - W\| = 0, \|W\| = 1$, and $0 < r(W) < 1$. The first systematic study of $\mathcal{N}(\mathcal{H})^-$ is apparently due to J. Hedlund [53], who exhibited a class of partial isometries in $\mathcal{N}(\mathcal{H})^-$ with spectra the closed unit disk; indeed, the operator

$$V = \sum_{k=1}^{\infty} \oplus q_k^{(\infty)}$$

belongs to this class (where q_k is the Jordan k-cell on $\mathbb{C}^k$).

Hedlund also studied the normal operators in $\mathcal{N}(\mathcal{H})^-$ and suggested that there are no nonzero normals in $\mathcal{N}(\mathcal{H})^-$. Using Hedlund's own results, however, D. Herrero [54] deduced the striking fact that $\mathcal{N}(\mathcal{H})^-$ does contain nonzero normal operators. (The complete description of such normal operators is given below in Theorem 2.6.)

To sketch the idea, let V be Hedlund's operator. Let $\epsilon > 0$ and let $\sigma_\epsilon(z)$ be the conformal mapping of the unit disk onto the interior of the ellipse with vertices $\{-1, 1, i\epsilon, -i\epsilon\}, \varphi_\epsilon(0) = 0$. Let $W_\epsilon = \varphi_\epsilon(V)$. Considerations with the series representation of φ_ϵ imply $W_\epsilon \in \mathcal{N}(\mathcal{H})^-$, $\|ReW_\epsilon\| = 1, \|ImW_\epsilon\| = \epsilon$, and $\sigma(W_\epsilon) = \varphi_\epsilon(\mathbb{D})^-$.

Let M be a normal operator with $\sigma(M) = [-1, 1]$. Upper semicontinuity of spectra implies

$$\lim_{\epsilon \to 0} \operatorname{dist}_H(\sigma(M), \sigma(Re(W_\epsilon))) = 0.$$

Proposition 1.28 implies $\lim_{\epsilon \to 0}(M, \mathcal{U}(Re(W_\epsilon))) = 0$, whence $\lim_{\epsilon \to 0} \operatorname{dist}(M, \mathcal{U}(W_\epsilon)) = 0$. Since $W_\epsilon \in \mathcal{N}(\mathcal{H})^-$, it follows that $M \in \mathcal{N}(\mathcal{H})^-$.

At least two other proofs of this result are known. The proof in [8] uses the fact that the Volterra operator possesses an nth root V_n such that $\lim \|Im(V_n)\|/\|Re(V_n)\| = 0$. The proof in [63] depends on Berg's orbit exchange technique.

Several approximation results rely on the fact that $\mathcal{N}(\mathcal{H})^-$ contains every normal operator whose spectrum is the closed unit disk:

Lemma 2.2. *If N is normal and $\sigma(N) = \mathbb{D}^-$, then $N \in \mathcal{N}(\mathcal{H})^-$.*

Proof: Let M be a normal operator with $\sigma(M) = [-1, 1]$; thus $M \in \mathcal{N}(\mathcal{H})^-$. Let f denote a continuous mapping of $[-1, 1]$ onto $\mathbb{D}$ such that $f(0) = 0$; we may uniformly approximate f by polynoimials $\{p_n\}$ without constant term. Since $M \in \mathcal{N}(\mathcal{H})^-$, each $p_n(M) \in \mathcal{N}(\mathcal{H})^-$, and thus $L \equiv f(M) = (\text{norm}) \lim p_n(M)$ is in $\mathcal{N}(\mathcal{H})^-$ and $\sigma(L) = \mathbb{D}^-$. Since $L \approx_a N$, the result follows.

Lemma 2.3. *Let $N \in \mathcal{L}(\mathcal{H})$ be a normal operator such that $\sigma_e(N)$ has an accumulation point λ, and let $Q \in \mathcal{N}(\mathcal{H})$; then $N \underset{\text{sim}}{\longrightarrow} N \oplus (\lambda + Q)$.*

Proof: Suppose Q is nilpotent of order $k \geq 1$, and consider the orthogonal decomposition

$$\mathcal{H} = \sum_{j=1}^{k} \oplus \mathcal{M}_j, \mathcal{M}_j = \ker Q^j \ominus \ker Q^{j-1} \quad (1 \leq j \leq k). \tag{2.1}$$

Let $\{\lambda_n\}_{n=1}^{\infty}$ be a sequence of distinct points of $\sigma_e(N)$ such that $\lambda_n \to \lambda$. Let

$$D_n = \sum_{j=1}^{k} \oplus \lambda_{n+j} 1_{\mathcal{M}_j}$$

relative to (2.1) ($n \geq 1$); since the operator matrix of Q relative to (2.1) is strictly upper triangular, $D_n + Q \sim D_n$ (Proposition 1.20). Since $\{\lambda_n\} \subset \sigma_e(N)$, (1.1) implies

$$N \approx_a N \oplus D_n \sim N \oplus (D_n + Q) \underset{\text{norm}}{\longrightarrow} N \oplus (\lambda + Q).$$

Let $K \subset \mathbb{C}$ be compact. We say that a compact set $L \subset \mathbb{C}$ is *obtained from K by disks* if there exists a finite collection of open disks $\{D_i\}_{i=1}^{p}$ (of possibly differing radii) such that each D_i is centered at a point of K, $K \subset \bigcup_{i=1}^{p} D_i$, and $L = \bigcup_{i=1}^{p} D_i^-$. Recall that K is *perfect* if K has no isolated points.

Lemma 2.4. *Suppose N is normal and $\sigma(N)$ is perfect. If M is a normal operator whose spectrum is obtained from $\sigma(N)$ by disks, then $N \underset{\text{sim}}{\longrightarrow} M$.*

Proof: There exists disks of the form $D_j \equiv D(\lambda_j, \epsilon_j), \lambda_j \epsilon \sigma(N), \epsilon_j > 0 (1 \leq j \leq p)$, such that $\sigma(N) \subset \bigcup_{j=1}^{p} D_j$ and $\sigma(M) = \cup_{j=1}^{p} D_j^-$. Let M_j be a normal operator such that $\sigma(M_j) = D_j^-$ and let

$$M' = \sum_{j=1}^{p} \oplus M_j; \sigma(M') = \sigma_e(M') = \sigma(M) = \sigma_e(M).$$

Let Q_j be a nilpotent operator. Since $\sigma(N)$ is perfect, each λ_j is an accumulation point of $\sigma_e(N)$, so Lemma 2.3 implies

$$N \approx_a \sum_{j=1}^{p} \oplus N \underset{\text{sim}}{\longrightarrow} \sum_{j=1}^{p} \oplus (N \oplus (\lambda_j + Q_j)) \approx_a N \oplus \sum_{j=1}^{p} \oplus (\lambda_j + Q_j).$$

Lemma 2.2 shows that for $\delta > 0, Q_j$ can be chosen so that $\|(\lambda_j + Q_j) - M_j\| < \delta$, and thus $N \underset{\text{sim}}{\longrightarrow} N \oplus M' \approx_a M$.

Lemma 2.5. *If M and N are normal operators such that $\sigma(N)$ is perfect, $\sigma(M) \supset \sigma(N)$, and each component of $\sigma(M)$ intersects $\sigma(N)$, then $N \underset{\text{sim}}{\longrightarrow} M$.*

Proof: Let $\epsilon > 0$. Since $\sigma(N)$ is perfect, $\sigma(N) \subset \sigma(M)$, and each component of $\sigma(M)$ intersects $\sigma(N)$, a compactness argument implies that there exist compact sets $K_0, K_1, \ldots, K_m, K_0 = \sigma(N)$, such that $\sigma(N) \subset K_1 \subset \cdots \subset K_m$, each K_{i+1} is obtained from K_i by disks, and $D_H(K_m, \sigma(M)) < \epsilon$. Now $\sigma(N)$ is perfect, and so is each K_i. Let N_i be a normal operator such that $\sigma(N_i) = K_i (1 \leq i \leq m)$. Successive application of Lemma 2.4 shows that

$$N \underset{\text{sim}}{\longrightarrow} N_1 \underset{\text{sim}}{\longrightarrow} \ldots \underset{\text{sim}}{\longrightarrow} N_m,$$

so $N \underset{\text{sim}}{\longrightarrow} N_m$. Since $d_H(K_m, \sigma(M)) < \epsilon$, then $\text{dist}(\mathcal{U}(N_m), M) < \epsilon$ (Proposition 1.28), so $\text{dist}(\mathcal{S}(N), M) \leq \text{dist}(\mathcal{U}(N_m), M) < \epsilon$ and the result follows.

The following characterization of the normal operators in $\mathcal{N}(\mathcal{H})^-$ is due to D. Herrero [54]; the following proof, also due to Herrero [63], differs from that in [54].

Theorem 2.6. *A normal operator M belongs to $\mathcal{N}(\mathcal{H})^-$ if and only if $\sigma(M)$ is connected and contains the origin.*

Proof: The necessity of the conditions is clear. For sufficiency, let $\epsilon > 0$ and let K_ϵ be obtained from $\sigma(M)$ by disks of radius $\epsilon/2$. Let M_ϵ be a normal operator such that $\sigma(M_\epsilon) = K_\epsilon$. Since $0 \in \sigma(M) \subset \text{interior}(K_\epsilon)$, there exists $\delta > 0$ such that $D(0,\delta)^- \subset \text{interior}(K_\epsilon)$. If N is a normal operator with $\sigma(N) = D(0,\delta)^-$, then $N \epsilon \mathcal{N}(\mathcal{H})^-$ (Lemma 2.2). Lemma 2.5 implies $M_\epsilon \in \mathcal{S}(N)^-$, so it follows that $M_\epsilon \in \mathcal{N}(\mathcal{H})^-$. Since $\text{dist}(M, \mathcal{U}(M_\epsilon)) < \epsilon$, then $\text{dist}(M, \mathcal{N}(\mathcal{H})) < \epsilon$ and the result follows.

Remark. We included Lemma 2.5 because it is needed in the sequel. A more direct proof of Theorem 2.6 starts with the fact that $\mathcal{N}(\mathcal{H})^-$ contains a normal operator N with $\sigma(N) = [-1.1]$. Since the operator M of Theorem 2.6 can be norm approximated by operators unitarily equivalent to finite direct sums $\sum \oplus p_i(N)$, where p_i is a polynomial without constant term, then $M \in \mathcal{N}(\mathcal{H})^-$ [8].

Recall that an operator A is algebraic if $p(A) = 0$ for some nonzero polynomial $p(z)$. If $\lambda_1, \ldots, \lambda_n$ are the distinct roots of the minimal polynomial of the algebraic operator A, then Proposition 1.20 or [49] implies that there exists an orthogonal decomposition

$$\mathcal{H} = \mathcal{H}_1 \oplus \ldots \oplus \mathcal{H}_n,$$

and nilpotent operators $Q_i \in \mathcal{L}(\mathcal{H}_i)(1 \le i \le n)$, such that

$$A \sim \sum_{i=1}^{n} \oplus(\lambda_i + Q_i).$$

Thus if N is normal, $\sigma(N)$ is perfect, and each $\lambda_i \in \sigma(N)(= \sigma_e(N))$, an induction argument based on Lemma 2.3 implies

$$N \underset{\text{sim}}{\longrightarrow} N \oplus \sum \oplus(\lambda_i + Q_i) \sim N \oplus A.$$

We next extend this result from algebraic operators to arbitrary operators with spectra contained in $\sigma(N)$.

Lemma 2.7. *Let N be a normal operator with perfect spectrum. If $T \in \mathcal{L}(\mathcal{H})$ and $\sigma(T) \subset \sigma(N)$, then $N \underset{\text{sim}}{\longrightarrow} N \oplus T$.*

Proof: Let $\epsilon > 0$ and let $\mathcal{U}_1, \mathcal{U}$ be analytic Cauchy domains such that $\sigma(N) \subset \mathcal{U}_1 \subset \mathcal{U}_1^- \subset \mathcal{U} \subset [\sigma(N)]_{\epsilon/2}$ and such that each component of $\mathcal{U}$ intersects $\sigma(N)$. Let M_1, M be normal operators such that $\sigma(M_1) = \mathcal{U}_1^-, \sigma(M) = \mathcal{U}^-$. Since $d_H(\sigma(N), \sigma(M)) < \epsilon$, Proposition 1.28 implies $\text{dist}(M, \mathcal{U}(N)) < \epsilon$, so it suffices to show that $N \underset{\text{sim}}{\longrightarrow} T \oplus M$. (1.1) and (1.6) imply that

$$M \approx_a M \oplus [(M \oplus M_1) \oplus M(\Gamma)^{(\infty)}], \text{ where } \Gamma = \text{bdry}(\mathcal{U}_1).$$

Now

$$M \oplus M_1 \oplus M(\Gamma)^{(\infty)} \approx M \oplus M_1 \oplus \begin{bmatrix} M_-(\Gamma)^{(\infty)} & 0 \\ Z(\Gamma)^{(\infty)} & M_+(\Gamma)^{(\infty)} \end{bmatrix}$$
$$\approx S \equiv \begin{pmatrix} M \oplus M_1 \oplus M_-(\Gamma)^{(\infty)} & 0 \\ Z' & M_+(\Gamma)^{(\infty)} \end{pmatrix}.$$

It thus suffices to prove that $N \underset{\text{sim}}{\longrightarrow} T \oplus M \oplus S$.

Lemma 2.5 implies

(i) $N \underset{\text{sim}}{\longrightarrow} M \oplus M_1$, and the above remarks imply

(ii) $M \underset{\text{sim}}{\longrightarrow} M \oplus A$ for every algebraic operator A with $\sigma(A) \subset \sigma(M)$. Moreover, Corollary 1.26 implies

(iii) $M_1 \underset{\text{sim}}{\longrightarrow} M_1 \oplus M_+(\Gamma)^{(\infty)} \oplus T$.

Thus (i)-(iii) imply

(iv) $N \underset{\text{sim}}{\longrightarrow} M \oplus A \oplus M_1 \oplus M_+(\Gamma)^{(\infty)} \oplus T \approx_a M \oplus T \oplus A \oplus M_+(\Gamma)^{(\infty)}$ (A algebraic, $\sigma(A) \subset \sigma(M)$).

Let us assume for a moment that there exists a sequence of algebraic operators, $\{A_k\}$, such that

(v) $\sigma(A_k) \subset \sigma(M)$ $(k \geq 1)$;

(vi) $\|A_k - M \oplus M_1 \oplus M_-(\Gamma)^{(\infty)}\| \to 0$ (as $k \to \infty$);

(vii) $\sigma(A_k) \cap \sigma(M_+(\Gamma)^{(\infty)}) = \phi$ $(k \geq 1)$.

Proposition 1.20 and (vii) imply

$$A_k \oplus M_+(\Gamma)^{(\infty)} \sim \begin{pmatrix} A_k & 0 \\ Z' & M_+(\Gamma)^{(\infty)} \end{pmatrix}$$

so from (iv)-(vi) we see that

$$N \underset{\text{sim}}{\longrightarrow} M \oplus T \oplus S, \text{ as desired.}$$

To complete the proof it remains to exhibit a sequence $\{A_k\}$ as above. Note that $\mathcal{U} \backslash \mathcal{U}_1^-$ has finitely many components, say $\mathcal{O}_1, \ldots, \mathcal{O}_p$. Let $\lambda_i \in \mathcal{O}_i$, let M_i be a normal operator with $\sigma(M_i) = \mathcal{O}_i^-$, and let $M' = M_1 \oplus \ldots \oplus M_p$. Theorem 2.6 implies that there exist nilpotent operators $Q_{i,k}$ such that

$$\|M_i - (\lambda_i + Q_{i,k})\| < \frac{1}{k}, 1 \leq i \leq p, 1 \leq k.$$

Let

$$B_k = \sum_{i=1}^{p} \oplus(\lambda_i + Q_{i,k});$$

clearly, B_k is algebraic, $\sigma(B_k) \subset \mathcal{U}\backslash\mathcal{U}_1^-$, and $\lim \|B_k - M'\| = 0$; in particular

$$\sigma(B_k) \subset \sigma(M)\backslash\sigma\left(M_+(\Gamma)^{(\infty)}\right).$$

Since each component of $\sigma(M)$ intersects $\sigma(M')$, Lemma 2.5 implies $M' \underset{\text{sim}}{\longrightarrow} M$, so via Corollary 1.26,

$$M' \underset{\text{sim}}{\longrightarrow} M \approx_a M \oplus M_1 \underset{\text{sim}}{\longrightarrow} M \oplus M_1 \oplus M_-(\Gamma)^{(\infty)}.$$

Since $\lim \|B_k - M'\| = 0$, it follows that there exist operators $A_k \sim B_k$ satisfying (v)-(vii). The proof is complete.

Proof of Theorem 2.1. To complete the proof of Theorem 2.1, suppose $A \in \mathcal{L}(\mathcal{H})$ satisfies (i)-(iii) of the hypothesis and let $\epsilon > 0$. Since $\sigma_0(A) = \phi$, Theorem 1.27 and (iii) imply that there exists $K_1 \in \mathcal{K}(\mathcal{H}), \|K_1\| < \epsilon/4$, such that $A_1 \equiv A - K_1$ satisfies $\sigma(A_1) = \sigma_{\text{lre}}(A_1)(= \sigma_{\text{lre}}(A) = \sigma_e(A))$. Thus $\sigma(A_1)$ is connected and contains the origin.

Let ρ be a faithful unital *-represenation of $C^*(\widetilde{A}_1)$ on a separable Hilbert space $\mathcal{H}_\rho$, and let $A_2' = \rho(\widetilde{A}_1)$. Let $A_2 = (A_2')^{(\infty)}$, so that $\sigma_{\text{lre}}(A_2) = \sigma(A_2) = \sigma(A_1)$. Voiculescu's Theorem implies that there exists $K_2 \in \mathcal{K}(\mathcal{H}), \|K_2\| < \epsilon/4$, such that

$$A_3 \equiv A_1 - K_2 \approx A_1 \oplus A_2.$$

Let N be a normal operator such that $\sigma(N) = \sigma_{\text{lre}}(A_1)(= \sigma(A_1))$; an application of Proposition 1.17 implies that there exist compact operator L_1, L_2, both of norm less than $\epsilon/4$, such that

$$A_1 - L_1 \approx \begin{pmatrix} N & C_1 \\ 0 & B_1 \end{pmatrix} \text{ and } A_2 - L_2 \approx \begin{pmatrix} B_2 & C_2 \\ 0 & N \end{pmatrix}$$

If $K = K_1 + K_2 + (L_1 \oplus L_2)$, then $\|K\| < \epsilon$ and

$$\text{2.2} \qquad A - K \approx (A_1 - L_1) \oplus (A_2 - L_2) \approx \begin{pmatrix} N & 0 & C_1 & 0 \\ 0 & B_2 & 0 & C_2 \\ 0 & 0 & B_1 & 0 \\ 0 & 0 & 0 & N \end{pmatrix}.$$

Note that if Q_1 and Q_2 are nilpotent, then so is

$$\begin{pmatrix} Q_1 & C_1 \oplus C_2 \\ 0 & Q_2 \end{pmatrix}.$$

It thus suffices to verify that $N \oplus B_i \in \mathcal{N}(\mathcal{H})^-(i = 1, 2)$. Since $\sigma(N) = \sigma(A_1)$ (a set that is connected and contains the origin), Theorem 2.6 implies $N \in \mathcal{N}(\mathcal{H})^-$. Now $\sigma(B_i) \subset \sigma(N)$, so Lemma 2.7 implies $N \underset{\text{sim}}{\longrightarrow} N \oplus B_i$; thus $N \oplus B_i \in \mathcal{N}(\mathcal{H})^-$ and the proof is complete.

With the techniques now at hand we may also describe the norm closure of $\text{Alg}(\mathcal{H})$, the set of algebraic operators in $\mathcal{L}(\mathcal{H})$. This result, due to D. Voiculescu [78], was used in the original proof of Theorem 2.1 in [8]. We obtain it as a corollary of this result.

Theorem 2.8. *The norm closure of* $\text{Alg}(\mathcal{H})$ *is the set of those operators* $A \in \mathcal{L}(\mathcal{H})$ *for which* $\text{ind}(A - \lambda) = 0 (\lambda \in \rho_{SF}(A))$. *(From Theorem 1.18, the index condition on* A *is equivalent to the hypothesis that* A *is biquasitriangular.)*

Proof: Suppose $\lim \|T - A_n\| = 0$ with each A_n algebraic. If $T - \lambda$ is semi-Fredholm, then (for large n) $A_n - \lambda$ is semi-Fredholm. Since A_n is algebraic, $\sigma(A_n)$ is finite, so $\text{ind}(A_n - \lambda) = 0$. Continuity of the index now implies $\text{ind}(T - \lambda) = 0$.

Suppose A satisfies the index condition. Proceeding as in the proof of Theorem 2.1, we may assume (up to small-norm compact perturbation) that A has a matrix form as in (2.2), where N is normal and $\sigma(N) = \sigma_e(N) = \sigma_e(A) = \sigma(A) \supset \sigma(B_j)(j = 1, 2)$.

Since N is normal, it is a limit of normal operators with finite spectra, whence $N \in \text{Alg}(\mathcal{H})^-$. Lemma 2.7 implies $N \underset{\text{sim}}{\longrightarrow} N \oplus B_j$, whence $N \oplus B_j \in \text{Alg}(\mathcal{H})^-(j = 1, 2)$. Observe that if A_1 and A_2 are algebraic, then so is

$$\begin{pmatrix} A_1 & C_1 \oplus C_2 \\ 0 & A_2 \end{pmatrix}.$$

It is now clear from (2.2) that $A \in \text{Alg}(\mathcal{H})^-$.

The same techniques used in the proof of Theorem 2.1 also yield the following description of the closure of the similarity orbit of a normal operator with perfect spectrum.

Theorem 2.9. (D. Herrero [57]) *Let* N *be a normal operator such that* $\sigma(N)$ *is perfect. An operator* $A \in \mathcal{L}(\mathcal{H})$ *belongs to* $\mathcal{S}(N)^-$ *if and only if it satisfies the following conditions:*

(i) $\sigma(A) \supset \sigma(N)$ and each component of $\sigma(A)$ intersects $\sigma(N)$.

(ii) $\sigma_e(A) \supset \sigma_e(N)(= \sigma(N))$ and each component of $\sigma_e(A)$ intersects $\sigma_e(N)$.

(iii) $\text{ind}(A - \lambda) = 0$ for every $\lambda \in \rho_{SF}(A)$.

By considering the *essential* hypotheses in the preceeding results, we can formulate analogues for elements of the Calkin algebra.

Theorem 2.10. [63] *The closure of the set of nilpotent elements of the Calkin algebra is the set of $\widetilde{A} \in \mathcal{A}(\mathcal{H})$ satisfying:*

(i) *$\sigma_e(A)$ is connected and contains the origin;*

(ii) $\text{ind}(A - \lambda) = 0$ $(\lambda \in \rho_{SF}(A))$.

Proof: Necessity of the conditions is clear. If A satisfies the hypotheses, then it follows from Theorem 1.27 that $A + K$ satisfies the hypotheses of Theorem 2.1 for some $K \in \mathcal{K}(\mathcal{H})$. Thus $A + K \in \mathcal{N}(\mathcal{H})^-$ and so $\widetilde{A}$ is the limit of nilpotents in the Calkin algebra.

The analogues of Theorem 3.8 and 3.9 may be formulated and proven in much the same fashion.

3. Closed Similarity Orbits

In this lecture we describe the operators and elements of the Calkin algebra with closed similarity orbits. For $T \in \mathcal{L}(\mathcal{H})$, let $[T]$ denote the equivalence class of T with respect to the equivalence relation # (asymptotic similarity). A partial ordering on $\mathcal{L}(\mathcal{H})/\#$ may be defined as follows: $[S] < [T]$ if $\mathcal{S}(S)^- \subset \mathcal{S}(T)^-$; clearly, if $\mathcal{S}(T)$ is closed, then $[T]$ is a minimal element of $\mathcal{L}(\mathcal{H})/\#$. We begin our considerations with normal operators.

Proposition 3.1. *A normal operator has a closed similarity orbit if and only if its spectrum is finite.*

Proof: Let D be a normal operator with finite spectrum and let $z_1, \ldots, z_n$ denote the distinct elements of $\sigma(D)$. Let $p(z) = (z - z_1) \cdots (z - z_n)$, the minimal polynomial of D. Suppose $D \underset{\text{sim}}{\longrightarrow} T$ and $q(z)$ is a polynomial such that $q(T) = 0$, then $\{0\} = \sigma(q(T)) = q(\sigma(T)) \supset q(\sigma(D)) = q(\{z_1, \ldots, z_n\})$, so $p|q$ and it follows that p is the minimal polynomial of T. Thus T is similar to a normal operator N with $\sigma(N) = \sigma(D)$([49] and Proposition 1.19). For $1 \leq i \leq n$, there is a polynomial $q_i(z)$ such that $q_i(D) = P_{\ker(D-z_i)}$ and $q_i(N) = P_{\ker(N-z_i)}$; since $q_i(D) \underset{\text{sim}}{\longrightarrow} q_i(N)$, it follows that $\dim\ker(D - z_i) = \dim\ker(N - z_i)(1 \leq i \leq n)$, whence $T \sim N \sim D$.

Conversely, suppose N is normal, $\sigma(N)$ is infinite, and let λ denote an accumulation point of $\sigma(N)$. If $\lambda \notin \sigma_p(N)$, then by (1.1) $N \oplus \lambda \in \mathcal{U}(N)^- \backslash \mathcal{S}(N)$; similarly, if $\lambda \in \sigma_p(N)$, then $N|(\mathcal{H} \ominus \ker(N - \lambda)) \in \mathcal{U}(N)^- \backslash \mathcal{S}(N)$; thus $\mathcal{S}(N)$ is not closed.

We next proceed to the analogue of Proposition 4.1 for normal elements of the Calkin algebra. Recall a result of Calkin [24] which implies that a finite system of pairwise orthogonal self-adjoint projections in $\mathcal{A}(\mathcal{H})$ lifts to a system of pairwise orthogonal self-adjoint projections in $\mathcal{L}(\mathcal{H})$; consequently, if $\widetilde{T}$ is normal and $\sigma(\widetilde{T})$ is finite, then there is a normal operator $N, \sigma(N) = \sigma(\widetilde{T})$, such that $\widetilde{N} = \widetilde{T}$.

Lemma 3.2. *If $\widetilde{T}$ is a normal element of $\mathcal{A}(\mathcal{H})$ with finite spectrum, then $\mathcal{S}(\widetilde{T})$ is closed.*

Proof: By the preceding observation, we may assume T is normal, with $\sigma(T) = \sigma_e(T)$. Since $\sigma(T)$ is finite, the minimal polynomial of $T, p(z)$, has simple roots. Exactly as in the proof of Proposition 3.1, we see that if $\widetilde{T} \underset{\text{sim}}{\longrightarrow} \widetilde{S}$, then p is the minimal polynomial of $\widetilde{S}$, whence $\sigma_e(S) = \sigma_e(T)$. By Olsen's Theorem [65], there exists compact operator K such that $p(S + K) = 0$. Thus, as before, there is a normal operator N similar to $S + K$. Since N and T are normal, $\sigma_e(N) = \sigma_e(S) = \sigma_e(T)$, and these sets are finite, it is elementary that $N \approx_K T$; thus $\widetilde{S} \sim \widetilde{N} \approx \widetilde{T}$ and the result follows.

Let us digress briefly to observe that if $\widetilde{T}$ is any normal element of the Calkin algebra, then $\mathcal{S}(\widetilde{T})$ can be lifted to $\mathcal{S}(T)$ up to compact perturbation. For T in $\mathcal{L}(\mathcal{H})$, let

$$\mathcal{S}_K(T) = \{X^{-1}TX + K : S \text{ is invertible}, K \in \mathcal{K}(\mathcal{H})\}(= \mathcal{S}(T) + \mathcal{K}(\mathcal{H})),$$

and let

$$\mathcal{S}_F(T) = \{XTY + K : \widetilde{X}^{-1} = \widetilde{Y}, K \in \mathcal{K}(\mathcal{H})\}.$$

It is easy to see that the following properties are equivalent:

(i) $\mathcal{S}_K(T) = \mathcal{S}_F(T)$;

(ii) $\mathcal{S}(\widetilde{T}) = \pi(\mathcal{S}(T))$ (where π is the Calkin map);

(iii) $\pi^{-1}(\mathcal{S}(\widetilde{T})) = \mathcal{S}(T) + \mathcal{K}(\mathcal{H})$.

Proposition 3.3. (cf. [34]) *If $\widetilde{T}$ is normal, then $\mathcal{S}(\widetilde{T}) = \pi(\mathcal{S}(T))$.*

Proof: From the eiquvalence of (i) and (ii) above, it suffices to verify that $\mathcal{S}_F(T) \subset \mathcal{S}_K(T)$. Let $S = XTY + K$, where $\widetilde{X} = \widetilde{Y}^{-1}$ and $K \in \mathcal{K}(\mathcal{H})$. Thus Y is Fredholm, and we consider here only the case $\text{ind}(Y) < 0$. In this case, there is an operator L and a finite rank operator G such that $L(Y+G) = 1$. Let $\mathcal{M} = \text{Ran}(Y+G)$. Then $\mathcal{M}$ is a subspace of finite codimension in $\mathcal{H}$ and thus $XTY - LP_M T(Y+G)$ is compact. Let $T_{\mathcal{M}}$ in $\mathcal{L}(\mathcal{M})$ denote the compression of T to $\mathcal{M}, T_{\mathcal{M}} = P_{\mathcal{M}}T|\mathcal{M}$, and note that $T_{\mathcal{M}}$ is similar to $LP_{\mathcal{M}}T(Y+G)$. Indeed, $Y+G : \mathcal{H} \to \mathcal{M}$ is invertible, with inverse $L|\mathcal{M} : \mathcal{M} \to \mathcal{H}$, and

$$LP_{\mathcal{M}}T(Y+G) = (L|\mathcal{M})T_{\mathcal{M}}(Y+G).$$

Since $\mathcal{M}$ has finite codimension in $\mathcal{M}$, it is easy to check that $T_{\mathcal{M}}$ is essentially normal, $\sigma_e(T_{\mathcal{M}}) = \sigma_e(T)$, and $\text{ind}(T-z) = \text{ind}(T_{\mathcal{M}} - z)$ for every $z \in \rho_F(T_{\mathcal{M}})$. An application of Theorem 1.5 shows that $T_{\mathcal{M}} \approx_K T$. Thus $S \approx_K LP_{\mathcal{M}}T(Y+G) \sim T_{\mathcal{M}} \approx_K T$, whence $S \in \mathcal{S}_K(T)$ and the proof is complete.

It turns out that if $\widetilde{T}$ is not algebraic, then

$$\mathcal{S}(\widetilde{T})^- = [\pi(\mathcal{S}(T)]^- = [\pi(\mathcal{S}(T))]^- [5].$$

However, we cannot expect that $\mathcal{S}(\widetilde{T}) = \pi(\mathcal{S}(T))$ in general. Indeed, if $Q = q_2^{(\infty)} \oplus q_4^{(\infty)}$, then $S \equiv q_1 \oplus Q \notin \mathcal{S}_K(Q)$ although $\widetilde{S} \sim \widetilde{Q}$ [5, page 57]. It is a conjecture of D. Herrero that if T is not similar to a compact perturbation of a Jordan operator (defined below), then $\mathcal{S}(\widetilde{T}) = \pi(\mathcal{S}(T))$.

Question 3.4. For which operators T is $\mathcal{S}(\widetilde{T}) = \pi(\mathcal{S}(T))$?

To prove the converse of Lemma 3.2 we cannot rely exclusively on closures of unitary orbits, as we did in proving the *only if* direction of Proposition 3.1. The obstacle to such a course is Proposition 1.7, which shows that $\mathcal{U}(\widetilde{T})$ is closed for $\widetilde{T}$ normal. The converse of Proposition 3.2 depends instead on some bona fide similarities, as follows.

Lemma 3.5. [34] *If N is a normal operator whose essential spectrum is infinite and M is a normal opertor whose spectrum is connected and contains $\sigma(N)$, then $N \underset{\text{sim}}{\longrightarrow} M$; in particular, $\mathcal{S}(N)^- \not\subset \mathcal{S}_K(N)$.*

Proof: We may assume that the origin is an accumulation point of $\sigma_e(N)$. Lemma 3.3 implies $N \underset{\text{sim}}{\longrightarrow} N \oplus Q$ for each nilpotent operator Q. Since $\sigma(M)$ is connected and $0 \in \sigma(N) \subset \sigma(M)$, Theorem 2.6 implies $M \in \mathcal{N}(\mathcal{H})^-$, whence $N \underset{\text{sim}}{\longrightarrow} N \oplus M \approx_a M$. By choosing M so that $\sigma_e(M) \neq \sigma_e(N)$ we conclude that $\mathcal{S}(N)^- \neq \mathcal{S}_K(N)$.

Now we may obtain the analogue of Proposition 3.3.

Proposition 3.6. *If $\widetilde{T}$ is normal, $\mathcal{S}(\widetilde{T})$ is closed if and only if $\sigma(\widetilde{T})$ is finite.*

Proof: In view of Lemma 3.2, it suffices to consider the case when $\sigma(\widetilde{T})$ is infinite. Let N be a normal operator such that $\sigma(N) = \sigma_e(N) = \sigma(\widetilde{T})$ and let M be a normal operator whose spectrum is connected and properly contains that of N. Theorem 1.5 implies $T \approx_K T \oplus N$ and Lemma 3.5 implies $N \underset{\text{sim}}{\longrightarrow} M$. Thus

$$\widetilde{T} \underset{\text{sim}}{\longrightarrow} \widetilde{T} \oplus \widetilde{M},$$

and since $\sigma(\widetilde{T}) \neq \sigma(\widetilde{T} \oplus \widetilde{M})$, it follows that $\mathcal{S}(\widetilde{T})$ is not closed.

We next extend Propositions 3.1 and 3.6 to the non-normal case.

Lemma 3.7. *Let σ be a compact and infinite subset of $\sigma_{le}(T)$. Let N be a normal operator such that $\sigma(N)$ is connected and contains σ. Then $\mathcal{S}(T)^-$ contains an operator of the form $N \oplus R$ (up to unitary equivalence).*

Proof: From Proposition 1.17, there exists an operator S in $\mathcal{U}(T)^-$ unitarily equivalent to

$$\begin{pmatrix} D & S_{12} \\ 0 & S_{22} \end{pmatrix},$$

where D is a normal operator with $\sigma(D) = \sigma_e(D) = \sigma$. Lemma 1.21 shows $S \underset{\text{sim}}{\longrightarrow} D \oplus S_{22}$; since Lemma 3.5 implies $D \underset{\text{sim}}{\longrightarrow} N$, then

$$T \approx_a S \underset{\text{sim}}{\longrightarrow} D \oplus S_{22} \underset{\text{sim}}{\longrightarrow} N \oplus S_{22},$$

and the result follows.

Lemma 3.8. *If $\sigma_{le}(T)$ is infinite, then $\mathcal{S}(T)$ is not closed in $\mathcal{L}(\mathcal{H})$ and $\mathcal{S}(\widetilde{T})$ is not closed in $\mathcal{A}(\mathcal{H})$; moreover, $\mathcal{S}(T)^- \not\subset \mathcal{S}_K(T)$.*

Proof: Let τ be a compact connected set that properly contains $\sigma(T)$. Lemma 3.7 implies that $\mathcal{S}(T)^-$ contains an operator S with $\sigma_e(S) \supset \tau$; since $\sigma_e(S) \neq \sigma_e(T)$, all the desired conclusions follow.

Lemma 3.9. [6] *Let $T \in \mathcal{L}(\mathcal{H}), \sigma(\widetilde{T}) = \{0\}$. Then $\mathcal{S}(T)^-$ contains a compact normal operator D such that $\sigma(D)\backslash\{0\} = \sigma(T)\backslash\{0\}$.*

Proof: We only consider here the case when $\sigma(T)\backslash\{0\}$ is infinite, with distinct points $\{\lambda_i\}_{i=1}^{\infty}$, so that $\lim \lambda_i = 0$. Let $\mathcal{H}_\lambda$ denote the (finite dimensional) Riesz spectral subspace for T corresponding to $\{\lambda\} \subset \sigma(T)\backslash\{0\}$. Let $\{e_1^i, \ldots, e_{n_i}^i\}$ denote an orthonormal basis for $\mathcal{H}_{\lambda_i}$ relative to which the matrix of $T|\mathcal{H}_{\lambda_i}$ is upper triangular.

Let

$$\mathcal{M} = \bigvee_{i=1}^{\infty} \mathcal{H}_{\lambda_i}; \text{ if } \mathcal{B} = \cup_{i=1}^{\infty} \mathcal{B}_i$$

is ordered as

$$\ldots, e_1^i, \ldots, e_{n_i}^i, e_1^{i+1}, \ldots, e_{n_{i+1}}^{i+1}, \ldots,$$

then an orthonormalization of $\mathcal{B}$ produces an orthonormal basis $\{e_n\}$ for $\mathcal{M}$ with the following properties:

(i) The matrix (T_{ij}) of $T|\mathcal{M}$ with respect to $\{e_n\}$ is upper triangular and every $\mathcal{H}_{\lambda_i}$ is spanned by a finite subset of $\{e_n\}$.

(ii) Every diagonal entry of (T_{ij}) is a λ_i, and each λ_i appears on the diagonal only finitely many times; thus $\lim T_{ii} = 0$.

Let $P_n = P_{<e_n>}$ and $Q_n = \sum_{k=1}^{n} P_k$; (ii) implies that $D \equiv \sum_{k=1}^{\infty} P_k T P_k$ is norm-convergent to a compact normal operator with $\sigma(D)\backslash\{0\} = \sigma(T)\backslash\{0\}$, and we claim that $D \in \mathcal{S}(T)^-$. From Lemma 1.21

(iii) $T \underset{\text{sim}}{\longrightarrow} Q_n T Q_n + (1 - Q_n)T(1 - Q_n)$, and Lemma 1.21 also implies

(iv) $Q_n T Q_n \longrightarrow_{\text{sim}} \sum_{k=1}^{n} P_k T P_k$. The finite dimensionality of the $\mathcal{H}_\lambda$'s implies that

$$R_n \equiv (1 - Q_n)T(1 - Q_n)|(1 - Q_n)\mathcal{H}$$

satisfies $\sigma(\widetilde{R}_n) = \{0\}$, and (i) and (ii) imply

$$\sigma(R_n)\backslash\{0\} \subset \sigma(T)\backslash\{0\}$$

and

$$\lim r(T_n) = 0 (r(\cdot) = \text{ spectral radius}).$$

Using Corollary 1.23, we may choose $R_n' \sim R_n$ such that

$$\|R_n'\| \leq r(R_n) + 1/n.$$

Let

$$S_n = \sum_{k=1}^{n} P_k T P_k + R_n'(1 - Q_n).$$

From (iii) and (iv), $T \underset{\text{sim}}{\longrightarrow} S_n$, and

$$\begin{aligned}\|S_n - D\| &\leq \|S_n - \sum_{k=1}^{n} P_k T P_k\| + \left\|\sum_{k=1}^{n} P_k T P_k - D\right\| \\ &\leq r(R_n) + 1/n + \left\| \sum_{k=n+1}^{\infty} P_k T P_k \right\| \rightarrow 0 (n \rightarrow \infty).\end{aligned}$$

Thus $T \underset{\text{sim}}{\longrightarrow} D$.

We may now prove the main results of this section.

Theorem 3.10. [6,61] *$\mathcal{S}(T)$ is closed if and only if T is similar to a normal operator with finite spectrum.*

Proof: If $\mathcal{S}(T)$ is closed, then Lemma 3.8 implies $\sigma_{le}(T)$ is finite, whence $\sigma_e(T)$ is finite [43]. If $\sigma_e(T)$ has n distinct points, then Proposition 1.20 implies

$$T \sim T_1 \oplus \ldots \oplus T_n, \text{ where } \sigma(T_i) \cap \sigma(T_j) = \emptyset \text{ for } i \neq j$$

and $\sigma_e(T_i)$ is a singleton. Lemma 3.9 implies $T_i \underset{\text{sim}}{\longrightarrow} D_i$, where D_i is normal, so $\mathcal{S}(T)^-$ contains $D \equiv D_1 \oplus \ldots \oplus D_n$. Since $\mathcal{S}(T)$ is closed, Proposition 3.1 implies $T \sim D$ and $\sigma(D)$ is finite. The converse follows from Proposition 3.1.

Theorem 3.11. *$\mathcal{S}(\widetilde{T})$ is closed if and only if $\widetilde{T}$ is similar to a normal element with finite spectrum.*

Proof: If $\mathcal{S}(\widetilde{T})$ is closed, then exactly as in the preceding proof we see that $T \underset{\text{sim}}{\longrightarrow} D$, where D is a normal operator with finite essential spectrum. Thus $\widetilde{D} \in \mathcal{S}(\widetilde{T})^- = \mathcal{S}(\widetilde{T})$. The converse follows from Lemma 3.2.

Note from the remarks preceding Lemma 3.2, and from Proposition 3.3, that $\widetilde{T}$ is similar to a normal element with finite spectrum if and only if T is similar to a compact perturbation of a normal operator with finite spectrum.

4. Universal Quasinilpotent Operators

Having just considered closed similarity orbits, we turn our attention to some orbits whose closures are as large as possible (within the constraints of (1.2)-(1.5)). For a quasinilpotent operator $Q, \mathcal{S}(Q)^- \subset \mathcal{Q}(\mathcal{H})^-$. It is a remarkable fact, discovered by D. Herrero [59], that $\mathcal{Q}(\mathcal{H})^-$ (as described by Theorem 2.1) can actually be realized as

$\mathcal{S}(Q)^-$ for certain quasinilpotent operators Q; our goal in this lecture is to describe this phenomenon.

Let $\mathcal{A}$ denote a complex Banach algebra with identity and let $\mathcal{Q}(\mathcal{A})$ denote the set of all quasinilpotent elements of $\mathcal{A}$; $Q \in \mathcal{Q}(\mathcal{A})$ is a *universal quasinilpotent* if $\mathcal{S}(Q)^- \supset \mathcal{Q}(\mathcal{A})$. Let $\mathcal{J}$ denote a nonzero closed 2-sided ideal in $\mathcal{A}$. A quasinilpotent $Q \in \mathcal{J}$ is $\mathcal{J}$-universal if $\mathcal{S}(Q)^- \supset \mathcal{J} \cap \mathcal{Q}(\mathcal{A})$. (When $\mathcal{A} = \mathcal{L}(\mathcal{H})$ and $\mathcal{J} = \mathcal{K}(\mathcal{H}), \mathcal{J} \cap \mathcal{Q}(\mathcal{H})$ is closed, so in this case, Q is a compact universal quasinilpotent if and only if $\mathcal{S}(Q)^- = \mathcal{K}(\mathcal{H}) \cap \mathcal{Q}(\mathcal{H})$). Note that if $\mathcal{J} \cap \mathcal{Q}(\mathcal{A})$ contains a non-nilpotent quasinilpotent, then no nilpotent Q in $\mathcal{J}$ is $\mathcal{J}$-universal; indeed, $Q \underset{\text{sim}}{\longrightarrow} R$ implies $Q^n \underset{\text{sim}}{\longrightarrow} R^n$ for $n \geq 1$. The next three results characterize the universal quasinilpotents in $\mathcal{L}(\mathcal{H})$ and $\mathcal{A}(\mathcal{H})$ and the compact universal quasinilpotents in $\mathcal{K}(\mathcal{H})$.

Theorem 4.1. (C. Apostol [3], D. Herrero [60]) *A quasinilpotent operator $Q \in \mathcal{Q}(\mathcal{H})$ is a universal quasinilpotent if and only if Q^n is not compact for every $n \geq 1$ (equivalently, $\widetilde{Q}$ is not nilpotent).*

Corollary 4.2. *A quasinilpotent element $\widetilde{T} \in \mathcal{A}(\mathcal{H})$ is a universal quasinilpotent if and only if $\widetilde{T}$ is not nilpotent.*

Theorem 4.3. [3] *A compact quasinilpotent operator $K \in \mathcal{K}(\mathcal{H})$ is a compact universal quasinilpotent if and only if K is not nilpotent.*

Since there exist quasinilpotent operators Q and K such that $\widetilde{Q}$ is not nilpotent and K is compact but not nilpotent, the necessity of the *not nilpotent* condition is clear in each result. The proof of sufficiency depends on some preliminary results.

A nilpotent operator $Q \in \mathcal{N}_{\mathcal{H}}$ is a *Jordan nilpotent* if and only if it is unitarily equivalent to an operator of the form

$$q_{k_1}^{(\alpha_1)} \oplus \ldots \oplus q_{k_n}^{(\alpha_n)}, k_i \geq 1, 1 \leq \alpha_i \leq \infty (1 \leq i \leq n).$$

Jordan nilpotents provide convenient models for arbitrary nilpotents in several different ways. For example, C. Apostol, R. Douglas, and C. Foias [4] proved that every nilpotent is *quasisimilar* to an essentially unique Jordan nilpotent. About the same time, C. Apostol and J. Stampfli [12], L. Gray [44], and L. Williams [84] independently proved that a nilpotent operator T is similar to a Jordan nilpotent if and only if $\text{Ran}(T^j)$ is closed for every $j \geq 1$. In [44], Gray also obtained the following density theorem. Let $\mathcal{N}_k(\mathcal{H})$ denote the set of nilpotents of order k.

Theorem 4.4. *If $T \in \mathcal{N}_k(\mathcal{H})$, then there exists $\{T_n\} \subset \mathcal{N}_k(\mathcal{H})$ such that $\lim \|T - T_n\| = 0$ and T_n is similar to a Jordan nilpotent for every $n \geq 1$.*

The foregoing definitions and results extend to algebraic operators. An algebraic operator is a *Jordan operator* if it is of the form

$$(\lambda_1 + N_1) \oplus \ldots \oplus (\lambda_k + N_k),$$

where $\lambda_1, \ldots, \lambda_k$ are scalars and $N_1, \ldots, N_k$ are Jordan nilpotents. Although our interest in Jordan operators is due to their importance for the study of similarity orbits, they were first studied by C. Apostol and J. Stampfli [12] in connection with inner derivations, and we wish to digress briefly to mention some results in this area. For $A \in \mathcal{L}(\mathcal{H})$, let $\delta_A : \mathcal{L}(\mathcal{H}) \to \mathcal{L}(\mathcal{H})$ denote the inner derivation on $\mathcal{L}(\mathcal{H})$ induced by $A, \delta_A(X) = AX - XA$.

Theorem 4.5. (C. Apostol [2]) *For $A \in \mathcal{L}(\mathcal{H})$ the following are equivalent:*

(i) $\mathrm{Ran}(\delta_A)$ *is closed in* $\mathcal{L}(\mathcal{H})$;

(ii) A is similar to a Jordan model;

(iii) A is algebraic and $\mathrm{Ran}(p(A))$ *is closed for every polynomial $p(z)$.*

Let $\delta_{\widetilde{A}}$ denote the inner derivation on $\mathcal{A}(\mathcal{H})$ induced by $\widetilde{A}$.

Theorem 4.6. (L. Fialkow and D. Herrero [41]) *For $A \in \mathcal{L}(\mathcal{H})$, the following are equivalent:*

(i) $\mathrm{Ran}\delta_{\widetilde{A}}$ *is closed in* $\mathcal{A}(\mathcal{H})$;

(ii) A is similar to a compact perturbation of a Jordan operator;

(iii) $\mathrm{Ran}\delta_{A+K}$ *is closed in $\mathcal{L}(\mathcal{H})$ for some compact operator K.*

Apart from the range, many other aspects of δ_A have been studied. The norm of δ_A was computed by J. Stampfli [75]; more generally, for the generalized derivation T_{AB} defined by $T(X) = AX - XB$, Stampfli's formula is that

$$\|T_{AB}\| = \inf_{\lambda \in \mathbb{C}} \|A - \lambda\| + \|B - \lambda\|.$$

Beginning with Rosenblum's Theorem [71], spectral properties of generalized derivations were studied by many authors (cf. R. Harte [52], C. Davis and P. Rosenthal [30]). Many of these results have been extended to the *elementary operator* R_{AB} defined by

$$R(X) = A_1 X B_1 + \ldots + A_n X B_n \text{ [35]}.$$

Approximation techniques based on Voiculescu's Theorem have been used to describe $\sigma_e(R), \rho_{SF}(R), \text{ind}(R-\lambda)$ [39]. Can these techniques be used to compute $\|R_{AB}\|$?

δ_A has at least one obvious invariant subspace, namely $\mathcal{K}(\mathcal{H})$. J. Froelich has observed that $\delta_{\widetilde{A}}$ has an invariant subspace since $\mathcal{A}(\mathcal{H})$ is nonseparable.

Question 4.7. Does $\delta_A|\mathcal{K}(\mathcal{H})$ have an invariant subspace?
(Question 4.7 was suggested to the author by P. R. Halmos. Several variants suggest themselves: replace $\mathcal{K}(\mathcal{H})$ by an arbitrary *norm ideal* in $\mathcal{L}(\mathcal{H})$; replace δ_A by R_{AB}.)

There is an interesting connection between inner derivations and local structures of similarity orbits. An operator $T \in \mathcal{L}(\mathcal{H})$ has a *local similarity cross section* if the mapping $X \to X^{-1}TX$ (X invertible) admits a norm continuous local cross section. Which operators have this property? It turns out that if T has a local similarity cross section, then $\text{Ran}(\delta_T)$ is norm-closed in $\mathcal{L}(\mathcal{H})$ [36], and thus, by Theorem 4.5, T is similar to a Jordan operator. Perhaps surprisingly, the converse is *false*; only certain Jordan models have local similarity cross sections:

Theorem 4.8. (L. Fialkow and D. Herrero [40]) *For T in $\mathcal{L}(\mathcal{H})$, the following are equivalent:*

(i) T has a local similarity cross section;

(ii) T is similar to a nice Jordan operator, i.e., to an operator of the form

$$J = \bigoplus_{i=1}^{n} [\lambda_i 1_{\mathcal{H}_i} + \bigoplus_{j=1}^{m_i} q_{k_{ij}}^{(\alpha_{ij})}],$$

where $\lambda_1, \ldots, \lambda_n$ are distinct complex numbers, and for each $i, k_{ij} \geq 1, \alpha_{ij} \geq 1, \bigoplus_{j=1}^{m_i} q_{k_{ij}}^{(\alpha_{ij})}$ acts on a Hilbert space $\mathcal{H}_i$, and $\alpha_{ij} = \infty$ for at most one value of $j (j = 1, \ldots, m_i)$.

The analogue for elements of the Calkin algebra is as follows:

Theorem 4.9. [5] *For $\widetilde{T}$ in $\mathcal{A}(\mathcal{H})$, the following are equivalent:*

(i) $\widetilde{T}$ has a local similarity cross section relative to the mapping $\widetilde{X} \to \widetilde{X}^{-1}\widetilde{T}\widetilde{X}$;

(ii) T is similar to a compact perturbation of a nice Jordan operator.

We also note that the concept of *global cross sections* has been studied by M. Pecuch Herrero [68], who has characterized global unitary and similarity cross sections in $\mathcal{L}(\mathcal{H})$ (including the finite dimensional case) and global similarity cross sections for elements of the Calkin algebra. These results are as follows:

T has a global similarity cross section if and only if

$$T \approx \left(\lambda_1 + q_{k_1}^{(\infty)}\right) \oplus \ldots \oplus (\lambda_n + q_{k_n}^{(\infty)})$$

for a sequence of distinct scalars $\lambda_1, \ldots, \lambda_n$;
T has a global unitary cross section if and only if

$$T \approx A^{(\infty)}, A \in \mathcal{L}(\mathbb{C}^k);$$

$\widetilde{T}$ has a global similarity cross section if and only if

$$\widetilde{T} = \lambda, \lambda \in \mathbb{C}.$$

The proofs of these results require topological methods that are beyond the scope of these lectures; however, one aspect of their proofs is relevant to our main theme. In [18], J. Barria and D. Herrero presented a comprehensive analysis of the closure of the similarity orbit of a Jordan nilpotent, and their results are essential to the theorems on similarity cross sections. From this study we cite one particular result which is a key to understanding universal quasinilpotents.

Proposition 4.10. [18] *If $k > 1$, then*

$$\mathcal{S}\left(q_1^{(\infty)} \oplus q_k^{(\infty)}\right)^- \supset \mathcal{N}_{k-1}(\mathcal{H}).$$

We now proceed towards the proof of Theorem 4.1. Let Q be a non-nilpotent quasinilpotent operator in $\mathcal{L}(\mathcal{H})$. Suppose $n \geq 1, x \in \mathcal{H}, \|x\| = 1$, and $Q^n x \neq 0$. Let $\mathcal{H}_n(x)$ denote the span of $A_x \equiv \{A^{k-1}x\}_{k=1}^{n+1}$ and let $B_x \equiv \{e_k(x)\}_{k=1}^{n+1}$ denote the orthonormalization of A_x using the Gram-Schmidt process. The compression $T_n(x)$ of Q to $\mathcal{H}_n(x)$ has a matrix relative to B_x of the form

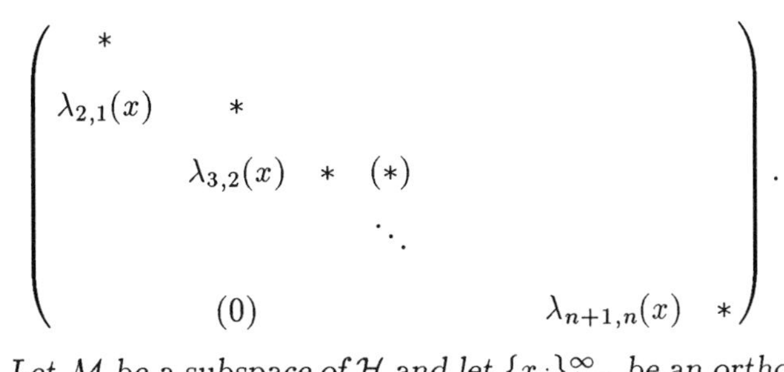

$$\begin{pmatrix} * & & & & & \\ \lambda_{2,1}(x) & * & & & & \\ & \lambda_{3,2}(x) & * & (*) & & \\ & & & \ddots & & \\ & (0) & & & \lambda_{n+1,n}(x) & * \end{pmatrix}.$$

Lemma 4.11. *Let $\mathcal{M}$ be a subspace of $\mathcal{H}$ and let $\{x_j\}_{j=1}^{\infty}$ be an orthonormal sequence in $\mathcal{M}$ such that $\mathcal{H}_n(x_j) \perp \mathcal{H}_n(x_k)$ for all $j \neq k$. if $Q^n|\mathcal{M}$ is bounded below, then*

$$\delta \equiv \inf \left\{ \left|\lambda_{k+1,k}(x_j)\right| : 1 \leq k \leq n, 1 \leq j < \infty \right\} > 0.$$

Proof: Assume to the contrary that $\delta = 0$; then there exist $k_0, 1 \leq k_0 \leq n$, and a subsequence $\{x_{j_m}\}_{m=1}^{\infty}$ of $\{x_j\}$ such that

$$\lim |\lambda_{k_0+1,k_0}(x_{j_m})| = 0.$$

Passing, if necessary, to a further subsequence, we may assume in the following argument that for $1 \leq k, m \leq n+1$, $\{\lambda_{k,m}(x_j)\}_{j=1}^{\infty}$ is convergent, say to $\lambda_{k,m}$; thus $\lambda_{k_0+1,k_0} = 0$.

Define $T_{n,j}$ to be the operator on $\mathcal{H}_n(x_j)$ whose matrix relative to B_{x_j} is

$$(\lambda_{k,m})_{1 \leq k,m \leq n+1}.$$

Let $P_n(x_j)$ denote the orthogonal projection of $\mathcal{H}$ onto $\mathcal{H}_n(x_j)$ and set

$$T = Q + \sum_{j=1}^{\infty} (T_{n,j} - T_n(x_j)) P_n(x_j).$$

Since $\lambda_{k_0+1,k_0} = 0$, it is easy to see that $\mathcal{N} \equiv \operatorname{span}\{e_1(x_j), \ldots, e_{k_0}(x_j)\}_{j=1}^{\infty}$ is T-invariant and that $T|\mathcal{N}$ is unitarily equivalent to an orthogonal direct sum of countably many copies of $L \equiv (\lambda_{k,m})_{1 \leq k,m \leq k_0}$. Clearly, $T - Q$ is compact, so T is essentially quasinilpotent, and so also is $T|\mathcal{N}$. Since $T|\mathcal{N} \approx L^{(\infty)}$, we conclude that $L^{k_0} = 0$, whence $T^n|\mathcal{N} = (T|\mathcal{N})^n \approx L^{n(\infty)} = 0$. Finally, since $(T^n - Q^n)|\mathcal{N}$ is compact, it now follows that $Q^n|\mathcal{N}$ is compact, contradicting the assumption that $Q^n|\mathcal{M}$ is bounded below.

Lemma 4.12. *Let $\mathcal{M}$ be an infinite dimensional subspace of $\mathcal{H}$ such that $Q^n|\mathcal{M}$ is bounded below (Q as above). Then $\mathcal{S}(Q)^-$ contains an operator unitarily equivalent to $q_{n+1}^{(\infty)} \oplus q_1^{(\infty)}$.*

Proof: By Rota's Corollary we may assume $\|Q\| < \epsilon$. Since $\mathcal{M}$ is infinite dimensional, we can inductively produce an orthogonal sequence $\{x_j\}_{j=1}^{\infty} \subset \mathcal{M}$ such that

$$\mathcal{H}_n(x_j) \perp \mathcal{H}_n(x_m), j \neq m,$$

and

$$\dim \left(\mathcal{H} \ominus \sum_{n=1}^{\infty} \mathcal{H}_n(x_j) \right) = \infty.$$

From Lemma 4.11, $\delta \equiv \inf\{|\lambda_{k+1,k}(x_j)| : 1 \leq k \leq n, 1 \leq j < \infty\} > 0$.

For $j \geq 1$, define $q_{n+1}(x_j), A_n(x_j) \in \mathcal{L}(\mathcal{H}_n(x_j))$ by matrices (relative to B_{x_j}) of the form

$$q_{n+1}(x_j) = \begin{pmatrix} 0 & & & & \\ 1 & 0 & & & \\ & 1 & & & \\ & & \vdots & & \\ & & & 1 & 0 \end{pmatrix},$$

$$A_n(x_j) = \begin{pmatrix} 1 & & & \\ & a_1(x_j) & & \\ & & \ddots & \\ & & & a_n(x_j) \end{pmatrix},$$

where $a_1(x_j) = \lambda_{2,1}(x_j), a_2(x_j) = \lambda_{21}(x_j)\lambda_{32}(x_j), \ldots$. Since $|\lambda_{k+1,k}(x_j)| \leq \|Q\| < 1$, then $\|A_n(x_j)\| = 1$. A calculation shows that $q_{n+1}(x_j) - A_n(x_j)^{-1}T_n(x_j)A_n(x_j)$ is an upper triangular matrix whose nonzero entries are entries from $T_n(x_j)$ multiplied by certain products of the $\lambda_{k+1,k}(x_j)$; thus

$$\|q_{n+1}(x_j) - A_n(x_j)^{-1}T_n(x_j)A_n(x_j)\| < (n+1)^2\|Q\|.$$

Let $P = 1 - \sum_{j=1}^{\infty} P_n(x_j)$ and define $A = \sum_{j=1}^{\infty} A_n(x_j)P_n(x_j) + P$; since $\delta > 0$, A is invertible. Relative to the decomposition

$$\mathcal{H} = \sum_{j=1}^{\infty} \mathcal{H}_n(x_j) \oplus P\mathcal{H},$$

the matrix of Q is of the form

$$\begin{pmatrix} Q_1 & Q_2 \\ Q_3 & Q_4 \end{pmatrix},$$

where

$$Q_1 = \left(\sum_{j=1}^{\infty} T_n(x_j)P_n(x_j)\right) |(1-P)\mathcal{H}.$$

Thus $A^{-1}QA$ is of the form

$$\begin{pmatrix} \sum_{j=1}^{\infty} A_n(x_j)^{-1}T_n(x_j)A_n(x_j)P_n(x_j) & \left[\sum_{j=1}^{\infty} A_n(x_j)P_n(x_j)\right]Q_2 \\ Q_3\left[\sum_{j=1}^{\infty} A_n(x_j)P_n(x_j)\right] & Q_4 \end{pmatrix},$$

and so

$$\left\| \sum_{j=1}^{\infty} q_{n+1}(x_j)P_n(x_j) - A^{-1}QA \right\| < (n+1)^2\|Q \mid + 3\|Q\|.$$

Note that $\sum_{j=1}^{\infty} q_{n+1}(x_j)P_n(x_j)$ is unitarily equivalent to $q_{n+1}^{(\infty)} \oplus q_1^{(\infty)}$, and thus

$$\operatorname{dist}(q_{n+1}^{(\infty)} \oplus q_1^{(\infty)}, \mathcal{S}(Q)) < (3 + (n+1)^2)\epsilon;$$

since $\epsilon > 0$ is arbitrary, it follows that $q_{n+1}^{(\infty)} \oplus q_1^{(\infty)} \in \mathcal{S}(Q)^-$.

Proof of Theorem 4.1. Assume Q is a quasinilpotent such that $\widetilde{Q}$ is not nilpotent. In view of Theorem 2.1, to prove $\mathcal{S}(Q)^- \supset \mathcal{Q}(\mathcal{H})$, is suffices to prove $\mathcal{S}(Q)^- \supset \mathcal{N}(\mathcal{H})$. Let $k \geq 1$; since Q^k is not compact, there is an infinite dimensional subspace $\mathcal{M}$ such that $Q^k|\mathcal{M}$ is bounded below. Lemma 4.12 implies $Q \underset{\text{sim}}{\longrightarrow} q_{k+1}^{(\infty)} \oplus q_1^{(\infty)}$, so it follows from Proposition 4.10 that $\mathcal{S}(Q)^- \supset \mathcal{N}_k(\mathcal{H})$. Thus $\mathcal{S}(Q)^- \supset \bigcup_{k=1}^{\infty} \mathcal{N}_k(\mathcal{H}) = \mathcal{N}(\mathcal{H})$ and the proof is complete.

Proof of Corollary 4.2. This follows from Theorem 4.1 and the fact that each quasinilpotent element of $\mathcal{A}(\mathcal{H})$ lift to a quasinilpotent operator [82].

Before proving Theorem 4.3 we require a lemma due to R. G. Douglas (see [48]).

Lemma 4.13. *Every compact quasinilpotent is the norm limit of finite rank nilpotents.*

Proof: Suppose K is compact and quasinilpotent. Let $\{P_n\}$ be an increasing sequence of finite rank projections converging strongly to the identity. Then $K_n \equiv P_nKP_n$ satisfies $\lim \|K - K_n\| = 0$. By upper semicontinuity of spectra, we may assume $\sigma(K_n) \subset D(0, 1/n)$. If we put K_n in upper triangular form, then each diagonal entry has modulus less than $1/n$. Let L_n be the operator obtained from K_n by replacing each diagonal entry by 0. Then L_n is a finite rank nilpotent and $\|L_n - K_n\| < 1/n$.

Proof of Theorem 4.3. If K is not nilpotent, then by a simplification of Lemma 4.12, $\mathcal{S}(K)^-$ contains $q_{n+1} \oplus q_1^{(\infty)} (n \geq 1)$. By [17], $\mathcal{S}(q_{n+1})^-$ contains every nilpotent operator on $\mathbb{C}^{n+1}$. It follows that $\mathcal{S}(Q)^-$ contains every finite rank nilpotent and thus every compact quasinilpotent.

5. The Similarity Orbit Theorem

We previously suggested that the Similarity Orbit Theorem characterizes the closures of the similarity orbits of all operators, but this is not quite the case. If T is quasinilpotent and $\widetilde{T}$ is not nilpotent, then $\mathcal{S}(T)^- = Q(\mathcal{H})^-$ (Theorem 4.1). However, there are certain essential nilpotents T for which $\mathcal{S}(T)^-$ has not been completely characterized; indeed, the entire problem of characterizing closures of similarity orbits has been reduced to an intractable class of essential nilpotents. To discuss these issues we need to develop additional invariants for $\underset{\text{sim}}{\longrightarrow}$ beyond (1.2)-(1.5). Some of these invariants are suggested by the finite dimensional case that we mentioned briefly in Lecture 1.

Suppose $\mathcal{H} \approx \mathbb{C}^k, T \in \mathcal{L}(\mathcal{H})$, and $T \underset{\text{sim}}{\longrightarrow} S$. Let $\{X_n\}$ denote a sequence of invertible opertors such that $\lim \|S - X_n^{-1}TX_n\| = 0$. For every polynomial $q(z)$ we have $\lim \|q(S) - X_n^{-1}q(T)X_n\| = 0$, and since rank $X_n^{-1}q(T)X_n =$ rank $q(T)$ for each n, the *lower semicontinuity of rank* implies

$$\text{rank } q(S) \le \text{rank } q(T).$$

(Since $T^* \underset{\text{sim}}{\longrightarrow} S^*$, this condition is equivalent to $\dim\ker q(T) \le \dim\ker q(S)$.) A careful and quite nontrivial analysis with Jordan canonical forms serves to show that the preceding condition is the complete $\underset{\text{sim}}{\longrightarrow}$ invariant:

Theorem 5.1. (J. Barria and D. Herrero [17]) *Let $T \in \mathcal{L}(\mathbb{C}^k)$ and let $p(z)$ denote the minimal monic polynomial of T. Then $S \in \mathcal{S}(T)^-$ if and only if*

$$\text{rank } q(S) \le \text{rank } q(T)$$

for every monic polynomial $q(z)$ that divides $p(z)$.

Theorem 5.1 suggests $\underset{\text{sim}}{\longrightarrow}$ invariants concerning isolated points of the spectrum in the infinite dimensional case. Suppose $T \in \mathcal{L}(\mathcal{H})$ and $\lim \|X_n^{-1}TX_n - A\| = 0$ for some $A \in \mathcal{L}(\mathcal{H})$. If $f(z)$ is any function analytic in a neighborhood of $\sigma(A) \cup \sigma(T)$, then

$$\lim \|f(X_n^{-1}TX_n) - f(A)\| = 0. \tag{5.1}$$

Suppose λ is an isolated point of $\sigma(A)$. Upper semi-continuity of the separate parts of the spectrum implies that λ is isolated in $\sigma(T)$; thus f_λ, the characteristic function of $\{\lambda\}$ relative to $\sigma(T) \cup \sigma(A)$, satisfies $f_\lambda(A) \equiv P(\lambda, A) =$ Riesz indempotent for A corresponding to $\{\lambda\}$, and $f_\lambda(X_n^{-1}TX_n) = X_n^{-1}P(\lambda,T)X_n$, where $P(\lambda,T)$ is the Riesz idempotent for T corresponding to $\{\lambda\}$. Since these operators are idempotents and $\lim \|X_n^{-1}P(\lambda,T)X_n - P(\lambda,A)\| = 0$, it follows that

$$\dim \mathcal{H}(\lambda, A) = \dim \mathcal{H}(\lambda, X_n^{-1}TX_n)$$

for large n, whence

5.2 $$\dim \mathcal{H}(\lambda, A) = \dim \mathcal{H}(\lambda, T);$$

in particular,

5.3 $$\sigma_0(A) \subset \sigma_0(T).$$

Let $k \geq 1$ and let $f_k(z) = z^k f_\lambda(z)$; an application of (5.1) shows that

$$\lim \|X_n^{-1} T^k P(\lambda, T) X_n - A^k P(\lambda, A)\| = 0,$$

so lower semicontinuity of rank implies

5.4 $$\operatorname{rank} A^k | \mathcal{H}(\lambda, A) \leq \operatorname{rank} T^k | \mathcal{H}(\lambda, T).$$

Let us next consider $\underset{\text{sim}}{\longrightarrow}$ invariants related to isolated points of the essential spectrum. Suppose λ is an isolated point of $\sigma_e(T)$ (not necessarily isolated in $\sigma(T)$). If $\rho : \mathcal{A}(\mathcal{H}) \to \mathcal{L}(\mathcal{H}_\rho)$ is a faithful $*$-representation, then via the Riesz Decomposition Theorem,

5.5 $$\rho(\widetilde{T}) \sim (\lambda + Q_\lambda(\widetilde{T})) \oplus R_\lambda(\widetilde{T}),$$

where $Q_\lambda \equiv Q_\lambda(\widetilde{T})$ is quasinilpotent and $\lambda \notin \sigma(R_\lambda(\widetilde{T}))$. For $\lambda \in \mathbb{C}$, define

$$k(\lambda, \widetilde{T}) = \begin{cases} 0, & \text{if } \lambda \text{ is not an isolated point of } \sigma_e(T) \\ n, & \text{if } \lambda \text{ is an isolated point of } \sigma_e(T) \text{ and } Q_\lambda \text{ is nilpotent of order n,} \\ \infty, & \text{if } \lambda \text{ is an isolated point of } \sigma_e(T) \text{ and } Q_\lambda \text{ is not nilpotent.} \end{cases}$$

The definition of $k(\lambda, \widetilde{T})$ is independent of the representation ρ. Let $\sigma_{ne}(T) = \{\lambda \in \sigma_e(T) : 0 < k(\lambda, \widetilde{T}) < \infty\}$, the set of *essentially nilpotent* points of $\sigma(T)$.

Suppose $T \underset{\text{sim}}{\longrightarrow} A$ (with $\{X_n\}$ as above). Upper semicontinuity of the separate parts of the essential spectrum implies

$$[\sigma_e(A)]_{\text{isol}} \subset [\sigma_e(T)]_{\text{isol}}\,.$$

If $\lambda \in [\sigma_e(A)]_{\text{isol}}$ and $\widetilde{f}_\lambda$ denotes the characteristic function of $\{\lambda\}$ relative to $\sigma_e(A) \cup \sigma_e(T)$, then for $k \geq 1$ and $\beta \neq 0$, let

$$g_\lambda(z) = (z - \lambda)^k \widetilde{f}_\lambda(z) + \beta(1 - \widetilde{f}_\lambda(z)).$$

Then

$$\text{5.6} \qquad \lim \|\rho(\widetilde{X}_n^{-1} g_\lambda(\widetilde{T})\widetilde{X}_n) - \rho(g_\lambda(\widetilde{A}))\| = 0.$$

Since $g_\lambda(\widetilde{T}) \sim Q_\lambda(\widetilde{T})^k \oplus \beta, g_\lambda(\widetilde{A}) \sim Q_\lambda(\widetilde{A})^k \oplus \beta$, from (5.6) it is clear that $Q_\lambda(\widetilde{T})^k = 0 \Rightarrow Q_\lambda(\widetilde{A})^k = 0$; thus

$$\text{5.7} \qquad k(\lambda, \widetilde{A}) \leq k(\lambda, \widetilde{T}) \quad (\lambda \in \mathbb{C}),$$

and

$$\text{5.8} \qquad \sigma_{ne}(A) \subset \sigma_e(T)_{\text{isol}}.$$

If $\lambda \in [\sigma_e(T)]_{\text{isol}}$ and $k = k(\lambda, T)$ satisfies $2 \leq k < \infty$, we wish to distinguish the case when $Q_\lambda^{k-1} + Q_\lambda^*$ is invertible (e.g., consider $Q_\lambda \approx q_k^{(\infty)}$). In this case the description of $\mathcal{S}(Q_\lambda)^-$ or $\mathcal{S}(T)^-$ tends to be very complicated and to date has not been completely solved. The following result of C. Apostol and D. Voiculescu [14] provides a useful alternate characterization of this case.

Proposition 5.2. *If $Q \in \mathcal{L}(\mathcal{H})$ and $\widetilde{Q}$ is nilpotent of order $k (k \geq 2)$, then the following are equivalent:*

(i) $\widetilde{Q}^{k-1} + \widetilde{Q}^*$ is invertible;

(ii) $Q \sim J + K$, where J is a nice Jordan nilpotent and K is compact.

The significance of Proposition 5.2 is that it shows that the invertibility of $\widetilde{Q}^{k-1} + \widetilde{Q}^*$ is a *similarity* invariant for essential nilpotents of order k.

Suppose now that $T \underset{\text{sim}}{\longrightarrow} A$ and $\lambda \in [\sigma_e(A)]_{\text{isol}}$. If $Q_\lambda(\widetilde{A})$ is nilpotent of order k and $Q_\lambda(\widetilde{A})^{k-1} + Q_\lambda(\widetilde{A})^*$ is invertible, then (5.6) and Proposition 5.2 imply that either

$$k(\lambda, \widetilde{T}) > k,$$

or

$$k(\lambda, \widetilde{T}) = k \text{ and } Q_\lambda(\widetilde{T})^{k-1} + Q_\lambda(\widetilde{T})^*$$

is invertible. For $\lambda \in \mathbb{C}$ we now define

$$k^+(\lambda, \widetilde{T}) = \begin{cases} 0, & \text{if } \lambda \text{ is not an isolated point of } \sigma_e(T); \\ 1, & \text{if } \lambda \in \sigma_{ne}(T) \text{ and } Q_\lambda = 0; \\ n, & \text{if } \lambda \in \sigma_{ne}(T), Q_\lambda \text{ has order } n \leq 2, \\ & \quad \text{and } Q_\lambda^{n-1} + Q_\lambda^* \text{ is not invertible;} \\ n + 1/2, & \text{if } \lambda \in \sigma_{ne}(T), Q_\lambda \text{ has order } n \geq 2, \\ & \quad \text{and } Q_\lambda^{n-1} + Q_\lambda^* \text{ is invertible;} \\ \infty, & \text{if } \lambda \text{ is an isolated point of } \sigma_e(T) \\ & \quad \text{but } Q_\lambda \text{ is not nilpotent.} \end{cases}$$

Let $\sigma_{me}(T) = \{\lambda \in \mathbb{C} : k^+(\widetilde{T}) = n + 1/2 \text{ for some } n \geq 1\}$

The preceding discussion implies that if $T \underset{\text{sim}}{\longrightarrow} A$, then

$$k^+(\lambda, \widetilde{A}) \leq k^+(\lambda, \widetilde{T}) \quad (\lambda \in \mathbb{C}).$$

We now have a collection of *algebraic* invariants for $\underset{\text{sim}}{\longrightarrow}$ based on dimensions of Riesz subspaces and orders of nilpotency. We may summarize some of these invariants for $T \underset{\text{sim}}{\longrightarrow} A$ as follows:

(A) $\dim \mathcal{H}(\lambda, A) = \dim \mathcal{H}(\lambda, T)$ for all $\lambda \in \sigma_o(A)$; $k^+(\lambda, \widetilde{A}) \leq k^+(\lambda, \widetilde{T}) \quad (\lambda \in \mathbb{C})$. If $\lambda \in \sigma_{ne}(A)$ is isolated in $\sigma(A)$, then

$$\text{rank } (A - \lambda)^k | \mathcal{H}(\lambda, A) \leq \text{rank } (T - \lambda)^k | \mathcal{H}(\lambda, T) \ (k \geq k(\lambda, \widetilde{A})).$$

Other $\underset{\text{sim}}{\longrightarrow}$ invariants arise from semi-Freholm behavior. Suppose $T \underset{\text{sim}}{\longrightarrow} A$ (with $\{X_n\}$ as above) and $\lambda \in \rho_{SF}(A)$. Since $\lim \|X_n^{-1}(T - \lambda)^k X_n - (A - \lambda)^k\| = 0$, then $\dim \ker(A - \lambda)^k \geq \limsup_{n \to \infty} \dim \ker X_n^{-1}(T - \lambda)^k X_n = \dim \ker(T - \lambda)^k$, $\dim \ker(A - \lambda)^{k^*} \geq \limsup_{n \to \infty} \dim \ker(X_n^{-1}(T - \lambda)^k X_n)^* = \dim \ker(T - \lambda)^{k^*}$, and for large n, $X_n^{-1}(T - \lambda) X_n$ is semi-Fredholm with

$$\text{ind}(A - \lambda) = \text{ind}(X_n^{-1}(T - \lambda) X_n) (= \text{ind}(T - \lambda)).$$

Thus, if $T \underset{\text{sim}}{\longrightarrow} A$, then [64] imples

$$(F) \qquad \rho_{SF}(A) \subset \rho_{SF}(T), \text{ind}(A - \lambda) = \text{ind}(T - \lambda) (\lambda \in \rho_{SF}(A)),$$

and

$$\text{min. ind.}(A-\lambda)^k \geq \text{min. ind.}(T-\lambda)^k (\lambda \in \rho_{SF}(A), k \geq 1).$$

Note that since $\sigma_{ne}(A)$ is a set of isolated points of $\sigma_e(A)$, $\sigma_{lre}(A)\backslash\sigma_{ne}(A)$ is compact, and each component of this set is a component of $\sigma_{lre}(A)$. This observation, upper semicontinuity of the separate parts of the spectrum and essential spectrum, and considerations (using (F)) about the possible *spectral pictures* for A imply that if $T \underset{\text{sim}}{\longrightarrow} A$, then

(S) $\qquad \sigma_0(A) \subset \sigma_0(T)$ and every componet of $\sigma_{lre}(A)\backslash\sigma_{ne}(A)$

intersects

$$\sigma_e(T)\backslash\sigma_{ne}(T).$$

For $T, A \in \mathcal{L}(\mathcal{H})$, we say that T *spectrally dominates* $A(T >_{sp} A)$ if T and A satisfy (S), (F), and (A). (It may seem surprising that the definition of $>_{sp}$ does not explicitly include all the necessary conditions for dominance that we found in the first lecture; however, it is a relatively straightforward (though not trivial) exercise to verify that $T >_{sp} A$ implies (1.2)-(1.5).) It is now easy to state the main results of C. Apostol, D. Herrero, and D. Voiculescu [10,5].

Similarity Orbit Theorem (Reduced Form). If

$$T \in \mathcal{L}(\mathcal{H}) \text{ and } \sigma_{me}(T) \subset \text{interior } \sigma(T),$$

then

$$\mathcal{S}(T)^- = \{A \in \mathcal{L}(\mathcal{H}) : T >_{sp} A\}.$$

Similarity Orbit Theorem (General Form). Let $T \in \mathcal{L}(\mathcal{H})$. An operator A belongs to $\mathcal{S}(T)^-$ if and only if the following conditions are satisfied:

(1) $T >_{sp} A$;
(2) For every $\lambda \in \sigma_{me}(A) \cap \text{bdry } \sigma(A)$ such that $k^+(\lambda, \widetilde{T}) = k^+(\lambda, \widetilde{A})$, and for every closed-and-open subset σ_λ of $\sigma(A)$ such that $\sigma_\lambda \cap \sigma_e(A) = \{\lambda\}, \sigma_\lambda$ is closed-and-open in $\sigma(T)$, and $T|\mathcal{H}(\sigma_\lambda, T) \underset{\text{sim}}{\longrightarrow} A|\mathcal{H}(\sigma_\lambda, A)$.

Unfortunately, the characterization of $\mathcal{S}(T)^-$ essentially nilpotent operators corresponding to points of $\sigma_{me}(\cdot)$ is incomplete; we address this issue briefly in the sequel.

The Calkin algebra analogue of the orbit theorem comes from looking at the *essential* ingredients of $>_{sp}$.

Similarity Orbit Theorem (Calkin Algebra). Let $T, A \in \mathcal{L}(\mathcal{H})$. Then $\widetilde{A} \in \mathcal{S}(\widetilde{T})^-$ if and only if the following conditions hold:

(s) Each component of $\sigma_{lre}(A) \backslash \sigma_{ne}(A)$ intersects $\sigma_e(T) \backslash \sigma_{ne}(T)$;

(f) $\rho_{SF}(A) \subset \rho_{SF}(T)$ and $\text{ind}(A - \lambda) = \text{ind}(T - \lambda)(\lambda \in \rho_{SF}(A))$;

(a) $k^+(\lambda, \widetilde{A}) \leq k^+(\lambda, \widetilde{T})(\lambda \in \mathbb{C})$.

The proofs of the orbit theorems require all the techniques we have previously discussed, and several new ones concerning, for example, canonical #-models and one-sided resolvents. As a first indication of the scope of these results, consider again the main results of Lectures 2–4. Let $\{e_n\}$ denote an orthonormal basis for $\mathcal{H}$ and let K denote the unilateral weighted shift defined by $Ke_n = (1/n)e_{n+1}$. Let $Q = K^{(\infty)}$; then Q is quasisnilpotent and Q^k is not compact for every $k \geq 1$. For $m \geq 1$, let K_m denote the operator obtained from K by replacing the weight 1/jm by $0 (j = 1, 2, \ldots)$, and let $Q_m = K_m^{(\infty)}$. Cleraly, Q_m is nilpotent and $\lim \|Q - Q_m\| = 0$. Thus $\mathcal{S}(Q) \subset \mathcal{N}(\mathcal{H})^-$. Now suppose $A \in \mathcal{L}(\mathcal{H})$ satisfies the hypotheses of Theorem 2.1: $\sigma(A)$ and $\sigma_e(A)$ are connected and contain 0, and $\text{ind}(A - \lambda) = 0 (\lambda \in \rho_{SF}(A))$. It is trivial to verify that $Q >_{sp} A$, and since $\sigma_{me}(Q) = \phi$, the Similarity Orbit Theorem implies $Q \underset{\text{sim}}{\longrightarrow} A$. Thus $A \in \mathcal{S}(Q)^- \subset \mathcal{N}(\mathcal{H})^-$. This argument proves Theorem 2.1.

Let Q be any quasinilpotent operator such that Q^n is noncompact for every $n \geq 1$. It is clear that if R is quasinilpotent, then $Q >_{sp} R$, whence $Q \underset{\text{sim}}{\longrightarrow} R$ by the Similarity Orbit Theorem; this proves Theorem 4.1.

Considerations with the possible types of *spectral pictures* show that if $\sigma(T)$ is infinite, then the Similarity Orbit Theorem implies $\mathcal{S}(T)$ is not closed. On the other hand, if $\sigma(T) = \{0\}$ and $\mathcal{S}(T)$ is closed, then the conditions of the theorem readily imply $T = 0$; this essentially proves one direction in Theorem 3.10.

The orbit theorems can be used to give short proofs to many other results in approximation theory. We cite a typical example from [5].

Theorem 5.3. (C. Apostol and B. Morrell [11]) *Let σ be a nonempty subset of $\mathbb{C}$ and let $\mathcal{S}(\sigma) = \{T \in \mathcal{L}(\mathcal{H}) : \sigma(T) \subset \sigma\}$. Then $\mathcal{S}(\sigma)^-$ is the set of all operators A in $\mathcal{L}(\mathcal{H})$ such that:*

(1) Every component of $\sigma_0(A) \cup \sigma_e(A)$ intersects σ^-; and

(2) $\{\lambda \in \rho_{SF}(A) : \text{ind}(A - \lambda) \neq 0\} \subset \sigma$.

To conclude, we wish to indicate a few recent developments related to the similarity orbit theorems. The *open* part of the orbit theorem concerns operators T with the following property:

(5.9) There exists $k \geq 2$ such that T^k is an infinite rank compact operator and $\widetilde{T}^{k-1} + \widetilde{T}^*$ is invertible.

This class of essentially nilpotent opertors has been studied by C. Apostol and C. Herrero, who completely characterize $\mathcal{S}(T)^-$ in case T is an algebraic operator satisfying (5.9) for $k = 2$ [9]. Surprisingly, the results of [9] suggest that a general solution to the orbit problem for essential nilpotents may require at least some understanding of closures of joint similarity orbits of n-tuples of operators. Two n-tuples $T = (T_1, \ldots, T_n)$ and $A = (A_1, \ldots, A_n)$ are *jointly similar* if there exists an invertible operator X such that $A_i = X^{-1}T_iX (1 \leq i \leq n)$; the relations $\approx, \approx_a, \approx_K, \underset{\text{sim}}{\longrightarrow}$ can be similarly formulated for tuples of operators. Closures of joint similarity orbits have been studied recently by R. Curto and D. Herrero [27]; among the results of [27] are the following:

(i) *Generalized Rota models for n-tuples of commuting operators* (cf. J. Ball [15]);
(ii) *Corollary:* If Q is a commutative n-tuple of quasinilpotent operators, then $Q \underset{\text{sim}}{\longrightarrow} 0$.
(iii) *Normal limits of nilpotent tuples:* For a normal n-tuple N the following are equivalent:

 (a) N is a limit of n-tuples of commuting nilpotent operators.
 (b) N is a limit of n-tuples of commuting quasinilpotent operators.
 (c) $\sigma_T(N)$ (the Taylor spectrum of N) is connected and contains the origin.

(iv) *Existence of compact universal quasinilpotent tuples:* There exists a compact, quasinilpotent, commutative n-tuple K^u such that if K is a commutative n-tuple of compact quasinilpotents, then $K \in \mathcal{S}(K^u)^-$.
(v) *Closed Orbits:* (commutative case). For a commutative n-tuple T the following are equivalent:

 (a) $\mathcal{S}(T)$ is closed;
 (b) $\mathcal{S}(T_j)$ is closed $(1 \leq j \leq n)$;
 (c) T is similar to a normal n-tuple with finite Taylor spectrum.

The orbit theorems have rather diverse applications; we cite some recent examples. Recall the concept of a *finite operator* [83] introduced by J. P. Williams: $T \in \mathcal{L}(\mathcal{H})$ is finite if $\|TX - XT - 1\| \geq 1$ for every $X \in \mathcal{L}(\mathcal{H})$. The set of finite operators in $\mathcal{L}(\mathcal{H})$

contains the set of quasidiagonal operators as a proper subset. In [83], Williams asked for a characterization of the operators each of whose similarities is finite.

Theorem 5.4. (L. Fialkow and D. Herrero [42]) *For $T \in \mathcal{L}(\mathcal{H})$ the following are equivalent:*

(1) Every operator similar to T is finite;

(2) Every operator similar to T is quasidiagonal;

(3) $\widetilde{T}$ satisfies a nonzero linear or quadratic polynomial

The equivalence of (2) and (3) is due to D. Herrero [62]; the remaining equivalence depends on the Similarity Orbit Theorem.

In a different direction, recall that operators T and S are *quasisimilar* ($T \sim_{qs} S$) if there exist operators X and Y, each one-to-one with dense range, such that $TX = XS$ and $YT = SY$ [77]. Several classes of operators, e.g. cyclic subnormal operators [25] and algebraic operators [4], [85], have nice models up to quasisimilarity. Some very basic questions about quasisimilarity remain open: it is known that if $T \sim_{qs} S$, then every component of $\sigma(S)$ intersects $\sigma(T)$, but it is unknown whether every component of $\sigma_e(S)$ intersects $\sigma_e(T)$. Although this question has been affirmatively answered in the case when $T = V$, the unilateral shift (see [38]), nevertheless the shift does provide a focal point for several other quasisimilarity questions. It would thus be of interest to characterize $(V)_{qs}$, the quasisimilarity orbit of V. The Similarity Orbit Theorem can be used to show that

$$(V)_{qs} \subset \mathcal{S}(V)^- \text{ [38].}$$

More generally, if T is a cyclic hyponormal unilateral weighted shift or a cyclic subnormal operator quasisimilar to V, then $(T)_{qs} \subset \mathcal{S}(T)^-$. Moreover, a normal operator T has this property if and only if $[\sigma_e(T)]_{\text{isol}} \subset [\sigma(T)]_{\text{isol}}$. Which other operators T satisfy $(T)_{qs} \subset \mathcal{S}(T)^-$? There are also connections between quasisimilarity and asymptotic similarity which depend on the Similarity Orbit Theorem:

Theorem 5.5. [38]

(1) Quasisimilar cyclic subnormal operators are asymptotically similar (cf. [69]).

(2) If T and S are quasisimilar bilateral weighted shifts and 0 is not an isolated point of $\sigma_e(T)$, then T and S are asymptotically similar.

A very recent application of approximation theory and the Similarity Orbit Theorem concerns roots and logarithms of operators. let

$$\mathcal{R}_p = \{A^p : A \in \mathcal{L}(\mathcal{H})\}, p = 1, 2, \ldots,$$

and

$$\mathcal{E} = \{\exp(A) : A \in \mathcal{L}(\mathcal{H})\}.$$

In [26] J. Conway and B. Morrel characterized the norm closures of these sets as follows: $T \in \mathcal{R}_p^-$ if and only if the set $\{\lambda \in \mathbb{C} : T - \lambda$ is Fredholm and $\operatorname{ind}(T - \lambda)$ is neither 0 nor divisible by $p\}$ does not separate 0 from ∞; $T \in \mathcal{E}^-$ if and only if the set

$$\{\lambda \in \mathbb{C} : \lambda \in \rho_{SF}(T) \text{ and } \operatorname{ind}(T - \lambda) \neq 0\}$$

does not separate 0 from ∞.

We have already demonstrated (in at least a limited way) the role of Apostol's Lemma in approximation; the setting for this result is bounded operators on Hilbert space. By modifying the proof a bit, it is possible to prove an analogue for n-tuples of commuting operators with respect to Taylor spectra. This result provides a key tool in describing the spectral picture of a system of left and right multiplication operators.

Theorem 5.6. (R. Curto and L. Fialkow [28]) *Let A and B each be commutative n-tuples of operators on $\mathcal{H}$ and let L_A and R_B denote the corresponding n-tuples of left and right multiplication operators on $\mathcal{L}(\mathcal{H})$.*

(1) $\sigma_T(L_A, R_B) = \sigma_T(A) \times \sigma_T(B)$.
(2) $\sigma_{Te}(L_A, R_B) = \sigma_T(A) \times \sigma_{Te}(B) \cup \sigma_{Te}(A) \times \sigma_T(B)$.
(3) If A and B are Fredholm, then (L_A, R_B) is Fredholm and $\operatorname{ind}(L_A, R_B) = (-1)^n \operatorname{ind}(A)\operatorname{ind}(B)$.

This result extends a long line of research emanating from Rosenblum's Theorem on linear operator equations and from the Brown-Pearcy Theorem $\sigma(A \otimes B) = \sigma(A)\sigma(B)$ [21]. The pleasing and sometimes surprising interplay between modern approximation theory and classical operator theory can be expected to be a source of rich mathematical development in the future.

References

1. Apostol, C., The correction by compact perturbations of the singular behavior of operators, Rev. Roum. Math. Pures Appl. **21** (1976), 155-175.
2. ———, Inner derivations with closed range, Rev. Roum. Math. Pures Appl **21** (1976), 249-265.
3. ———, Universal quasinilpotent operators, Rev. Roum. Math. Pures Appl. **25** (1980), 135-138.
4. ———, R. Douglas, and C. Foias, Quasi-similar models for nilpotent operators, Trans. Amer. Math. Soc. **224** (1976), 407-415.
5. ———, L. Fialkow, D. Herrero, and D. Voiculescu, Approximation of Hilbert space operators. Vol. II, Research Notes in Mathematics **102** (1984), Pitman.
6. ——— and C. Foias, Operators with closed similarity orbits, Rev. Roum. Math. Pures Appl. **22** (1977), 13-15.
7. ———, C. Foias, and D. Voiculescu, Some results on nonquasitriangular opertors IV, Rev. Roum. Math. Pures Appl. **18** (1973), 487-514.
8. ———, C. Foias, and D. Voiculescu, On the norm-closure of nilpotents II, Rev. Roum. Math. Pures Appl **19** (1974), 549-577.
9. ——— and D. A. Herrero, On the closures of similarity orbits of essential nilpotent operators of order 2, Integral Equations and Operator Theory **8** (1985), 437-461.
10. ———, D. A. Herrero, and D. Voiculescu, The closure of the similarity orbit of a Hilbert space operator, Bull. Amer. Math. Soc. **6** (1982), 421-426.
11. ——— and B. B. Morrel, On uniform approximation of operators by simple models, Indiana Univ. Math. J. **26** (1977), 427-442.
12. ——— and J. Stampfli, On derivation ranges, Indiana Univ. Math. J. **25** (1976), 857-869.
13. ——— and D. Voiculescu, On a problem of Halmos, Rev. Roum. Math. Pures Appl. **19** (1974), 283-284.
14. ——— and D. Voiculescu, Closure of similarity orbits of nilpotent and quasinilpotent operators, preprint, 1977.
15. Ball, J. A., Rota's theorem for general functional Hilbert spaces, Proc. Amer. Math. Soc. **64** (1977), 55-61.
16. Barria, J. and D. A. Herrero, Closures of similarity orbits of Hilbert space operators IV. Normal operators, J. London Math. Soc. **17** (1978), 525-536.

17. ______ and D. A. Herrero, Closure of similarity orbits of nilpotent operators I. Finite rank operators, J. Operator Theory **1** (1979), 177-185.

18. ______ and D. A. Herrero, Closure of similarity orbits of nilpotent operators II, preprint, 1977.

19. Berg, I. D., An extension of the Weyl-von Neumann Theorem to normal operators, Trans. Amer. Math. Soc. **160** (1971), 365-371.

20. ______, On approximation of normal operators by weighted shifts, Michigan Math. J. **21** (1974), 377-383.

21. Brown, A. and C. Pearcy, Spectra of tensor products of operators, Proc. Amer. Math. Soc. **17** (1966), 162-169.

22. ______ and C. Pearcy, Compact restrictions of operators, Acta Sci. Math. (Szeged) **32** (1971), 271-282.

23. Brown, L. G., R. G. Douglas, and P. A. Filmore, Unitary equivalence modulo the compact operators and extensions of C^*-algebras, Lecture Notes in Math. Vol. 345 (1973), 58-128, Springer-Verlag.

24. Calkin, J. W., Two-sided ideals and congruences in the ring of bounded operators in Hilbert space, Ann. of Math. **42** (1941), 839-873.

25. Clary, W. S., Quasisimilarity and subnormal operators, Ph.D. Dissertation, Univ. of Michigan, 1973.

26. Conway, J. and B. Morrel, Roots and logarithms of bounded operators on Hilbert space, preprint, 1986.

27. Curto, R. and D. Herrero, On closures of joint similarty orbits, Integral Equations and Operator Theory **8** (1985), 489-556.

28. ______ and L. Fialkow, The spectral picture of (L_A, R_B), Journal Funct. Anal. **71** (1987), 371-392.

29. Davidson, K., The distance between unitary orbits of normal elements in the Calkin algebra, preprint, 1984.

30. Davis, C. and P. Rosenthal, Solving linear operator equations, Canad. J. Math. **26** (1974), 1384-1389.

31. Deckard, D. and L. Fialkow, Characterization of Hilbert space operators with unitary cross sections, J. Operator Theory **2** (1979), 153-158.

32. Douglas, R. and C. Pearcy, A note on quasitriangular operators, Duke Math. J. **37** (1970), 177-188.

33. ——— and C. Pearcy, Invariant subspaces of nonquasitriangular operators, Lecture Notes in Math. **345** (1973), 13-57, Springer-Verlag.

34. Fialkow, L., The similarity orbit of a normal operator, Trans. Amer. Math. Soc. **210** (1975), 129-137.

35. ———, A note on unitary cross sections for operators, Canadian J. Math. **30** (1978), 1215-1227.

36. ———, Similarity cross sections for operators, Indiana Univ. Math. J. **28** (1979), 71-86.

37. ———, Spectral properties of elementary operators II, Trans. Amer. Math. Soc. **290** (1985), 415-429.

38. ———, Quasisimilarity and closures of similarity orbits of operators, J. Operator Theory **14** (1985), 215-238.

39. ———, The index of an elementary operator, Indiana Univ. Math. J. **73** (1986), 73-102.

40. ——— and D. Herrero, Characterization of Hilbert space operators with local similarity cross sections, preprint, 1978.

41. ——— and D. Herrero, Inner derivations with closed range in the Calkin algebra, Indiana Univ. Math. J. **33** (1984), 185-211.

42. ——— and D. Herrero, Finite operators and similarity orbits, Proc. Amer. Math. Soc. **93** (1985), 601-609.

43. Fillmore, P. A., J. G. Stampfli, and J. P. Williams, On the essential spectrum, the essential numerical range and a problem of Halmos, Acta Sci. Math. (Szeged) **33** (1972), 179-192.

44. Gray, L., Jordan representation for a class of operators, Indiana Univ. Math. J. **26** (1977), 57-64.

45. Hadwin, D. W., Closures of unitary equivalence classes, Dissertation, Indiana University, 1975.

46. ———, An operator valued spectrum, Indiana Univ. Math. J. **26** (1977), 329-340.

47. Halmos, P. R., Quasitriangular operators, Acta Sci. Math. (Szeged) **29** (1968), 283-293.

48. ———, Ten problems in Hilbert space, Bull. Amer. Math. Soc. **76** (1970), 887-933.

49. ———, Capacity in Banach algebras, Indiana Univ. Math. J. **20** (1971), 855-863.

50. ———, Limits of shifts, Acta Sci. Math. (Szeged) **34** (1973), 131-139.

51. ———, Spectral approximation of normal operators, Proc. Edinburgh Math. Soc. **19** (1974), 51-58.

52. Harte, R., Tensor products, multiplication operators and the spectral mapping theorem, Irish Acad. Sect. A. **73** (1973), 285-302.

53. Hedlund, J. H., Limits of nilpotent and quasinilpotent operators, Mich. Math. J. **19** (1972), 249-255.

54. Herrero, D. A., Normal limits of nilpotent operators, Indiana Univ. Math. J. **23** (1974), 1097-1108.

55. ———, Toward a spectral characterization of the set of norm limits of nilpotent operators, Indiana Univ. Math. J. **24** (1975), 847-864.

56. ———, Clausura de las orbitas de similaridad de operadores en espacios de Hilbert, Rev. Un. Mat. Argentina **27** (1976), 244-260.

57. ———, Closure of similarity orbits of Hilbert space operators II. Normal operators, J. London Math. Soc **13** (1976), 299-316.

58. ———, A Rota universal model for operators with multiply connected spectrum, Rev. Roum. Math. Pures Appl. **21** (1976), 15-23.

59. ———, Universal quasinilpotent operators, Acta Sci. Math. **38** (1976), 291-300.

60. ———, Almost every quasinilpotent Hilbert space operator is a universal quasinilpotent, Proc. Amer. Math. Soc. **71** (1978), 212-216.

61. ———, Closure of similarity orbits of Hilbert space operators III, Math. Ann. **232** (1978), 195-204.

62. ———, Quasidiagonality, similarity and approximation by nilpotent operators, Indiana Univ. Math. J. **30** (1981), 199-233.

63. ———, Approximation of Hilbert space operators, Volume I, Research Notes in Math. **72**, Pitman, 1982.

64. Kato, T., *Perturbation Theory for Linear Operators*. Springer-Verlag, New York, 1966.

65. Olsen, C. L., A structure theorm for polynomially compact opertors, Amer. J. Math. **93** (1971), 686-698.

66. Pearcy, C. and N. Salinas, Compact perturbations of seminormal operators, Indiana Univ. Math. J. **22** (1973), 789-793.

67. ——— and N. Salinas, Operators with compact self-commutator, Canadian J. Math. **26** (1974), 115-120.

68. Pecuch-Herrero, M., Global cross sections of unitary and similarity orbits of Hilbert space operators, J. Operator Theory **12** (1984), 265-283.
69. Raphael, M., Quasisimilarity and essential spectra for subnormal operators, Indiana Univ. Math. J. **31** (1982), 243-246.
70. Riesz, F. and B. Sz.-Nagy, *Functional Analysis*, Ungar, New York, 1955.
71. Rosenblum, M., On the operator equation $BX - XA = Q$, Duke Math J. **23** (1956), 263-269.
72. Rota, G.-C., On models for linear operators, Comm. Pure Appl. Math. **13** (1960), 469-472.
73. Salinas, N., Reducing essential eigenvalues, Duke Math. J. **40** (1973), 561-580.
74. Sikonia, W., The von Neumann converse of Weyl's theorem, Indiana Univ .Math. J. **21** (1971), 121-123.
75. Stampfli, J., The norm of a derivation, Pac. J. Math. **33** (1970), 737-747.
76. ———, Compact perturbations, normal eigenvalues and a problem of Salinas, J. London Math. Soc. (2) **9** (1974), 165-175.
77. Sz.-Nagy, B. and C. Foias, *Harmonic Analysis of operators on Hilbert space*. North Holland–American Elsevier, 1970.
78. Voiculescu, D., Norm-limits of algebraic operators, Rev. Roum. Math. Pures Appl. **19** (1974), 371-378.
79. ———, A non-commutative Weyl-von Neumann theorem, Rev. Roum. Math. Pures Appl. **21** (1976), 97-113.
80. Von Neumann, J., *Charakterisierung des Spektrums eines Integraloperators*. Hermann, Paris, 1935.
81. West, T.,
82. Weyl, H., Uber beschrankte quadratische Formen deren Differenz vollstetig ist, Ren. Circ. Mat. Palermo **27** (1909), 373-392.
83. Williams, J. P., Finite operators, Proc. Amer. Math. Soc. **26** (1970), 129-136.
84. Williams, L. R., Similarity invariants for a class of nilpotent operators, Acta Sci. Math. (Szeged) **38** (1976), 423-428.
85. ———, A quasisimilarity model for algebraic operators, Acta Sci. Math. **40** (1978), 185-188.

Lawrence A. Fialkow
Department of Mathematics and Computer Science
State University of New York at New Paltz
New Paltz, New York, 12561, U.S.A.

Acknowledgment. The author is grateful to Professor John B. Conway for the invitation to present the contents of these notes to the Asymmetric Algebras and Invariant Subspaces Conference at Indiana University in March, 1986. The author also thanks Professors K. Davidson, C. Foias, R. Harte, and D. Herrero for helpful comments and references concerning these notes. The author was partially supported by a National Science Foundation Grant.

Toward a Theory of K-Spectral Sets

by

Vern I. Paulsen

These expository notes are intended to accompany a series of lectures given at the conference on *C^*-algebras and Single Operators* at Indiana University.

The theories of spectral and K-spectral sets are partially an attempt to develop model theories for operators on Hilbert space and can also be viewed as an attempt to develop a representation theory for the simplest kinds of commutative operator algebras.

We attempt to survey the central results and the main unanswered questions.

I. Introduction

In 1951 J. von Neumann [18] proved that an operator T on a Hilbert space is a contraction if and only if $\|p(T)\| \leq \sup\{|p(z)| : |z| \leq 1\}$ for all polynomials. This result has come to be known as *von Neumann's inequality.* In 1953, B. Sz.-Nagy [30] proved that if T is a contraction on a Hilbert space H, then there is a unitary operator U on a Hilbert space K containing H as a subspace such that

$$(^*)\ T^n = P_H U^n|_H, \text{ for all integers } n \geq 0,$$

which implies von Neumann's inequality. Sarason [24] observed that for relation $(^*)$ to hold, $H = M \ominus N$, where M and N are invariant subspaces for U, and any such subspace H is called a *semi-invariant subspace.* Thus, Sz.-Nagy's theorem shows that restrictions of unitary operators to semi-invariant subspaces serve as models for all contraction operators.

The theory of spectral sets is an attempt to generalize this correspondence to more general normal operators.

Let X be a compact set in the complex plane $\mathbb{C}$, and let ∂X denote its boundary. We'll let $C(\partial X)$ denote the continuous complex-valued functions on ∂X, equipped with the sup norm topology and let $R(X)$ denote the subalgebra consisting of all quotients of polynomials p/q where the $0'$s of q lie off X.

If T is an operator on H with spectrum, $\sigma(T)$, contained in X, then $f(T)$ is defined for all f in $R(X)$ and the map $\rho(f) = f(T)$ defines a homomorphism from $R(X)$ into $L(H)$, the algebra of bounded, linear operators on H.

We call X a *spectral set* for T if $\|f(T)\| \leq \|f\|$ for all f in $R(X)$, i.e., if $\|\rho\| \leq 1$.

We say that T has a *normal ∂X-dilation*, if there exists a Hilbert space K containing H and a normal operator N on K with $\sigma(N)$ contained in ∂X, such that,

$$f(T) = P_H f(N)|_H, \tag{1}$$

for all f in $R(X)$.

It is easy to see that if T has a normal ∂X-dilation, then X is a spectral set for T. The question of whether the converse of this statement is true is discussed in Sz.-Nagy-Foias [31], and is still largely unresolved.

Problem 1: If X is a spectral set for T, then must T have a normal ∂X-dilation?

If we are given that X is a spectral set for T, then that is equivalent to being given a contractive homomorphism ρ of $R(X)$ with $\rho(z) = T$. Analogously, if we are given a normal operator N, with $\sigma(N)$ contained in ∂X, then that is equivalent to being given a $*$-homomorphism π of $C(\partial X)$ with $\pi(Z) = N$. In this notation formula (1) becomes

$$\rho(f) = P_H \pi(f)|_H. \tag{1'}$$

Thus, we see that Problem 1 is equivalent to the problem of whether or not every unital, contractive homomorphism ρ of $R(X)$ is of the form (1′). Problem 1 can be viewed as a question about the representation theory of $R(X)$. There are no known counterexamples to Problem 1, and only a few instances where the anwser is known to be affirmative. The answer is yes when:

(i) $X = \mathbb{D}^-$ the closed unit disc (Sz.-Nagy's dilation Theorem),

(ii) $R(X) + \overline{R(X)}$ is dense in $C(\partial X)$ (Berger [4], Foias [10], Lebow [15]),

(iii) T is a 2×2 matrix with 1 eigenvalue (Misra [16]),

(iv) T is a 2×2 matrix ([23]),

(v) X is an annulus (Agler [1]),

(vi) if and only if X is a complete spectral set (Arveson [3]),

(vii) if X is finitely connected, then there is an operator similar to T which has a normal ∂X-dilation (Douglas-P [8]).

The question is still open for sets with 2 holes, and when T is a 3×3 Jordan block and X is arbitrary.

In Section 2 we discuss the role that positivity plays in the study of this question. In Section 3, we discuss complete positivity and touch briefly on the multi-dimensional case. Section 4 introduces K-spectral sets, completely bounded maps, and similarity theory. Section 5 develops a theory, up to similarity for finitely connected sets. Finally, Section 6 returns to the multi-dimensional theory.

2. Spectral Sets and Positivity

Generally, one proves Sz.-Nagy's dilation theorem by actually geometrically constructing a larger Hilbert space and a unitary transformation on it, which satisfies the conclusions of the theorem. The inequality of von Neumann can then be deduced as a corollary. It is fairly well-known that if one starts with von Neumann's inequality, then one can also prove Sz.-Nagy's dilation theorem.

In order to illustrate the central role of positivity considerations, we want to show how to deduce von Neumann's inequality from first principles, using positivity considerations. Most of the ideas of this section can be found in Arveson [3].

The key concept for using positivity is an *operator system*. Let B be any C^*-algebra with unit, then any subspace S of B containing 1 and satisfying $a^* \varepsilon S$ whenever $a \varepsilon S$ is called an operator system. We write $p \geq 0$ to indicate that p is a positive element of S.

It is not hard to check that an operator system has sufficiently many positive elements such that if $a \varepsilon S$, then $a = (p_1 - p_2) + i(p_3 - p_4)$ where $p_i \geq 0$ are positive elements of S and $\|p_i\| \leq \|a\|, i = 1, 2, 3, 4$. A linear map $\phi : S \to L(H)$ is called *positive*, if $\phi(p) \geq 0$ whenever $p \geq 0$.

Lemma 2.1. *Let S be an operator system and $\phi : S \to L(H)$ positive, then ϕ is bounded.*

Proof: Note that if $-1 \leq p \leq 1$ then $-\phi(1) \leq \phi(p) \leq \phi(1)$ and so $\|\phi(p)\| \leq \|\phi(1)\|$. Thus, for $p = p^*$, $\|\phi(p)\| \leq \|p\| \cdot \|\phi(1)\|$, and consequently, $\|\phi(a)\| \leq 2\|a\| \cdot \|\phi(1)\|$, since $a = h + ik$ with $h = h^*, k = k^*$ and $\|h\|, \|k\| \leq \|a\|$. Actually, with slightly more care, one can show that 2 is sharp [3].

Lemma 2.2. *Let $P_i \in L(H), P_i \geq 0$ and let $\lambda_i \in \mathbb{C}, |\lambda_i| \leq 1, i = 1, \ldots, n$. If $Q = P_1 + \ldots + P_n$, then $\|\lambda_1 P_1 + \ldots + \lambda_n P_n\| \leq \|Q\|$.*

Proof: Fix $x, y \in H$, then

$$\begin{aligned} | < (\lambda_1 P_1 + \ldots + \lambda_n P_n)x, y > | &\leq | < \lambda_1 P_1^{1/2} x, P_1^{1/2} y > | + \ldots + |\lambda_n P_n^{1/2} x, P_n^{1/2} y > | \\ &\leq \|P_1^{1/2} x\| \cdot \|P_1^{1/2} y\| + \ldots + \|P_n^{1/2} x\| \cdot \|P_n^{1/2} y\| \\ &\leq \|Qx\|^{1/2} \|x\|^{1/2} \|Qy\|^{1/2} \|y\|^{1/2} \\ &\leq \|Q\| \cdot \|x\| \; \|y\|. \end{aligned}$$

Lemma 2.3. *Let Y be a compact, Hausdorff space and let $\phi : C(Y) \to L(H)$ be positive, then $\|\phi\| = \|\phi(1)\|$.*

Proof: Given $f \in C(Y), \|f\| \leq 1$ use a partition of unity argument to approximate f uniformly by a sum of the form $\lambda_1 p_1(y) + \ldots + \lambda_n p_n(y)$, where $\lambda_i = f(y_i)$ and $p_i \geq 0$, with $p_1 + \ldots + p_n = 1$. The image of this sum under ϕ is, $\lambda_1 \phi(p_1) + \ldots + \lambda_n \phi(p_n)$, which by Lemma 2.2 has norm at most $\|\phi(1)\|$. By Lemma 2.1, ϕ is continuous and so $\|\phi(f)\| \leq \|\phi(1)\|$. ■

Lemma 2.3 is a special case of the Russo-Dye Theorem.

These are the only general positivity considerations needed to prove von Neumann's inequality.

Theorem 2.4. *Let $\|T\| \leq 1$, let $\mathbb{T}$ denote the unit circle and let $S = \{p + \overline{q} :$ p, q are polynomials$\} \subseteq C(\mathbb{T})$. Then the map $\phi : S \to L(H)$ given by $\phi(p + \overline{q}) = p(T) + q(T)^*$ is positive.*

Proof: We may assume $\|T\| < 1$. Set

$$P(t, T) = (1 - e^{-it} T)^{-1} + (1 - e^{-it} T^*)^{-1} - 1,$$

the operator-valued Poisson kernel for T.

It is straight-forward to check that,

$$\phi(p + \overline{q}) = \frac{1}{2\pi} \int_0^{2\pi} (p(e^{it}) + \overline{q(e^{it})}) P(t, T) dt$$

and that $P(t, T) \geq 0$ for all t. Hence, the map ϕ is positive.

Corollary 2.5. (von Neumann) *Let $\|T\| \leq 1$, then $\|p(T)\| \leq \|p\|$ for every polynomial p.*

Proof: The space S is dense in $C(\mathbb{T}), \phi$ is bounded by Lemma 2.1 and so extends to a positive map on $C(\mathbb{T})$. But then by Lemma 2.3, $\|\phi\| \leq 1$. ■

Given the above theorem, it is possible to deduce Sz.-Nagy's dilation theorem from a theorem of M. A. Naimark. In 1943, Naimark [17] proved that positive operator valued measures can be dilated to spectral measures. There is a natural correspondence between operator-valued measures and linear maps on the space of continuous functions given by integration. Using this corespondence, we can state Naimark's theorem as:

Theorem 2.6. (Naimark) *Let Y be a compact, Hausdorff space and let $\phi : C(Y) \to L(H)$ be positive. Then there exists a Hilbert space K, a bounded operator $V : H \to K$ and a unital *-homomorphism, $\pi : C(Y) \to L(K)$ such that $\phi(f) = V^*\pi(f)V$ for all $f \in C(Y)$.*

Note that when $\phi(1) = 1$, then V is an isometry and after identifying VH with H, the formula becomes $\phi(f) = P_H\pi(f)|_H$.

Corollary 2.7. (Sz.-Nagy) *Let $T \in L(H), \|T\| \leq 1$, then there exists unitary operator U on a Hilbert space $K, H \leq K$, such that $T^n = P_H U^n|_H, n = 1, 2, \ldots$.*

Proof: The map ϕ of Theorem 2.4 extends to a positive map on $C(\mathbb{T})$. Apply Naimark's Theorem to this map. ■

Remark 2.8. If the numerical radius of T is less than 1, then $Q(t,T) = P(t,T)+1 \geq 0$ for all t. This implies that the map defined by

$$\psi(p+\overline{q}) = \frac{1}{4\pi}\int_0^{2\pi}(p(e^{it}) + \overline{q(e^{it})})Q(t,T)dt = \frac{1}{2}(p(T) + q(T)^*) + \frac{1}{2}(p(0) + q(0))I,$$

is positive. Applying Naimark's Theorem to this map yields

$$\frac{1}{2}T^n = \psi(Z^n) = P_H U^n|_H, n = 1, 2, \ldots,$$

which is Berger's *strange dilation Theorem* [5].

The full picture of the relationship between positivity and spectral sets can now be given.

Proposition 2.9. [3] *Let B be a C^*-algebra with unit, M a subspace, $1 \in M$ and let $\phi : M \to L(H)$ be a unital, contractive mapping. Then there is a well-defined positive map $\phi : M + M^* \to L(H)$ given by $\phi(a + b^*) = \phi(a) + \phi(b)^*$.*

The converse of the above proposition is not true in general [3], however for the case in which we are interested, it is.

Proposition 2.10. [22] *Let B be a C^*-algebra with unit and let A be a subalgebra, $1 \in A$. Then $\phi : A \to L(H)$ is a unital, contractive map if and only if $\phi : A + A^* \to L(H)$ is a unital, positive map.*

Combining these two results, we have:

Theorem 2.11. *Let $T \in L(H)$ with $\sigma(T) \subseteq X$. Then X is a spectral set for T if and only if the map $\widetilde{\rho} : R(X) + \overline{R(X)} \to L(H)$ given by $\widetilde{\rho}(f + \overline{g}) = f(T) + g(T)^*$ is positive.*

Theorem 2.12. [3] *Let X be a spectral set for T. Then T has a normal ∂X-dilation if and only if the positive map $\widetilde{\rho}$ extends to a positive map on $C(\partial X)$.*

Proof: Apply Naimark's dilation Theorem to the extension of ρ.

Corollary 2.13. (Berger [4], Foias [10], Lebow [15]) *Let X be a spectral set for T. If $R(X) + \overline{R(X)}$ is dense in $C(\partial X)$, then T has a normal ∂X-dilation.*

Let X be an analytic, Cauchy domain, that is, X is a connected, compact set whose boundary consists of $n+1$ disjoint, analytic, Jordan curves. If we let S denote the closure of $R(X)) + \overline{R(X)}$, then by a Theorem of Walsh [34], S has codimension n in $C(\partial X)$. In fact, if we choose one point in each bounded component of the complement of X, say $z_i, i = 1, \ldots, n$, then $C(\partial X)$ is the span of S and $\log |z - z_i|, i = 1, \ldots, n$. When we have a standard annulus, $X = \{z : r \leq |z| \leq R\}$, then $C(\partial X)$ is the span of S and a single function $h(z)$, which we can take to be either $\log |z|, |z|$, or $|z|^2$.

If our standard annulus is a spectral set for some operator T, then to prove that T has a normal ∂X-dilation, we need only extend $\widetilde{\rho}$ to a positive map ϕ on $C(\partial X)$. This amounts to choosing a positive operator $B = \phi(h)$, which to preserve the positivity needs to satisfy:

$$g(T) + g(T)^* \leq B \leq f(T) + f(T)^*,$$

whenever $f, g \in R(X)$ with $g + \overline{g} \leq h \leq f + \overline{f}$ on ∂X.

Problem 2: Prove Agler's Theorem via this approach.

Problem 3: Use Agler's proof to find B.

To see the difficulty with this approach, we recall [8] that for the annulus, there does exist a unital, positive map $\psi : R(X) + \overline{R(X)} \to M_2$, which has no positive extension. By Agler's result, $\psi \neq P_K \widetilde{\rho}|_K$ for any operator $T \in L(H)$ having X as a spectral set, and any two dimensional subspace, K contained in H.

3. Complete Positivity

In 1969 Arveson [3] gave necessary and sufficient conditions for the positive maps we considered in the previous section to have the desired positive extensions. The condition is that the map be completely positive. This concept was first introduced by Stinespring [28].

Given a C^*-algebra B, we define $M_n(B)$ to be the C^*-algebra of all n by n matrices with entries from B. If M is a subspace of a C^*-algebra A, then we identify $M_n(M)$ with a subspace of $M_n(A)$. This endows $M_n(M)$ with a norm and determines positive elements. If $\phi : M \to B$, then we define $\phi_n : M_n(M) \to M_n(B)$ via $\phi_n((a_{ij})) = (\phi(a_{ij}))$. If S is an operator system $\phi : S \to B$, and ϕ_n is a positive map for all n, then we say ϕ is *completely positive.*

If $\sup_n \|\phi_n\|$ is finite, we call ϕ completely bounded, and denote this supremum by $\|\phi\|_{cb}$. When $\|\phi\|_{cb} \leq 1$, we say that ϕ is *completely contractive.*

Proposition 3.1. [28] *Let Y be a compact, Hausdorff space and let $\phi : C(Y) \to L(H)$ be positive, then ϕ is completely positive.*

Theorem 3.2. (Stinespring [28]) *Let B be a C^*-algebra and let $\phi : B \to L(H)$ be completely positive. Then there exists a Hilbert space K, a bounded linear map $V : H \to K$ and a *-homomorphism $\pi : B \to L(K)$ such that $\phi(a) = V^*\psi(a)V$ for all $a \in B$.*

If $\phi(1) = 1$, then again V is an isometry and after identifying H with VH, we have $\phi(a) = P_H\pi(a)|_H$.

Combining Proposition 3.1 and Theorem 3.2 yields another proof of Naimark's dilation Theorem.

Proposition 3.1 also has a direct implication for spectral sets. If a positive map from an operator system in $C(Y)$ can be extended to a positive map on $C(Y)$, then the extended map, and hence the original map, is completely positive. Arveson proved a striking converse of this observation, namely, that completely positive maps always extend.

Theorem 3.3. (Arveson [3]) *Let A be C^*-algebra with unit, let $S \subseteq A$ be an operator system, and let $\phi : S \to L(H)$ be completely positive, then there exists a completely positive map, $\psi : A \to L(H)$ extending ϕ.*

Corollary 3.4. (Arveson [3]) *Let X be a spectral set for T and let $\rho : R(X) \to L(H)$ be given by $\rho(f) = f(T)$. The following are equivlent:*

(i) T has a normal ∂X-dilation,

(ii) $\tilde{\rho} : R(X) + \overline{R(X)} \to L(H)$ is completely positive,

(iii) $\rho : R(X) \to L(H)$ is completely contractive.

Whenever (iii) holds, we call X a *complete spectral set* for T. We see that Problem 1 can be re-stated as:

Problem 1′: Is every unital, contractive homomorphism $\rho : R(X) \to L(H)$, completely contractive?

There are examples of unital, contractive homomorphisms of commutative algebras which are not completely contractive, but these occur for multi-variable algebras. To discuss these, it will be necessary to state the full version of Corollary 3.4.

Let A be a subalgebra of the C^*-algebra B containing the unit. A map $\rho : A \to L(H)$ with $\rho(1) = 1$ has a *B-dilation*, if there is a Hilbert space K containing H and a *-homomorphism $\pi : B \to L(K)$ such that $\rho(a) = P_H \pi(a)|_H$.

Theorem 3.5. (Arveson [3]) *Let $\rho : A \to L(H)$ with $\rho(1) = 1$. Then ρ has a B-dilation if and only if ρ is completely contractive.*

We begin with the major result in 2 variables.

Theorem 3.6. (Ando [2], Sz.-Nagy-Foias [31]) *Let T_1, T_2 be contractions on H. Then there exists a Hilbert space K, containing H as a subspace, and two commuting unitary operators U_1, U_2 on K such that*

$$p(T_1, T_2) = P_H p(U_1, U_2) \big|_H,$$

for all polynomials $p(z, w)$ in two variables.

Let $P(\mathbb{D}^2)$ denote the algebra of all polynomials in 2 variables, regarded as a subalgebra of $C(\mathbb{T}^2)$, where $\mathbb{T}^2$ denotes the torus. Theorem 3.6 is equivalent to the following two statements:

(1) Every pair of commuting contractions define a unital contractive homomorphism $\rho : P(\mathbb{D}^2) \to L(H)$.

(2) Every unital contractive homomorphism on $P(\mathbb{D}^2)$ is completely contractive.

The first statement is generally expressed by saying that von Neumann's inequality holds in two variables, and is also equivalent to the statement that the map $\widetilde{\rho}$ on $P(\mathbb{D}^2) + \overline{P(\mathbb{D}^2)}$ is positive. The second statement is equivalent to $\widetilde{\rho}$ having a positive extension to $C(\mathbb{T}^2)$, i.e., that ρ is completely positive.

Problem 4: Can a proof of (1) be given by using positivity considerations?

Problem 5: Can a proof of (2) be given by using complete positivity? In particular, is every unital positive map on $P(\mathbb{D}^2) + \overline{P(\mathbb{D}^2)}$ completely positive?

We expect that every positive map on $P(\mathbb{D}^2) + \overline{P(\mathbb{D}^2)}$ is not completely positive.

These questions are of more than pedagogical interest. Statements (1) and (2) both fail for 3 or more variables. Crabbe and Davie [6] and also Varopulos [32] have given examples of 3 commuting, contractions for which von Neumann's inequality fails, that is, for which the induced homomorphism of $P(\mathbb{D}^3)$ is not contractive. In fact, their examples are finite matrices. Parrott [19] has given an example of 3 commuting contractions, which induce a unital contractive homomorphism of $P(\mathbb{D}^3)$ but this homomorphism is not completely contractive. Again, in Parrott's example, the 3 operators can be taken to be finite matrices.

The failure of (1) for 3 variables is a property of Hilbert spaces, and in some sense does not say much about the algebra $P(\mathbb{D}^3)$. However, the failure of (2) for 3 variables is somehow intrinsically related to properties of the algebra $P(\mathbb{D}^3)$.

Problem 6: What is inherently different in the structure of the algebras $P(\mathbb{D}^2)$ and $P(\mathbb{D}^3)$, that allows unital contractive homomorphisms of $P(\mathbb{D}^3)$ to fail to be completely contractive? In particular, can a property of commutative algebras be found that explains this difference?

This property of algebras should really depend on the pair, the subalgebra and the C^*-algebra that it sits inside of, i.e., on the complete system of norms. This, hopefully, explains somewhat the relevance of Problem 5. An independent proof of the Ando-Sz.-Nagy-Foias result, which used positivity or complete positivity would at least be a starting point to study Problem 6.

We close this section with a related result that is not very well-known, but that could contribute to solving these problems.

Theorem 3.7. (Gaspar-Racz [11]) *Let $T_1, \ldots, T_n$ be cyclically commuting, contractions on H, then there is a Hilbert space K containing H as a subspace and cyclically* *commuting unitaries $U_1, \ldots, U_n$ on K such that*

$$T_{i_1}^{k_1} \ldots T_{i_m}^{k_n} = P_H U_{i_1}^{k_1} \ldots U_{i_n}^{k_n} \mid_H$$

for all n-tuples of positive integers $k_1, \ldots, k_n$, where $i_1, \ldots, i_n$ is some cyclic permutation of the integers $l, \ldots, n$.

Of course, when $n = 2$ cyclically commuting and commuting are the same and so Theorem 3.6 generalizes Theorem 3.5, but when $n \geq 3$ these concepts are different. Theorem 3.6 implies that we can construct normed algebras, C_n in n cyclically commuting variables, with $C_2 = P(\mathbb{D}^2)$, but $P(\mathbb{D}^n)$ will be a quotient of C_n for $n \geq 3$. Again, by Theorem 3.6, the algebras C_n will have the property that every unital, contractive, homomorphism is completely contractive. If these algebras and their properties could be deduced in some C^*-algebra framework, independent of Theorem 3.6, this should add some insight into Problem 6.

4. K-spectral Sets

The theory of spectral sets is interesting and involves many deep connections between function theory and operator theory. However, if one views it as a theory that attempts to model certain classes of operators by normal operators, then one is faced with determining when an operator belongs to the class under consideration. For example, if an annulus $X = \{z : r \leq |z| \leq R\}$ is a spectral set for T, then by Agler's Theorem [1], T will have a normal, ∂X-dilation. But how does one characterize the operators which have X as a spectral set? Clearly, one needs that $\|T\| \leq R$ and $\|T^{-1}\| \leq r^{-1}$. However, these two conditions are not sufficient as has been noticed by Shields [25] and Misra [16]. To give a simple example of this phenomena, set $r = R^{-1}, R > \sqrt{3}$, and let

$$T = \begin{pmatrix} 1 & t \\ 0 & 1 \end{pmatrix}, T^{-1} = \begin{pmatrix} 1 & -t \\ 0 & 1 \end{pmatrix}.$$

To have $\|T\| = \|T^{-1}\| = R$ one needs, $t = (R - R^{-1})$, but then, $\|T - T^{-1}\| = 2t > \|z - z^{-1}\| = R + R^{-1}$.

In fact, it is instructive to attempt to calculate the largest value of t such that X is a spectral set for T as an explicit function of R. Misra [16] finds this value in terms of the solution of an extremal problem on X.

The difficulty here really lies with the holes in X. If X is simply connected with analytic boundary, then the Riemann mapping f yields a one-to-one bianalytic mapping from X to the closed unit disk, which is analytic in a neighborhood of X. Thus, X will be a spectral set for T if and only if $\sigma(T) \subseteq X$ and $\|f(T)\| \leq 1$.

Problem 7: If X is an analytic Cauchy domain, is there a finite set of functions such that X is a spectral set for T if and only if $\sigma(T) \subseteq X$ and $\|f(T)\| \leq 1$ for all the functions in the set?

We suspect the answer to the above problem is *no*. However, if one weakens the requirement that X be a spectral set, then one does get a workable theory. As an example, we cite:

Theorem 4.1. (Douglas-Paulsen [8]) *If* $\|T\| \leq R$ *and* $\|T^{-1}\| \leq r^{-1}$, *then there exists an invertible operator* S, *such that* $S^{-1}TS$ *has a normal* ∂X*-dilation, where* $X = \{z : r \leq |z| \leq R\}$.

J. Stampfli [27] has studied an analogous phenomena. If two disks, $\mathbb{D}_1$ and $\mathbb{D}_2$, are each a spectral set for an operator T, then their intersection need not be a spectral set for T. However, under certain technical conditions, there is an invertible operator S such that $S^{-1}TS$ has a normal $\partial(\mathbb{D}_1 \cap \mathbb{D}_2)$-dilation. Another motivation comes from the fact that spectral sets are not a similarity invariant.

Note that if $T_1 = S^{-1}TS$ has X as as spectral set, then

$$\|f(T)\| = \|Sf(T_1)S^{-1}\| \leq \|S\| \cdot \|S^{-1}\| \; \|f\|,$$

for any f in $R(X)$. In fact, $\|(f_{ij}(T))\| \leq \|S\| \cdot \|S^{-1}\| \cdot \|(f_{ij}(T_1))\|$, for any f_{ij} in $R(X)$. These observations motivate the following:

Definition: Let X be a compact subset of $\mathbb{C}$ and let $\sigma(T)$ be contained in X. Then X is called a K*-spectral set* for T, provided $\|f(T)\| \leq K\|f\|$ for all f in $R(X)$. We call X a *complete* K*-spectral set* for T if $\|(f_{ij}(T))\| \leq K\|(f_{ij})\|$ for every n by n matrix of functions in $R(X)$, where K is independent of n. In terms of the homomorphism $\rho : R(X) \to L(H)$, these two statements are the requirements that ρ be respectively, bounded or completely bounded with $\|\rho\| \leq K$ or $\|\rho\|_{cb} \leq K$.

The outstanding problem concerning these concepts is Problem 6 of Halmos' *Ten Problems in Hilbert Space* [12].

Problem 8: If the disk is a K-spectral set for T, then is T similar to a contraction?

There are two related problems, which we refer to as the weak and strong generalized versions of Halmos' problem, respectively.

Problem 9: If X is a K-spectral set for T, then is T similar to an operator for which X is a spectral set?

Problem 10: If X is a K-spectral set for T, then is T similar to an operator which has a normal, ∂X-dilation?

Clearly an affirmative anser to 10 implies an affirmative answer to 9, and an affirmative answer to 9 implies an affirmative answer to 8. Also, if Problem 1 has an affirmative answer, then 9 and 10 are equivalent.

In the positive direction, we do have the following:

Theroem 4.2. [21] *Let B be a unital, C^*-algebra and let A be a subalgebra containing the unit. Then $\rho : A \to L(H)$ is a completely bounded homomorphism if and only if ρ is similar to a completely contractive homomorphism. Moreover, there exists such a similarity with $\|S\| \cdot \|S^{-1}\| \leq \|\rho\|_{cb}$.*

Corllary 4.3. [20] *If X is a complete K-spectral set for T, then there exists an invertible operator S, such that $S^{-1}TS$ has a normal, ∂X-dilation. Moreover, one may choose S such that $\|S^{-1}\| \cdot \|S\| \leq K$.*

Corollary 4.4. [20] *An operator T is similar to a contraction if and only if the closed unit disk is a complete K-spectral set for T.*

Thus, we see that Problems 8 and 10 can be re-stated as:

Problem 8′: If the closed unit disk is a K-spectral set for T, then is it a complete K'-spectral set for T?

Problem 10′: If X is a K-spectral set for T, then is X a complete K'-spectral set for T?

To illustrate the use of this theorem, we show how to deduce a slightly weakened form of the generalized Rota model of Herrero and Voiculescu from it.

Theorem 4.5. (Herrero [13], Voiculescu [33]) *Let X be an analytic, Cauchy domain and let T be an operator with $\sigma(T)$ contained in the interior of X, then there exists an invertible operator S such that $S^{-1}TS$ has a normal ∂X-dilation.*

Proof: We need only prove that X is a complete K-spectral set for T. Let $f_{ij}, i, j = 1, \ldots, n$ be in $R(X)$, then by the Riesz functional calculus,

$$(f_{ij}(T)) = \frac{1}{2\pi i} \int_{\partial X} (f_{ij}(z)) \operatorname{Diag}((z - T)^{-1}) dz,$$

where $\mathrm{Diag}((z-T)^{-1})$ is the diagonal matrix with $(z-T)^{-1}$ for its diagonal entries. Thus,

$$\|(f_{ij}(T))\| \le \frac{1}{2\pi} \int_{\partial X} \|(z-T)^{-1}\| \cdot |dz| \cdot \|(f_{ij}(z))\|.$$

The reason that the above result is a slightly weakened form of Herrero-Voiculescu is that they are able to conclude that $S^{-1}TS$ dilates to a particular normal operator. Namely, the operator of multiplication by z on the Hilbert space of square-integrable H-valued functions, where the measure is arc length measure on ∂X. This more delicate result can be obtained from the consideration of completely bounded maps, but one needs to use von Neumann algebra versions. A several variable theory follows similarly, see Curto-Herrero [7].

It is also straightforward to show that if T is an operator of class $C_\rho (\rho > 1)$ in the sense of Sz.-Nagy and Foias [31], then the closed unit disk is a complete $(2\rho - 1)$-spectral set for T and so there exists an invertible S such that $S^{-1}TS$ is a contraction, $\|S\| \cdot \|S^{-1}\| \le (2\rho - 1)$.

Let $\rho : R(X) \to L(H), \rho(f) = f(T)$, and set $K(T) = \|\rho\|, K'(T) = \|\rho\|_{cb}$. When T is a matrix, $K(T)$ finite implies $K'(T)$ is finite, but what's important is whether or not the ratio $K'(T)/K(T)$ is bounded independent of n.

Proposition 4.5. (Smith [26]) *If T is n by n, then $K'(T) \le nK(T)$.*

Proposition 4.7. (Holbrook [14]) *Let $X = \mathbb{D}^-$,*

(1) If T is 2 by 2, then $K'(T) = K(T)$,

(2) There are finite matrices with $K'(T) \ne K(T)$.

Proposition 4.8. [23] *For arbitrary X and T 2 by 2, $K'(T) = K(T)$.*

It is still unknown for $X = \mathbb{D}^-$ and $T_t = \lambda I + tJ$ whether or not $K(T_t) = K'(T_t)$ or any estimate on $K'(T_t)/K(T_t)$ independent of n, where J is the $n \times n$ backwards shift.

5. Model Theory for Finitely-Connected Domains

Let X be a closed subset of the complex plane, and let $\widetilde{X}$ denote the closure of X in the complex sphere, i.e., in the one-point compactification of $\mathbb{C}$. We let $R(X)$ denote the rational functions which are bounded on X and hence extend to be continuous on $\widetilde{X}$. We regard $R(X)$ as a subalgebra of $C(\partial\widetilde{X})$. If T is a bounded operator with $\sigma(T)$ contained in X, then as before we have a homomorphism $\rho : R(X) \to L(H)$. If $\|\rho\|$ or

$\|\rho\|_{cb}$ is finite, then we call X, respectively, a K-spectral or complete K-spectral set for T. In particular, $\{z : |z| \geq r\}$ is a spectral set for T if and only if $\|T^{-1}\| \leq r^{-1}$.

More generally, if ψ is a linear fractional map such that $\psi\widetilde{X} = \psi(X)^{-}$ is compact, then X is a (complete) K-spectral set for T if and only if $\psi(X)$ is a (complete) K-spectral set for $\psi(T)$.

Let X denote a connected, compact set whose boundary consists of $n+1$ disjoint, rectifiable, closed, Jordan curves, $\gamma_o, \ldots, \gamma_n$, where γ_o is the boundary of the unbounded component of $\mathbb{C}/X$. We let X_o denote the interior of γ_o, and X_i denote the exterior of $\gamma_i, i = 1, \ldots, n$. We have that $X = X_o \cap X_1 \ldots \cap X_n$ with each of the X_i's simply connected.

Theorem 5.1. (Douglas-P [8]) *Let X be as above. If each $X_i, i = 1, \ldots, n$ is a complete K_i-spectral set for T, then T is similar to an operator with a normal ∂X-dilation.*

Corollary 5.2. *Let X be as above. If the answer to Problem 8′ is affirmative, then the answer to Problem 10′ is also affirmative for any such X.*

Corollary 5.3. (Douglas-P [8]) *Let X be as above. If each $X_i, i = 1, \ldots, n$ is a spectral set for T, then T is similar to an operator with a normal ∂X-dilation.*

Proof: Since each X_i is simply connected, $R(X_i) + \overline{R(X_i)}$ is dense in $C(\partial \widetilde{X})$. By the same proof as of the Berger-Foias-Lebow result, we have that each X_i is a complete spectral set.

For X an analytic, Cauchy domain, we have one-to-one bianalytic mappings $f_i; \widetilde{X}_i \to \mathbb{D}^{-}, i = 0, 1 \ldots, n$ which are analytic in a neighborhood of $\widetilde{X}_i$. The statement that each X_i is a spectral set for T, is equivalent to requiring that $\|f_i(T)\| \leq 1$. Thus, this finite set of norm inequalities is enough to ensure that up to similarity T has a normal ∂X-dilation. This should be compared with Problem 7.

In fact, somewhat more is true.

Corollary 5.4. *Let X be an analytic Cauchy domain. If there exist invertible operators $S_i, i = 0, \ldots, n$, such that $\|S_i^{-1} f_i(T) S_i\| \leq 1$, then there is an invertible S, such that $S^{-1}TS$ has a normal ∂X-dilation.*

Note that the above operator S is a single operator such that $\|S^{-1} f_i(T) S\| \leq 1, i = 0, \ldots, n$ simultaneously. In particular, for $X = \{z : r \leq |z| \leq R\}, r < R$, we conclude that if there are two operators such that $\|S_o^{-1} T S_o\| \leq R$ and $\|S_1^{-1} T^{-1} S_1\| \leq r^{-1}$ then there is a single operator such that $\|S^{-1}TS\| \leq R$ and $\|S^{-1}T^{-1}S\| \leq r^{-1}$. This is proved in [8]. When $r = R$, this was already known and is due to Sz.-Nagy [29]. In fact, in the $r = R$ case, Sz.-Nagy proved that if $\|(T/R)^n\|$ is uniformly bounded for all integers n (positive and negative), then there is an invertible S such that $\|S^{-1}TS\| = \|S^{-1}T^{-1}S\|^{-1} = R$.

Problem 11: If $r < R$ and $\|(T/R)^n\|, \|(rT^{-1})^n\|$ are uniformly bounded for all positive integers, then does there exist an invertible such that $\|S^{-1}TS\| \leq R, \|S^{-1}T^{-1}S\| \leq r^{-1}$?

We expect that the answer to this question is no. However, if it is yes, then it implies the above result for the annulus. If $\|T^n\|$ is uniformly bounded for all *positive* integers n, then T is called *power bounded.* Foguel [9] gives an example of an operator that is power bounded, but not similar to a contraction. Note that if the answer to Problem 11 is yes, then any power bounded, invertible operator would be similar to a contraction. Foguel's operator is essentially invertible, that is, it has a compact perturbation which is invertible, but it is not clear if these can also be power bounded and not similar to a contraction. There exist compact perturbations of Foguel's example which are similar to contractions.

We close with a result true for more general sets than those considered above. Let X be a compact subset of the plane, if the closure of $R(X) + \overline{R(X)}$ is of codimension n in $C(\partial X)$, then we call X *hypo-Dirichlet of codimension n.* The nice n holed sets considered above, were all hypo-Dirichlet of codimension n. Usually, a compact set with n holes, will be hypo-Dirichlet of codimension $m, m \leq n$. For example, Figure 1 is hypo-Dirichlet of codimension 1.

Figure 1.

Theorem 5.5. (Douglas-P [8]) *Let X be hypo-Dirichlet of codimension n. If X is a spectral set for T, then there exists an invertible S such that $S^{-1}TS$ has a normal ∂X-dilation, with $\|S^{-1}\| \cdot \|S\| \leq (2n+1)$.*

6. Multi-variable Theory

We have seen earlier that if we are given two commuting contractions, then a 2-variable version of von Neumann's inequality is true, while for 3 commuting contractions, this fails. However, it is not known how badly von Neumann's inequality fails. Let $(T_1, \ldots, T_n)$ be an n-tuple of commuting contractions, and let $P(\mathbb{D}^n)$ denote the algebra of polynomials in n-variables, regarded as a subalgebra of $C(\mathbb{T}^n)$. We have a homomorphism $\rho : P(\mathbb{D}^n) \to L(H)$ given by $\rho(p) = p(T_1, \ldots, T_n)$. In general, for $n \geq 3, \rho$ is not contractive.

Problem 12: Is ρ either bounded or completely bounded?

See [32], [35], and [36] for more on this problem.

By Theorem 4.2, ρ is completely bounded if and only if there is an invertible operator S such that $(S^{-1}T_1S, \ldots, S^{-1}T_nS)$ has a commuting unitary dilation. Problem 12 is really a boundary problem. If $\sigma(T_i)$ is contained in the open unit disk, $i = 1, \ldots, n$, then we have a multi-variable Riesz functional calculus. Arguing as in the proof of the generalized Rota model, we can see that ρ is completely bounded. For a more general multi-variable Rota model, see Curto-Herrero [7].

In general, so little is known in the multi-variable case that we only indicate the problems. For a more complete treatment, see [22].

If $X_i, i = 1, \ldots, n$ are compact sets in the complex plane, then we let $X = X_1 \times \ldots \times X_n$ and set $\partial_d X = \partial X_1 \times \ldots \times \partial X_n$. We let $R_d(X)$ denote the subalgebra of $C(\partial X)$ generated by functions in $R(X_i), i = 1, \ldots, n$. In general, this will be a smaller algebra than $R(X)$, which we use to denote the quotients of polynomials with poles off X. But $R_d(X)$ is dense in $R(X)$, which can be seen by applying Runge's Theorem. The algebra $R_d(X)$ can be identified with the tensor product of $R(X_i), i = 1, \ldots, n$. If $T = (T_i, \ldots, T_n)$ are commuting operators with $\sigma(T_i) \subseteq X_i$, then there is a well-defined homomorphism $\rho : R_d(X) \to L(H)$ given by $\rho(f) = f(T)$. We say that X is a (complete) K-spectral set if $\|\rho\| \leq K$ (respectively, $\|\rho\|_{cb} \leq K$).

We say that $T = (T_i, \ldots, T_n)$ has a *normal ∂X-dilation* if there exists a commuting

n-tuple of normal operators $N = (N_1, \ldots, N_n)$ with $\sigma(N_i) \subseteq \partial X_i$ such that

$$f(T_1, \ldots, T_n) = P_H f(N_1, \ldots, N_n) \,|_H$$

for every f in $R_d(X)$. By Arveson's Theorem, this is equivalent to requiring that X is a complete spectral set for T. There exists an invertible S such that $(S^{-1}T_1S, \ldots, S^{-1} T_nS)$ has a normal ∂X-dilation if and only if X is a complete K-spectral set [22].

Problem 13: If X_i is a (complete) K_i-spectral set for $T_i, i = 1, 2$, and T_1 commutes with T_2, then is X a (complete) K_1K_2-spectral set for (T_1, T_2)?

The case $K_1 = K_2 = 1$ is of primary interest. We have already seen that we can't expect this result to be true for more than 2 variables. What we are really asking is how two different norms on $R(X_1) \otimes R(X_2)$ are related.

There are some cases where more is known and these are more compatible with C^*-algebra theory. Operators $T_1, \ldots, T_n$ are said to *doubly commute* if $T_iT_j^* = T_j^*T_i$ and $T_iT_j = T_jT_i$ for all $i \neq j$. Assume that X_i is a complete spectral set for T_i and set $S_i = R(X_i) + \overline{R(X_i)}, i = 1, \ldots, n$. The tensor product of these operator systems has two natural norms defined on it, called the max and min norms. When the operators $T = (T_1, \ldots, T_n)$ doubly commutes, then there is a completely positive map $\psi : S_1 \otimes_{\max} \ldots \otimes_{\max} S_n \to L(H)$ satisfying $\psi(f_1 \otimes \ldots \otimes f_n) = \tilde{\rho}_1(f_1) \ldots \tilde{\rho}_n(f_n)$.

The set X is a complete spectral set for T if and only if the same map ψ is completely positive in the min norm.

Problem 14: If X_i is a complete spectral set for $T_i, i = 1, \ldots, n$, and $T = (T_1, \ldots, T_n)$ doubly commutes, then is X a complete spectral set for T?

This is clear if S_i is dense in $C(\partial X_i)$ for $n - 1$ of the X_i's. Since $C(\partial X_i)$ is a nuclear C^*-algebra, the min and max norms on the tensor products agree. Thus, ψ will be completely positive in the min norm. In particular, we see that n doubly commuting contractions will always have a joint unitary dilation, which had been obtained by Sz.-Nagy and Foias [31].

Theorem 5.5 does have a several variable generalization.

Theorem 6.6. [22] *Let X_i be hypo-Dirichlet of codimension $k_i, i = 1, \ldots, n-1$. If X_i is a spectral set for $T_i, i = 1, \ldots, n$, and $(T_1, \ldots, T_n)$ double commutes, then there is an invertible S, such that $(S^{-1}T_1S, \ldots, S^{-1}T_nS)$ has a normal ∂X-dilation.*

The above result is proved by showing that the min and max norms are *completely* equivalent.

References

1. Agler, J., Rational dilation on an annulus, Ann. of Math. **121** (1985), 537-564.
2. Ando, T., On a pair of commutative contractions, Acta Sci. Math. **24** (1963), 88-90.
3. Arveson, W. B., Subalgebras of C^*-algebras, Acta. Math **123** (1969), 141-224.
4. Berger, C. A., Ph.D. Thesis.
5. ———, A strange dilation theorem, Notices Amer. Math. Soc. **12** (1965), 590, Abstract 625-152.
6. Crabbe, M. J. and A. M. Davie, von Neumann's inequality for Hilbert space operators, Bull. London Math. Soc. **7** (1975), 49-50. 31.
7. Curto, R. and D. Herrero, On closures of joint similarity orbits, Integral Equations and Operator Theory **8** (1985), 489-556.
8. Douglas, R. G. and V. I. Paulsen, Completely bounded maps and hypo-Dirichlet algebras, Acta. Sci. Math., to appear.
9. Foguel, S. F., A counter example to a problem of Sz.-Nagy, Proc. Amer. Math. Soc. **15** (1964), 788-790.
10. Foias, C., Some applications of spectral sets. I. Spectral Harmonic Measures (Rumanian), Studii Cerca. Math. **10** (1959), 365-401.
11. Gasper, D. and A. Racz, An extension of a theorem of T. Ando, Michigan Math. J. **16** (1969), 377-380.
12. Halmos, P. R., Ten problems in Hilbert space, Bull. Amer. Math. Soc. **76** (1970), 887-933.
13. Herrero, D. A., A Rota universal model for operators with multiply connected spectrum, Rev. Roum. Math. Pures et Appl. **21** (1976), 15-23.
14. Holbrook, J. A. R., Distortion coefficients for crypto-contractions, Lin. Alg. and Appl. **18** (1977), 229-256.
15. Lebow, A., On von Neumann's theory of spectral sets, J. Math. Anal. and Appl. **7** (1963), 64-90.
16. Misra, G., Curvature inequalities and extremal properties of bundle shifts, J. Op. Thy. **11** (1984), 305-318.
17. Naimark, M. A., On a representation of additive operator set functions, C.R. (Doklady) Acad. Sci. URSS **41** (1943), 359-361.
18. von Neumann, J., Eine spektraltheorie für allgemeine Operatoren eines unitären Raumes, Math. Nachr. **4** (1951), 258-281.

19. Parrott, S. K., Unitary dilations for commuting contractions, Pacific J. Math. **34** (1970), 481-490.
20. Paulsen, V. I., Every completely polynomially bounded operator is similar to a contraction, J. Funct. Anal. **55** (1984), 1-17.
21. ———, Completely bounded homomorphisms of operator algebras, Proc. Amer. Math. Soc. **92** (1984), 225-228.
22. ———, *Completely Bounded Maps and Dilations.* Pitman Press, to appear.
23. ———, K-spectral values for some finite matrices, preprint.
24. Sarason, D., On spectral sets having connected complement, Acta Sci. Math. **26** (1965), 289-299.
25. Shields, A., Weighted shift operators and analytic function theory, in *Topics in Operator Theory* (C. Pearcy, ed.). American Mathematical Society, Providence, 1974.
26. Smith, R. R., Private communication.
27. Stampfli, J. G., Surgery on spectral sets, J. Op. Thy., to appear.
28. Stinespring, W. F., Positive functions on C^*-algebras. Proc. Amer. Math. Soc. **6** (1955), 211-216. QA1.A521, 104.
29. Sz.-Nagy, B., On uniformly bounded linear transformations in Hilbert space, Acta Sci. Math. Szeged **11** (1947), 152-157.
30. ———, Sur les contractions de e'espace de Hilbert, Acta Sci. Math. **15** (1953), 87-92.
31. ——— and C. Foias, *Harmonic Analysis of Operators on Hilbert Space.* American Elsevier, New York, 1970.
32. Th. Varopoulos, N., On a inequality of von Neumann and an application of the metric theory of tensor products to operators theory, J. Funct. Anal. **16** (1974), 83-100.
33. Voiculescu, D., Norm-limits of algebraic operators, Rev. Roum. Math. Pures et Appl. **19** (1974), 371-378.
34. Walsh, J. L., The approximation of harmonic functions by harmonic polynomials and by harmonic rational functions, Bull. Amer. Math. Soc. **35** (1919), 499-544.
35. Dixon, P. G., The von Neumann inequality for polynomials of degree greater than two, J. London Math. Soc. (2) **14** (1976)¡ 369-375.
36. ——— and S. W. Drury, Unitary dilations, polynomial identities and the von Neumann inequality, Math. Proc. Camb. Phil. Soc. **99** (1986), 115-122.

Vern I. Paulsen
Department of Mathematics
University of Houston, University Park
Houston, Texas 77004, U.S.A.

Invariant Subspaces for Subnormal Operators

by

James E. Thomson

1. The Existence of Invariant Subspaces

Let $\mathcal{B}(\mathcal{H})$ denote the bounded linear operators on a separable, infinite-dimensional complex Hilbert space $\mathcal{H}$. An operator S in $\mathcal{B}(\mathcal{H})$ is *subnormal* if there exists a normal operator N in $\mathcal{B}(\mathcal{K})$ for some Hilbert space $\mathcal{K}$ containing $\mathcal{H}$ such that N leaves $\mathcal{H}$ invariant and S is the restriction of N to $\mathcal{H}$. For a compactly supported finite positive measure μ on $\mathbf{C}$ let $H^2(\mu)$ denote the closure in $L^2(\mu)$ of the polynomials in z. J. Bram [1] has shown that if a subnormal operator S is cyclic, then it is unitarily equivalent to multiplication by z, M_z on $H^2(\mu)$ for some measure μ.

Example 1.1. Let D be the open unit disc in $\mathbf{C}$ and let μ be Lebesgue measure on ∂D. Then $H^2(\mu)$ is the Hardy space H^2 and M_z on $H^2(\mu)$ is the unilateral shift.

Example 1.2. Let μ be area measure on D. Then $H^2(\mu)$ is the Bergman space.

In each example above the functions in $H^2(\mu)$ that vanish at a common point in D comprise an invariant subspace for M_z. An open question is whether a cyclic subnormal operator always has an invariant subspace of this kind if $H^2(\mu) \neq L^2(\mu)$. Let us state this more precisely.

A point λ in $\mathbf{C}$ is a bounded point evaluation (bpe) for $H^2(\mu)$ if there exists $c > 0$ such that

$$|p(\lambda)| \leq c\|p\|_2$$

for every polynomial p. If λ is a bpe, then the closure of the polynomials that vanish at λ is a nontrivial invariant subspace.

Question 1.3. Does $H^2(\mu) \neq L^2(\mu)$ imply the existence of bpe's?

If 2 is replaced by $p > 2$ above, then the question has an affirmative answer [2].

Theorem 1.4. (Brennan) *Let μ be a measure. If $p > 2$ and $H^p(\mu) \neq L^p(\mu)$, then there exist bpe's.*

Lemma 1.5. *Let $E = \operatorname{supp}\mu$ and let q be such that $(1/p)+(1/q) = 1$. If $G(z) \in L^q(\mu)$, then $G(z)(z-\lambda)^{-1} \in L^q(\mu)$ except for those λ in a set of zero area in $\mathbf{C}$.*

Proof: Choose $R > 0$ so that E is contained in $\{|z| < R\}$. For $z \in E$,

$$\int_{|\lambda|\le R} |z-\lambda|^{-q} d\text{Area}(\lambda) = \iint_{|z-\lambda|\le R} |\lambda|^{-q} d\text{Area}(\lambda)$$
$$\le \iint_{|z|\le 2R} |\lambda|^{-q} d\text{Area}(\lambda)$$
$$= \frac{2\pi}{2-q}(2R)^{2-q}.$$

Therefore,

$$\iint_E \left[\int_E \left|\frac{g(z)}{z-\lambda}\right|^q d\mu(z)\right] d\text{Area}(\lambda)$$
$$= \int_E \left[\iint_E |z-\lambda|^{-q} d\text{Area}(\lambda)\right] |g(z)|^q \, d\mu(z)$$
$$\le \int_E \left[\iint_{|\lambda|\le R} |z-\lambda|^{-q} d\text{Area}(\lambda) l\right] |g(z)^q \, d\mu(z)$$
$$\le \frac{2\pi}{2-q}(2R)^{2-q} \int_E |g(z)|^q \, d\mu(z)$$
$$< \infty.$$

The conclusion now follows easily. ∎

Proof of Theorem 1.4. Suppose $H^p(\mu) \neq L^p(\mu)$. By the Hahn-Banach and Riesz representation theorems there exists a nonzero function g in $L^q(\mu)$ such that for every polynomial f, $\int fg\, d\mu = 0$. Since g is not the zero function, its Cauchy transform

$$\widehat{g}(\lambda) = \int g(z)(z-\lambda)^{-1} \, d\mu(z)$$

is nonzero on a set of positive area [5, p. 316]. Applying Lemma 1.5, we see there exists a point w such that $\widehat{g}(w) \neq 0$ and $g(z)(z-w)^{-1} \in L^q(\mu)$. Now for every polynomial f

$$\int \frac{f(z)-f(w)}{z-w} g \, d\mu = 0.$$

Thus,

$$f(w) = \widehat{g}(w)^{-1} \int fg(z-w)^{-1} \, d\mu,$$

and, consequently,

$$|f(w)| \le |\widehat{g}(w)^{-1}| \, \|(z-w)^{-1} g\|_q \, \|f\|_p. \quad ∎$$

The theorem and proof above go through with $R^p(\mu)$ in place of $H^p(\mu)$, where $R^p(\mu)$ denotes the closure in $L^p(\mu)$ of the rational functions with no poles on the support of μ. There is an example where μ is area measure on a compact set and $R^2(\mu) \neq L^2(\mu)$ without any bpe's for $R^2(\mu)$ [7].

To prove the existence of invariant subspaces for subnormal operators, it is not necessary to establish the existence of bpe's for $H^2(\mu)$ spaces. As Scott Brown observed, it suffices to find a point λ, a vector x in $H^2(\mu)$, and a vector y in $L^2(\mu)$ such that for every polynomial p,

$$p(\lambda) =< px, y > .$$

When this happens, the closed linear span of $\{(z - \lambda)^n x\}_{n=1}^{\infty}$ is a nontrivial invariant subspace. Brown [3] proved that invariant subspaces of this kind exist whenever an $H^2(\mu)$-space has no bpe's. Thus every subnormal operator has a nontrivial invariant subspace. His proof relies on an understanding of $P^\infty(\mu)$, the weak-star closure in $L^\infty(\mu)$ of the polynomials, and we will give his proof later. We will now give Thomson's simple proof [13].

Theorem 1.6. (S.W. Brown) *Every subnormal operator has a nontrivial invariant subspace.*

Proof: By Bram's result it suffices to show that M_z on $H^2(\mu)$ has a nontrivial invariant subspace, where μ is an arbitrary finite measure on $\mathbf{C}$ with compact support. Clearly we may assume that $H^2(\mu) \neq L^2(\mu)$. Let p be such that $2 < p \leq 4$, and let q be such that $(1/p) + (1/q) = 1$. Now $H^p(\mu) \neq L^p(\mu)$, so by Theorem 1.4 there exists a point λ in $\mathbf{C}$ such that λ is a bpe for $H^p(\mu)$. By the Hahn-Banach theorem there exists a norm-preserving extension on $L^p(\mu)$ of evaluation at λ. This extension is represented by a function h in $L^q(\mu)$ with $\|h\|_q$ equal to the linear functional norm. Since the closed unit ball of $H^p(\mu)$ is weakly compact, there exists a function x in $H^p(\mu)$ with $\|x\|_p = 1$ and

$$\int xh \, d\mu = \|h\|_q.$$

Thus,

$$\|h\|_q = \int xh \, d\mu \leq \|x\|_p \; \|h\|_q = \|h\|_q.$$

The equality in Hölder's inequality above implies that

$$|x|^p = a|h|^q \quad \text{a.e.}(\mu)$$

for some positive constant a.

Let $y = h/x$ on the set where x is nonzero and zero elsewhere. The function y is in $L^2(\mu)$ because

$$\int |y|^2\, d\mu = \int |h|^2\, |x|^{-2}\, d\mu \leq b \int |h|^q\, d\mu < \infty,$$

where b is a positive constant. For any polynomial f we have

$$f(\lambda) = \int fh\, d\mu = \langle fx, y\rangle.$$

Thus the closed linear span of $\{(z-\lambda)^n x\}_{n=1}^{\infty}$ is a nontrivial invariant subspace. ∎

An easy modification of the preceding proof yields the following stronger theorem.

Theorem 1.7. (Thomson) *Let $\mathcal{H}$ be a closed subspace of $L^2(\mu)$ with $1 \in \mathcal{H}$. Let $\mathcal{A}$ be a subalgebra of $L^\infty(\mu)$ containing the function z such that $\mathcal{A}\mathcal{H} \subset \mathcal{H}$. Then there exists a nontrivial subspace $\mathcal{H}'$ of $\mathcal{H}$ such that $\mathcal{A}\mathcal{H}' \subset \mathcal{H}'$.*

Let us point out two consequences of this theorem. A subnormal operator S on a space $\mathcal{H}$ is rationally cyclic if there exists a vector x in $\mathcal{H}$ such that the set $\{r(S)x : r \in \mathrm{Rat}(\sigma(S))\}$ is dense in $\mathcal{H}$, where $\mathrm{Rat}(\sigma(S))$ denotes the algebra of rational functions with no poles on $\sigma(S)$.

For each such S, there exists a measure μ such that S is unitarily equivalent to M_z on $R^2(\sigma(S),\mu)$, the closure in $L^2(\mu)$ of $\mathrm{Rat}(\sigma(S))$ [5, p. 146]. Under this representation each operator that commutes with S is represented by multiplication by a function in $R^2(\sigma(S),\mu) \cap L^\infty(\mu)$, and conversely [5, p. 147]. A subspace invariant for every operator that commutes with S is called *hyperinvariant.* If we let $\mathcal{H} = R^2(\sigma(S),\mu)$ and $\mathcal{A} = \mathcal{H} \cap L^\infty(\mu)$ in Theorem 1.7, then we obtain a nontrivial hyperinvariant subspace.

Corollary 1.8. *Every rationally cyclic subnormal operator has a nontrivial hyperinvariant subspace.*

The following is a trivial consequence of Corollary 1.8.

Corollary 1.9. *Every subnormal operator has a nontrivial subspace invariant for the algebra $\{r(S) : r \in \mathrm{Rat}(\sigma(S))\}$.*

2. A Functional Calculus for Subnormal Operators

The main goal of these lectures is to outline a proof that every subnormal operator is reflexive. This is proved in [9] by Olin and Thomson. Before defining reflexive, we need to introduce some notation. Let $W(T)$ denote the weakly closed algebra generated by an operator T and let Alg Lat$T$ denote the algebra of operators that leave invariant each invariant subspace of T. An operator T is *reflexive* if $W(T) =$ Alg Lat$T$.

We will show first that if a subnormal operator has an abundance of analytic invariant subspaces, then it is reflexive. Second, we will show that a subnormal operator has an abundance of analytic invariant subspaces. The key to the second part is the invariant subspace technique of Scott Brown. Before proceeding to this agenda, we need to do some background material.

Today we will review briefly the functional calculus for subnormal operators that was developed by Conway and Olin [6], and really opened the door for studying subnormal operators. The cornerstone of their work is the celebrated theorem of Sarason [12] that characterizes the weak-star closure of the polynomials. An excellent reference for all our background material is the book by Conway [5].

Regarding $\mathcal{B}(\mathcal{H})$ as the dual of the space of trace class operators, we can give $\mathcal{B}(\mathcal{H})$ a weak-star topology. That is, a net $\{A_\alpha\}$ in $\mathcal{B}(\mathcal{H})$ converges weak-star to zero in $\mathcal{B}(\mathcal{H})$ if and only if trace $(A_\alpha T) \to 0$ for every trace class operator T. The following proposition is a standard exercise.

Proposition 2.1. *A net $\{A_\alpha\}$ in $\mathcal{B}(\mathcal{H})$ converges weak-star to zero if and only if for every pair of sequences $\{g_n\}$ and $\{h_n\}$ of vectors in $\mathcal{H}$ such that $\sum \|g_n\|^2 < \infty$ and $\sum \|h_n\|^2 < \infty$, $\sum\langle A_\alpha g_n, h_n\rangle$ converges to zero.*

Let S in $\mathcal{B}(\mathcal{H})$ be subnormal with minimal normal extension (mne) N in $\mathcal{B}(\mathcal{K})$. For convenience, suppose that N has a cyclic vector, so we can take N to be M_z on $L^2(\mu)$ for some measure μ. Let $\mathcal{A}(N)$ denote the weak-star closed algebra generated by N. By the previous proposition a net $\{p_\alpha(N)\}$ of polynomials in N converges to zero weak-star if and only if

$$\sum < p_\alpha(N)g_n, h_n > \to 0 \tag{2.2}$$

whenever $\{g_n\}$ and $\{h_n\}$ are sequences of vectors in $L^2(\mu)$ with $\sum \|g_n\|^2 < \infty$ and $\sum \|h_n\|^2 < \infty$. We see easily that (2.2) holds if and only if $\int p_\alpha w \, d\mu \to 0$ for each w in $L^1(\mu)$.

Let $P^\infty(\mu)$ denote the weak-star closure of the polynomials in $L^\infty(\mu)$. It turns out that $\mathcal{A}(N)$ and $P^\infty(\mu)$ are isometrically isomorphic as Banach algebras and homeomorphic with their weak-star topologies under a mapping that takes N to z. Furthermore, $\mathcal{A}(N)|\mathcal{H}$ and $\mathcal{A}(S)$ are isometrically isomorphic as Banach algebras and weak-star homeomorphic under the restriction map [6].

Next we will describe $P^\infty(\mu)$. A subalgebra $\mathcal{A}$ of $L^\infty(\mu)$ is *antisymmetric* if $1 \in \mathcal{A}$ and the only functions in $\mathcal{A}$ that are real-valued a.e. (μ) are the constant functions.

Example 2.3. If G is a bounded region in $\mathbf{C}$, then $H^\infty(G)$ is an antisymmetric subalgebra of $L^\infty(\text{Area}|G)$.

Proposition 2.4. *Let μ be a measure. Then there is a Borel partition $(\Delta_0, \Delta_1, \ldots)$ of* suppμ *such that if $\mu_n = \mu|\Delta_n$, then*

$$P^\infty(\mu) = L^\infty(\mu_0) \oplus \bigoplus_1^\infty P^\infty(\mu_n),$$

where each $P^\infty(\mu_n)$ for $n \geq 1$ is antisymmetric and infinite-dimensional. Except for the ordering of these summands, this decomposition is unique.

Applying this proposition to our subnormal operator S acting on a subspace $\mathcal{H}$ of $L^2(\mu)$, we see that $\chi_{\Delta_n}\mathcal{H}$ is a reducing subspace for S for each n. For the purpose of studying invariant subspaces it follows that it is sufficient to consider the case where $P^\infty(\mu)$ is antisymmetric. We make that assumption for the rest of this paper.

Let m denote normalized Lebesgue measure on ∂D. We will consider functions in the Hardy algebra H^∞ to be defined on D and a.e.(m) on ∂D. Thus, if supp$\mu \subset \bar{D}$ and $\mu|\partial D \ll m$, then there is a natural restriction map of H^∞ into $L^\infty(\mu)$. If this map is an isometric isomorphism and weak-star homeomorphism of H^∞ onto $P^\infty(\mu)$, then we say that $P^\infty(\mu) = H^\infty(D,\mu)$.

More generally, let G be a bounded simply connected region such that the Riemann map φ of D onto G is a weak-star generator of H^∞. (That is, the weak-star closed algebra generated by φ in $L^\infty(m)$ is H^∞.) Recall that harmonic measure for G on ∂G is $m \circ \varphi^{-1}$, so that we can use φ to define boundary values for functions in $H^\infty(G)$. As before, we obtain a natural restriction map of $H^\infty(G)$ into $L^\infty(\mu)$ if supp$\mu \subset \bar{G}$ and $\mu|\partial G$ is absolutely continuous with respect to harmonic measure on ∂G. If this map is an isometric isomorphism and weak-star homeomorphism of $H^\infty(G)$ onto $P^\infty(\mu)$, then we say that $P^\infty(\mu) = H^\infty(G,\mu)$.

Theorem 2.4. *If $P^\infty(\mu)$ is antisymmetric, then there exists a simply connected region G as above such that $P^\infty(\mu) = H^\infty(G,\mu)$.*

Thus we have reduced our study of invariant subspaces to the case where $P^\infty(\mu) = H^\infty(G,\mu)$ for some G as above. Let $S_1 = M_{\varphi-1}$. It is an easy exercise to show that $\mathcal{A}(S_1) = \mathcal{A}(S)$. Hence, for most invariant subspaces questions we may as well study S_1 instead of S. This means we may as well assume that $P^\infty(\mu) = H^\infty(D,\mu)$.

As an aid to understanding various algebras of subnormal operators, we will look at two examples. The method of construction is due to Hastings [8].

Example 2.5. We will define a measure μ by a limiting processing. Let μ_1 be area measure on $D \setminus \{|z - \frac{1}{2}| < \frac{1}{2}\}$. Let $X = \text{supp}\mu$ and let $E_n = \{z : 0 < \arg z < \frac{1}{n}\}$ for each positive n. We will define a sequence $\{\mu_j\}$ of measures by induction. Suppose $\mu_1, \ldots, \mu_n$ have been defined. By Runge's Theorem there exists a polynomial p_n such that

$$\left\| p_n - \frac{1}{z - \frac{1}{2}} \right\|_{X \setminus E_n} < \frac{1}{n}.$$

Choose $c_n > 0$ such that

$$c_n \int_{E_n \cap X} \left| p_n - \frac{1}{z - \frac{1}{2}} \right|^2 d\mu_n < \frac{1}{n}.$$

Let $\mu_{n+1} = \mu_n$ on $X \setminus E_n$ and $\mu_{n+1} = c_n \mu_n$ on $X \cap E_n$. Thus,

$$\int \left| p_n - \frac{1}{z - \frac{1}{2}} \right|^2 d\mu_{n+1} \leq \left(\frac{1}{n}\right)^2 \mu_1(X \setminus E_n) + \frac{1}{n}.$$

Let μ be the norm limit of the μ_n's. Since $\mu \leq \mu_n$ for all n, it follows that

$$p_n \to \frac{1}{z - \frac{1}{2}}$$

in $L^2(\mu)$.

Noting that μ and area measure are mutually boundedly absolutely continuous off any neighborhood of one, we see that every function in $H^2(\mu)$ is analytic in the interior of X. Recall that the commutant of M_z on $H^2(\mu)$ is $H^2(\mu) \cap L^\infty(\mu)$ [15]. Thus, the commutant contains $1/(z - \frac{1}{2})$ and it strictly contains $P^\infty(\mu)$ $(= H^\infty(D,\mu))$. A theorem of Olin and Thomson [9] shows that $H^2(\mu)$ contains an invariant subspace $\mathcal{H}$ such that $\sigma(M_z \mid \mathcal{H}) = \overline{D}$. In fact, although evaluation at the point one-half is unbounded, there exist x and y in $H^2(\mu)$ such that $x \otimes y$ equals evaluation at one-half. This example can be modified to ensure that Thomson's method [13] will not yield the x and y.

Example 2.6. We modify the previous example. Let μ_1 be area measure on D. Suppose $\mu_1, \ldots, \mu_n$ are defined. Let E_n be as in Example 2.5. Let p_n be a polynomial such that $|p_n - \sqrt{z}| < \frac{1}{n}$ on $\bar{D} \setminus E_n$ and let $c_n > 0$ be such that

$$c_n \int_{D \cap E_n} \left|p_n - \sqrt{z}\right|^2 \, d\mu_n < \frac{1}{n}.$$

Let μ_{n+1} be μ_n off E_n and $c_n \mu_n$ on $D \cap E_n$. Let μ be the norm limit of the μ_n's. This time the functions in $H^2(\mu)$ are analytic on $D \setminus [0,1)$ and $\sqrt{z}$ is in $H^2(\mu) \cap L^\infty(\mu)$. We still have $P^\infty(\mu) = H^\infty(D,\mu)$.

3. Analytic Subspaces and Reflexivity

First, we will review some results on bounded point evaluations. Let μ be a measure. For each bpe λ for $H^2(\mu)$ there exists a function k_λ in $H^2(\mu)$ such that for every polynomial p,

$$p(\lambda) = \langle p, k_\lambda \rangle. \tag{3.1}$$

Define $\widehat{f}(\lambda)$ to be $< f, k_\lambda >$ for each f in $H^2(\mu)$.

Lemma 3.2. *Let E be the set of bpe's for $H^2(\mu)$. Then $\widehat{f} = f$ a.e.(μ) on E for each f in $H^2(\mu)$.*

Proof: Let $f \in H^2(\mu)$. Let $\{p_n\}$ be a sequence of polynomials such that $p_n \to f$ in $L^2(\mu)$ and $p_n \to f$ a.e.(μ). The conclusion follows by taking limits in (3.1). ∎

Lemma 3.3. *If $f \in H^2(\mu) \cap L^\infty(\mu)$ and $g \in H^2(\mu)$, then $fg \in H^2(\mu)$ and $(fg)\widehat{\ } = \widehat{f}\widehat{g}$.*

Proof: Let $\{p_n\}$ and $\{q_m\}$ be sequences of polynomials converging to f and g, respectively, in $L^2(\mu)$-norm and a.e.(μ). For each m, $p_n q_m \to f q_m$, so $f q_m \in H^2(\mu)$. Also, for each bpe λ, we have

$$< p_n q_m, k_\lambda > \to < f q_m, k_\lambda >,$$

so $(f q_m)\widehat{\ } = \widehat{f} q_m$ a.e.(μ) by Lemma 3.2. Letting $m \to \infty$, we see that

$$fg \in H^2(\mu) \quad \text{and} \quad (fg)\widehat{\ } = \widehat{f}\widehat{g} \text{ a.e.}(\mu). \quad \blacksquare$$

A point λ in the interior of the set of bpe's is an analytic bpe (abpe) if the map $z \to \widehat{f}(z)$ is analytic in a neighborhood of λ for each f in $H^2(\mu)$. An easy way for this to happen is for $\sigma(S) \setminus \sigma(N)$ to be nonempty. Here S is M_z on $H^2(\mu)$ and N is M_z on $L^2(\mu)$. Recall that $\sigma(S)$ is the union of $\sigma(N)$ and some of the components of $\mathbf{C} \setminus \sigma(\mathbf{N})$.

Proposition 3.4. *If G is a component of $\mathbf{C} \setminus \sigma(\mathbf{N})$ and $G \subset \sigma(S)$, then every point of G is an abpe. Also, if $f \in H^2(\mu)$ and $\{p_n\}$ is a sequence of polynomials converging in norm to f, then $\widehat{p}_n \to \widehat{f}$ uniformly on compact subsets of G.*

Proof: Let $\lambda \in G$. Since $G \cap \text{supp}\mu = \varphi$, it follows that $S - \lambda$ is bounded below. Noting that the linear span of $(S - \lambda)H^2(\mu)$ and 1 is $H^2(\mu)$, we conclude that $S - \lambda$ is Fredholm and $i(S - \lambda) = -1$. If k_λ is the vector in $H^2(\mu) \ominus (S - \lambda)H^2(\mu)$ with $< 1, k_\lambda >= 1$, then $< p, k_\lambda >= p(\lambda)$ for every polynomial p. See [5, p. 171] to see that $z \to k_z$ is coanalytic in a neighborhood of λ and hence that $z \to < f, k_z >$ is analytic in a neighborhood of λ for each f in $H^2(\mu)$. Also, the k_z's are uniformly bounded in norm in a neighborhood of λ. Thus, the second conclusion of the proposition directly follows for that neighborhood. Since λ was an arbitrary point of G, the conclusions for all of G follow immediately. ■

T. Trent [14] has characterized the set of abpe's. We will not prove it here.

Theorem 3.5. *The set of abpe's for $H^2(\mu)$ equals $\sigma(S) \setminus \sigma_{\text{ap}}(S)$.*

We are now ready to look at reflexivity for subnormal operators. For this we can restrict our attention to the case where $P^\infty(\mu) = H^\infty(D, \mu)$. For convenience we are assuming that S is M_z on $\mathcal{H}$, a subspace of $L^2(\mu)$, and that its mne N is M_z on $L^2(\mu)$. To avoid certain technical difficulties we will look only at the case where $\mu(\partial D) = 0$. Let us begin with a special case that will be useful in the general case.

Theorem 3.6. *Suppose $\mathcal{H} = H^2(\mu)$ and the set of abpe's for $H^2(\mu)$ equals D. Then S is reflexive.*

Proof: Let $T \in \text{Alg Lat}S$. For each λ in D let k_λ be the unique vector in $H^2(\mu)$ with $p(\lambda) =< p, k_\lambda >$ for every polynomial p. Since k_λ is orthogonal to $(S - \lambda)H^2(\mu)$, it follows that k_λ is an eigenvector for T^*. Thus there exists a scalar $\varphi(\lambda)$ such that $T^* k_\lambda = \overline{\varphi(\lambda)} k_\lambda$. Since

$$< T1, k_\lambda >=< 1, T^* k_\lambda >= \varphi(\lambda),$$

it follows that $\varphi(\lambda)$ is analytic in D. For f in $H^2(\mu)$ we have $(Tf)\widehat{}(\lambda) =< Tf, k_\lambda >=< f, T^* k_\lambda >= \varphi(\lambda)\widehat{f}(\lambda)$. Recalling that $\widehat{g} = g$ a.e.(μ) for $g \in H^2(\mu)$, we see that $Tf = \varphi f$.

It remains to show that φ is bounded. But by the definition of $\varphi(\lambda)$ it is clear that $|\varphi(\lambda)| \leq \|T^*\|$. Thus $\varphi \in H^\infty$ and we are done. ■

Now let us return to our subspace $\mathcal{H}$ of $L^2(\mu)$. For $a \in \mathcal{H}$ let $\mathcal{H}_a$ denote the closed linear span of $\{S^n a\}_0^\infty$. Since $S \mid \mathcal{H}_a$ is a cyclic subnormal operator, there exists a measure ν such that $S|\mathcal{H}_a$ is unitarily equivalent to M_z on $H^2(\nu)$. If the collection of abpe's for $H^2(\nu)$ equals D, then $\mathcal{H}_a$ is a *full analytic subspace* of $\mathcal{H}$. The following theorem is the key to the proof of reflexivity and other results about the lattice of a subnormal operator [9, 10, 11].

Theorem 3.7. *Let $a \in L^2(\mu)$. Then there exists $x \in \mathcal{H}$ such that $\mathcal{H}_{a+x}$ is a full analytic subspace. Moreover, x can be chosen so that $|a+x| > 0$ a.e.(μ) on the set where $|a| > 0$.*

Before proving this theorem, we will show how reflexivity follows from it. First, we will consider the cyclic case.

Theorem 3.8. *Suppose $\mathcal{H} = \mathcal{H}_a$ for some $a \in \mathcal{H}$. Then S is reflexive.*

Proof: Let $T \in$ Alg Lat$S$. Applying Theorem 3.6 to the vector zero and the space $\mathcal{H}$, we obtain a vector b in $\mathcal{H}$ such that $\mathcal{H}_b$ is a full analytic subspace. By Theorem 3.7 applied to the vector a and the space $\mathcal{H}_b$, there exists a vector x in $\mathcal{H}_b$ such that $\mathcal{H}_{a+x}$ is a full analytic subspace and $|a+x| > 0$ a.e.(μ). (Because the mne is not necessarily M_z on $L^2(\mu)$, there is a technicality to be checked. The mne is M_z on $L^2(\nu)$, where $\nu = \mu \mid A$ for some Borel set A. Apply the theorem to $a \mid A$ and $\mathcal{H}_b$.) Because x is an element in a full analytic subspace, H_x is also a full analytic subspace.

By Theorem 3.6 there exist φ and ψ in $P^\infty(\mu)$ such that $T \mid \mathcal{H}_{a+x} = M_\varphi$ and $T \mid \mathcal{H}_x = M_\psi$. Thus, for each nonnegative integer n,

$$(TS - ST)z^n a = (TS - ST)Z^n(a+x) - (TS - ST)z^n x = 0 \ - 0.$$

Hence $TS - ST = 0$. By a theorem of Yoshino [15] there exists ϑ in $H^2(\mu) \cap L^\infty(\mu)$ such that $T = M_\vartheta$. Since $|a+x| > 0$ a.e.(μ), it follows that $\vartheta = \varphi$ a.e.(μ) and thus $T = M_\varphi$. ■

Theorem 3.9. *The operator S is reflexive.*

Proof: Let $T \in$ Alg Lat$S$. We claim there exists ϑ in $P^\infty(\mu)$ such that $T|\mathcal{H}_a = M_\vartheta$ for every a in $\mathcal{H}$ with $|a| > 0$ a.e.(μ). By a result of Chaumat [4], such a's are dense in $\mathcal{H}$, so we will be done once we prove the claim.

Let a and b be in $\mathcal{H}$ with $|ab| > 0$ a.e.(μ). For all except countably many λ in $\mathbf{C}$ we have $|a + \lambda b| > 0$ a.e.(μ). Replacing b by an appropriate λb, we may assume that $|a + b| > 0$ a.e.(μ). By Theorem 3.8 there exist functions f, g, and h in $P^\infty(\mu)$ such that $T|\mathcal{H}_a = M_f$, $T|\mathcal{H}_b = M_g$, and $T|\mathcal{H}_{a+b} = M_h$. Since $T(a+b) = Ta + Tb$, it follows that $h(a+b) = fa + gb$ or

$$(h - f)a = (g - h)b. \tag{3.10}$$

Now apply T to both sides above and use the equality above to obtain

$$f(h - f)a = g(h - f)a.$$

Since $|a| > 0$ a.e.(μ), it follows that

$$f(h - f) = g(h - f).$$

If $h - f$ is not the zero function, then $f = g$. If $h = f$, then $g = h$ by equation (3.10). In any event, $f = g$. Since a and b were arbitrary, it follows that there exists ϑ in $P^\infty(\mu)$ with $T = M_\vartheta$. ∎

4. Scott Brown's Technique and Full Analytic Subspaces

Now we turn to the problem of constructing full analytic subspaces. Because our method grew out of Scott Brown's work, we begin with a version of his technique.

Recall that S is M_z on $\mathcal{H}$, a subspace of $L^2(\mu)$, and its mne N is M_z on $L^2(\mu)$. Also, we are assuming that $P^\infty(\mu) = H^\infty(D, \mu)$ and $\mu(\partial D) = 0$. Let $P^\infty(\mu)_\perp$ denote the preannihilator of $P^\infty(\mu)$, the set of those functions f in $L^1(\mu)$ such that $\int fg\, d\mu = 0$ for each g in $P^\infty(\mu)$. Let $P^\infty(\mu)_* = L^1(\mu)/P^\infty(\mu)_\perp$, the predual of $P^\infty(\mu)$, and let $\| \|_*$ denote the norm on this space. If $x, y \in L^2(\mu)$, define $x \otimes y$ in $P^\infty(\mu)_*$ by $(x \otimes y)(f) = < fx, y >$ for each f in $P^\infty(\mu)$.

Theorem 4.1. (S. W. Brown) *Suppose* $\sigma(S) = \sigma_{\mathrm{ap}}(S)$. *If* $L \in P^\infty(\mu)_*$, *then there exist* $x, y \in \mathcal{H}$ *such that* $L = x \otimes y$.

Theorem 4.2. (Olin and Thomson) *If* $L \in P^\infty(\mu)_*$, *then there exist* $x, y \in \mathcal{H}$ *such that* $L = x \otimes y$.

Theorem 4.2 leads to a proof that $W(S) = \mathcal{A}(S)$ and that the spaces are homeomorphic. Note that the theorem implies that weak-star continuous linear functionals on $P^\infty(\mu)$ are weak operator topology continuous.

Before proving Brown's theorem, we need some lemmas. Let e_λ denote evaluation at λ, a weak-star continuous linear functional on $P^\infty(\mu)$ for λ in D.

Lemma 4.3. *The closed convex hull of* $\{we_\lambda : |w| = 1,\ \lambda \in \sigma(S) \cap D\}$ *equals the unit ball of* $P^\infty(\mu)_*$.

Proof: Suppose not. Then by a separation theorem there exists $L \in P^\infty(\mu)_*$ with $\|L\|_* \leq 1$ and a function $f \in P^\infty(\mu)$ such that

$$|L(f)| > \sup_{\lambda \in D \cap \sigma(S)} |f(\lambda)|. \tag{4.4}$$

But $D \cap \sigma(S) \supset D \cap \mathrm{supp}\mu$ and

$$\|f\| = \sup\{|f(\lambda)| : \lambda \in D \cap \mathrm{supp}\ \mu\}$$

because $P^\infty(\mu) = H^\infty(D, \mu)$. Putting (4.4) and (4.5) together, we see that $|L(f)| > \|f\|$, a contradiction.

Lemma 4.6. *Let* $\lambda \in D$. *Suppose* $\{x_n\}$ *is a sequence of unit vectors in* $\mathcal{H}$ *such that* $(S - \lambda)x_n \to 0$ *in norm. Then* $x_n \otimes x_n \to e_\lambda$ *in predual norm and* $\int_\Delta |x_n|^2\, d\mu \to 1$ *for every neighborhood* Δ *of* λ.

Proof: Let $f \in P^\infty(\mu)$.

$$\begin{aligned} |(x_n \otimes x_n)(f) - f(\lambda)| &= |< fx_n, x_n > - f(\lambda)| \\ &= |< (f - f(\lambda))x_n, x_n >| \\ &\leq \int \left| \frac{f - f(\lambda)}{z - \lambda} \right| |z - \lambda|\, |x_n|^2\, d\mu \\ &\leq 2\|f\|(1 - |\lambda|)^{-1}\, \|(S - \lambda)x_n\|. \end{aligned}$$

The last quantity goes to zero independently of f, so $x_n \otimes x_n \to e_\lambda$.

Let ε be the distance from λ to the complement of a neighborhood Δ. Then

$$\int_{D\backslash\Delta} |z - \lambda|^2\, |x_n|^2\, d\mu \geq \varepsilon^2 \int_{D\backslash\Delta} |x_n|^2\, d\mu.$$

Since $\|(S - \lambda)x_n\| \to 0$, it follows that $\int_{D\backslash\Delta} |x_n|^2\, d\mu \to 0$. ∎

Lemma 4.7. *Let* $\lambda, \tau \in D$, $\lambda \neq \tau$. *Suppose* $\{x_n\}$ *and* $\{y_m\}$ *are sequences of unit vectors in* $\mathcal{H}$ *such that* $(S - \lambda)x_n \to 0$ *and* $(S - \tau)y_m \to 0$. *Then* $\|x_n \otimes y_m\|_* \to 0$ *as* $n,m \to \infty$. *Also, if* $u \in L^2(\mu)$ *and* $u(\lambda) = 0$, *then* $\|x_n u\|_1 \to 0$.

Proof: Let Δ be a closed neighborhood of λ that does not contain τ. Then

$$\begin{aligned}\int |x_n y_m|\,d\mu &= \int_\Delta |x_n y_m|\,d\mu + \int_{D\setminus\Delta} |x_n y_m|\,d\mu \\ &\le \left(\int_\Delta |x_n|^2\,d\mu\right)^{1/2}\left(\int_\Delta |y_m|^2\,d\mu\right)^{1/2} \\ &\quad + \left(\int_{D\setminus\Delta} |x_n|^2\,d\mu\right)^{1/2}\left(\int_{D\setminus\Delta} |y_m|^2\,d\mu\right)^{1/2} \\ &\le \left(\int_\Delta |y_m|^2\,d\mu\right)^{1/2} + \left(\int_{D\setminus\Delta} |x_n|^2\,d\mu\right)^{1/2}.\end{aligned}$$

Each summand in the last line above goes to zero as $m, n \to \infty$. Since $\|x \otimes y\|_* \le \|xy\|_1$ for any $x, y \in L^2(\mu)$, the first conclusion is established. The second follows in a similar fashion. ■

Lemma 4.8. *Suppose $\sigma(S) = \sigma_{\mathrm{ap}}(S)$. Let $L \in P^\infty(\mu)_*$ and $a \in \mathcal{H}$ and $b \in L^2(\mu)$. Let $\delta, \varepsilon > 0$. Suppose*

$$\|L - a \otimes b\|_* < \delta^2.$$

Then there exist $x \in \mathcal{H}$ with $\|x\| < \delta$ and $y \in L^2(\mu)$ with $\|y\| < \delta$, such that

$$\|L - (a + x) \otimes (y + b)\|_* < \varepsilon.$$

Proof: By Lemma 4.3 there exist $\lambda_1, \ldots, \lambda_n \in D \cap \sigma_{\mathrm{ap}}(S)$ and $w_1, \ldots, w_n \in \mathbf{C}$ with $\sum |w_j|^2 < \delta^2$ such that

$$\left\|\sum w_j e_{\lambda_j} - (L - (a \otimes b))\right\|_* < \varepsilon.$$

For each $m = 1, \ldots, n$, let $\{x_{mj}\}_j$ be a sequence of unit vectors in $\mathcal{H}$ such that $(S - \lambda_m)x_{mj} \to 0$ as $j \to \infty$. Let $c_j = |w_j|^{1/2}$ and let $d_j = w_j/|w_j|$. We compute:

$$\begin{aligned}&\left\|L - \left(a + \sum_{m=1}^n d_m c_m x_{mj}\right) \otimes \left(b + \sum_{m=1}^n c_m x_m\right)\right\|_* \\ &\qquad\le \left\|L - (a \otimes b) - \sum d_m c_m^2 x_{mj} \otimes x_{mj}\right\|_* + \text{ small terms.}\end{aligned}$$

Since $x_{mj} \otimes x_{mj} \to e_{\lambda_m}$ for each m, it follows that for j sufficiently large the right-hand side above is less than ε. Also,

$$\begin{aligned}\left\|\sum_m d_m c_m x_{mj}\right\|^2 &= \sum_m |d_m c_m|^2 \left\|x_{mj}\right\|^2 + \text{ small terms} \\ &= \sum |c_m|^2 + \text{ small terms.}\end{aligned}$$

A similar computation holds for $\|\sum_m c_m x_m\|^2$. Picking j sufficiently large and setting $x = \sum_m d_m c_m x_{mj}$ and $y = \sum_m c_m x_{mj}$, we are done. ■

Proof of Theorem 4.1. Let $L \in P^\infty(\mu)_*$. By Lemma 4.8 there exist vectors $x_1 \in \mathcal{H}$ and $y_1 \in L^2(\mu)$ with $\|x_1\|^2 < \|L\|$ and $\|y_1\|^2 < \|L\|$ such that

$$\|L - x_1 \otimes y_1\| < 2^{-2}\|L\|.$$

Proceed by induction. Suppose $x_1, \ldots, x_n \in \mathcal{H}$ and $y_1, \ldots, y_n \in L^2(\mu)$ have been chosen satisfying $\|x_j\| < 2^{-j+1}\|L\|^{1/2}$, $\|y_j\| < 2^{-j+1}\|L\|^{1/2}$, and

$$\left\| L - \left(\sum_1^n x_j\right) \otimes \left(\sum_1^n y_j\right) \right\|_* < 2^{-2n}\,\|L\|.$$

Use Lemma 4.8 to choose x_{n+1} and y_{n+1}. Let $x = \sum_1^\infty x_j$, $y = \sum_1^\infty y_j$. It is easy to see that $L = x \otimes y$. ■

To do this factorization in the general case we need to look at points in $\sigma(S) \backslash \sigma_{\text{ap}}(S)$. Let $\lambda \in \sigma(S) \setminus \sigma_{\text{ap}}(S)$. Since $S - \lambda$ is one-to-one and has closed range, it must be the case that $S - \lambda$ is not onto. Let $x \in \mathcal{H}\theta(S-\lambda)\mathcal{H}$ with $\|x\| = 1$. If $f \in P^\infty(\mu)$, then

$$(x \otimes x)(f) =< fx, x >= f(\lambda) < x, x > + < (f - f(\lambda))x, x >= f(\lambda).$$

So $x \otimes x = e_\lambda$. Noting that for each natural number n the subset $(S-\lambda)^{n-1}\ \mathcal{H}$ is a subspace that $S - \lambda$ cannot map onto itself, we see that $(S-\lambda)^{n-1}\mathcal{H} \ominus (S-\lambda)^n\mathcal{H}$ is not the zero space. Replacing $\mathcal{H}$ by $(S-\lambda)^{n-1}\mathcal{H}$ above and choosing $v_n \in (S-\lambda)^{n-1}\mathcal{H} \ominus (S-\lambda)^n\mathcal{H}$ with $\|v_n\| = 1$, we get $e_\lambda = v_n \otimes v_n$ for each n. ■

Lemma 4.9. *If $a \in L^2(\mu)$, then*

$$\|v_n \otimes a\|_* \to 0 \quad \text{and} \quad \|a \otimes v_n\|_* \to 0.$$

Proof: Note first that $\{v_n\}$ is an orthonormal sequence so that $v_n \to 0$ weakly. To establish both conclusions above it suffices to show that if $\{e_n\}$ is a sequence of unit vectors in $L^2(\mu)$ that converges weakly to zero and $b \in L^2(\mu)$, then $\|e_n \otimes b\|_* \to 0$.

Let $\varepsilon > 0$. Choose $r > 1$ such that $\int_{|z|>r} |b|^2\, d\mu < \varepsilon$. Choose an integer M_1 such that

$$\left| \sum_{M_1}^\infty c_m z^m \right| < \frac{\varepsilon}{\|b\|}$$

if $|z| \le r$ and $\sum |c_m|^2 \le 1$. Since $e_n \to 0$ weakly, there exists a positive integer M_2 such that

$$\left| \int_{|z| \le r} z^m e_n \bar{b} \, d\mu \right| < \frac{\varepsilon}{M_1}$$

for $m = 1, \ldots, M_1$ and $n \ge M_2$.

Let $f = \sum c_m z^m$ belong to $P^\infty(\mu)$ with $\|f\| \le 1$. Then for any $n \ge M_2$,

$$\begin{aligned} |(e_n \otimes b)(f)| &= \left| \int f e_n \bar{b} \, d\mu \right| \\ &\le \left| \int_{|z|>r} f \, e_n \bar{b} \, d_\mu \right| + \left| \sum_1^{M_1 - 1} c_m \int_{|z| \le r} z^m e_n \bar{b} d_\mu \right| \\ &\quad + \left| \int_{|z| \le r} \left(\sum_{M_1}^\infty c_m z^m \right) e_n \bar{b} d_\mu \right| \\ &\le \varepsilon + \varepsilon + \varepsilon. \end{aligned}$$

■

Theorem 4.2 will follow from the next lemma in the same way that Theorem 4.1 followed from Lemma 4.8.

Lemma 4.9a. *Let $a, b \in L^2(\mu)$ and let $L \in P^\infty(\mu)_*$. Let $\delta, \varepsilon > 0$. If $\|L - a \otimes b\| < \delta^2$, then there exist $x \in \mathcal{H}$ and $y \in L^2(\mu)$ with $\|x\| < \delta$ and $\|y\| < \delta$ such that*

$$\|L - (a + x) \otimes (b + y)\|_* < \varepsilon.$$

Proof: By Lemma 4.3 there exist $\lambda_1, \ldots, \lambda_m$ in $\sigma(S) \setminus \sigma_a(S)$, $\lambda_{m+1}, \ldots, \lambda_M$ in $D \cap \sigma_{\text{ap}}(S)$, $c_1, \ldots, c_M > 0$ with $\sum c_i^2 < \delta^2$, and $d_1, \ldots, d_M$ with $|d_1| = 1$ such that

$$\left\| L - (a \otimes b) - \sum_i^M d_i c_i^2 e_{\lambda_i} \right\|_* < \varepsilon.$$

Now let α be a small positive number. Let v_{ik} be a unit vector in $(S - \lambda_1)^{k-i}\mathcal{H} \ominus (S - \lambda_1)^k \mathcal{H}$. For k sufficiently large, $|v_{ik}(\lambda_i)| < \alpha$ for all i. (The concern here is with i's greater than m and where $\mu(\{\lambda_i\}) > 0$.) Also, for sufficiently large k, $\|v_{1k} \otimes a\|_*$, $\|a \otimes v_{1k}\|_*$, $\|b \otimes v_{1k}\|_*$, and $\|v_{1k} \otimes b\|_*$ are each less than α. Let $v_1 = v_{1k}$ for a sufficiently large k.

We will choose $v_2, \ldots, v_m$ by induction. Suppose $v_1, \ldots, v_{n-1}$ have been chosen and satisfy

(i) $v_i \otimes v_i = e_{\lambda_i}$;

(ii) $\|v_i \otimes v_j\|_* < \alpha$ if $i \neq j, i, j < n$;

(iii) $\left\{\begin{array}{l} \|v_i \otimes a\|_*, \\ \|a \otimes v_i\|_*, \\ \|v_i \otimes b\|_*, \\ \|b \otimes v_i\|_*, \end{array}\right.$ are each less than α;

(iv) $|v_i(\lambda_j)| < \alpha$ for $1 \leq i \leq n-1$ and $m < j \leq M$.

Let v_{nk} be a unit vector in $(S - \lambda_n)^{k-1}\mathcal{H} \otimes (S - \lambda_n)^k \mathcal{H}$. Let $v_n = v_{nk}$ for some k sufficiently large so that (i), (ii), (iii), and (iv) will be satisfied for $v_1, \ldots, v_n$.

For $n = m+1, \ldots, M$, choose sequences $\{v_{nj}\}_j$ of unit vectors such that $(S - \lambda_n)v_{nj} \to 0$ as $j \to \infty$. For sufficiently large j, we have $\|v_{nj} \otimes v_{kj}\|_* < \alpha$ for $n,k > m$ and $n \neq k$ and $\|v_{nj} v_k\|_1 < \alpha$ for $n > m$ and $k \leq m$. Then

$$\left\| L - \left(a + \sum d_j c_j v_j\right) \otimes \left(b + \sum c_j v_j\right) \right\|_* < \varepsilon + c\alpha$$

for some constant c that depends only on m and M. A small α does the trick. ■

To obtain analytic subspaces, we need to improve upon Theorem 4.2. Let $A_r = \{z : r < |z| < 1\}$. By the maximum modulus theorem and our assumption that $P^\infty(\mu) = H^\infty(D, \mu)$,

$$\|f\|_{D \cap \sigma(S)} = \|f\|_{A_r \cap \sigma(S)} = \|f\|$$

for each f in $P^\infty(\mu)$. From this observation we obtain the following improvement of Lemma 4.3.

Lemma 4.10. *Fix r, $0 < r < 1$. The closed convex hull of*

$$\{w e_\lambda : |w| = 1, \ \lambda \in \sigma(S) \cap A_r\}$$

equals the closed unit ball of $P^\infty(\mu)_*$.

Theorem 4.11. *Let $L \in P^\infty(\mu)_*$ and let $r < 1$. Then there exist $x \in \mathcal{H}$ and $y \in L^2(\mu \mid A_r)$ such that $L = x \otimes y$.*

Idea of Proof. There are two basic modifications needed in the proof of Theorem 4.2. The first is to use Lemma 4.10 in place of Lemma 4.3. The second is to replace terms of the form $x \otimes y$ by terms of the form $x \otimes y\chi_{A_r}$. There are also technical modifications needed in terms coming from points in $\sigma(S) \setminus \sigma_a(S)$. The key fact for those changes is that representing measures for points near the unit circle have very little mass on the disc of radius r.

Corollary 4.12. *Let $r < 1$. There exist cyclic subspaces of $\mathcal{H}$ that consist of functions analytic on $|z| < r$.*

Proof: By the theorem there exist $x \in \mathcal{H}$ and $y \in L^2(\mu \mid A_r)$ such that $e_0 = x \otimes y$. We claim that the closed linear span of $\{z^n x\}_{n=0}^{\infty}$ is an analytic subspace on $|z| < r$. Let ν be the measure such that $d\nu = |x|^2 \, d\mu \mid A_r$. It is sufficient to show that $H^2(\nu)$ has abpe's throughout $|z| < r$. Since ν is carried by A_r, it suffices to show that zero is in the spectrum of M_z on $H^2(\nu)$ (see Theorem 3.5). But

$$< 1, y/x >_{H^2(\nu)} = 1$$

while

$$< z^n, y/x >_{H^2(\nu)} = 0 \qquad \text{for } n \geq 1,$$

so 1 is not in the closed linear span of $\{z^n\}_1^{\infty}$. Thus M_z on $H^2(\nu)$ is not onto. ■

Theorem 4.13. *There exist full analytic invariant subspaces.*

Proof: Recall that we have reduced to the case where $P^\infty(\mu) = H^\infty(D,\mu)$. Also, we are assuming that S is M_z on $\mathcal{H}$, a subspace of $L^2(\mu)$, its mne N is M_2 on $L^2(\mu)$, and $\mu(\partial D) = 0$.

It suffices to find $x \in \mathcal{H}$ and a sequence $\{y_n\}$ of vectors in $L^2(\mu)$ with $y_n = 0$ a.e.(μ) on $|z| < 1 - (1/n)$ such that $e_0 = x \otimes y_n$ for every n. The idea of the proof is simultaneous approximation. We will demonstrate the idea in the following theorem. We leave it as an exercise to extend the method and finish the proof of Theorem 4.13. ■

Theorem 4.14. *Suppose $\sigma(S) = \sigma_{\text{ap}}(S)$ and μ has no atoms. Let L_1, $L_2 \in P^\infty(\mu)_*$. Then there exist $x \in \mathcal{H}$ and y_1,$y_2 \in L^2(\mu)$ such that $L_i = x \otimes y_i$ for $i =$ 1,2.*

Proof: By Lemma 4.8 there exist $x_1 \in \mathcal{H}$ and $y_1 \in L^2(\mu)$ with $\|x_1\|^2 < \|L_1\|$ and $\|y_1\|^2 < \|L_1\|$ such that

$$\|L_1 - x_1 \otimes y_1\|_* < 2^{-4}.$$

Choose $x_2 \in \mathcal{H}$ and $y_2 \in L^2(\mu)$ with $\|x_2\|^2 < \|L_2\|_*$ and $\|y_2\|^2 < \|L_2\|_*$ such that

$$\left\|L_2 - (x_1 + x_2) \otimes y_2\right\|_* < 2^{-2}$$

and so that $\|x_1 \otimes y_2\|_* < 2^{-4}$. Thus,

$$\left\|L_i - (x_1 + x_2) \otimes y_i\right\|_* < 2^{-2}$$

for $i = 1, 2$.

Now choose $x_3 \in \mathcal{H}$ and $y_3 \in L^2(\mu)$ with $\|x_3\| < 1/2$ and $\|y_3\| < 1/2$ so that

$$\left\|L_1 - (x_1 + x_2 + x_3) \otimes (y_1 + y_3)\right\|_* < 2^{-4}$$

and

$$\left\|L_2 - (x_1 + x_2 + x_3) \otimes y_2\right\|_* < 2^{-2}.$$

Then choose $x_4 \in \mathcal{H}$ and $y_4 \in L^2(\mu)$ with $\|x_4\| < 1/2$ and $\|y_4\| < 1/2$ so that

$$\left\|L_2 - (x_1 + x_2 + x_3 + x_4) \otimes (y_2 + y_4)\right\|_* < 2^{-4}$$

and

$$\left\|L_1 - (x_1 + x_2 + x_3 + x_4) \otimes (y_1 + y_3)\right\|_* < 2^{-4}.$$

Continue process by induction. Let $x = \sum x_i$, $w_1 = \sum y_{2n-1}$ and $w_2 = \sum y_{2n}$. Then $L_i = x \otimes w_i$. ■

Brown's technique was also the starting point for some remarkable work by Apostol, Bercovici, Crevreau, Foias, Langsam, Pearcy, Sz.–Nagy, and others [16].

References

1. Bram, J. Subnormal operators, Duke Math. J. **22** (1955), 75–94.
2. Brennan, J. E., Invariant subspaces and rational approximation, J. Funct. Anal. **7** (1971), 285–310.
3. Brown, S. W., Some invariant subpsace for subnormal operators, Integral Eqs. Op. Theory **1** (1978), 310–333.
4. Chaumat, J., Adherence faible etoile d'algebras de fractions rationelles, Publ. Math. Orsay **147** (1975), 75–140.
5. Conway, J. B., *Subnormal Operators*, Pitman, Boston, 1981.
6. ——— and R. F. Olin, A functional calculus for subnormal operators II, Memoirs Amer. Math. Soc. **184** (1977).
7. Fernstrom, C., Bounded point evaluations and approximation in L^p by analytic functions, *Spaces of Analytic Functions*, Springer-Verlag Lecture Notes in Math. **512** (1976), 65–68.

8. Hastings, W. W., A construction of Hilbert spaces of analytic functions, Proc. Amer. Math. Soc. **74** (1979), 295–298.
9. Olin, R. F. and J. E. Thomson, Algebras of subnormal operators, J. Funct. Anal. **37** (1980), 271–301.
10. ———, Cellular-indecomposable subnormal operators, Integral Equations and Operator Theory **7** (1984), 392–430.
11. ———, Cellular-indecomposable subnormal operators II, Integral Equations and Operator Theory **9** (1986), 600-609.
12. Sarason, D., Weak-star density of polynomials, J. Reine Angew. Mth. **252** (1972), 1–15.
13. Thomson, J. E., Invariant subspaces for algebras of subnormal operators, Proc. Amer. Math. Soc. **96** (1986), 462–464.
14. Trent, T. T. $H^2(\mu)$ spaces and bounded point evaluations, Pacific J. Math. **80** (1979), 279–292.
15. Yoshino, T., Subnormal operators with a cyclic vector, Tohoku Math. **21** (1969), 47–55.
16. Bercovici, H., C. Foias, and C. Pearcy, Dual algebras with applications to invariant subspaces and dilation theory, CBMS Regional Conf. Ser. in Math. **56** Amer. Math. Soc., Providence, Rhode Island, 1985.

James E. Thomson
Department of Mathematics
Virginia Polytechnic Institute and State University
Blacksburg, Virginia 24061